Molecular hydrogen is the most abundant molecule in the Universe. In recent years, advances in theory and laboratory experiments coupled with breakthrough observations with important new telescopes and satellites have revolutionized our understanding of molecular hydrogen in space. It is now possible to address the question of how molecular hydrogen formed in the early Universe and the role it played in the formation of primordial structures. This timely volume presents articles from a host of experts who reviewed this new understanding at an international conference in Paris.

This book provides the first multi-disciplinary synthesis of our new understanding of molecular hydrogen. It covers the theory of the physical processes and laboratory experiments, as well as the latest observations. It will therefore be an invaluable reference for all students and researchers in astrophysics and cosmology.

CAMBRIDGE CONTEMPORARY ASTROPHYSICS

Molecular Hydrogen in Space

CAMBRIDGE CONTEMPORARY ASTROPHYSICS

Series editors
José Franco, Steven M. Kahn, Andrew R. King and Barry F. Madore

Titles available in this series

Gravitational Dynamics, *edited by O. Lahav, E. Terlevich and R. J. Terlevich* (ISBN 0 521 56327 5)

High-sensitivity Radio Astronomy, *edited by N. Jackson and R. J. Davis* (ISBN 0 521 57350 5)

Relativistic Astrophysics, *edited by B. J. T. Jones and D. Marković* (ISBN 0 521 62113 5)

Advances in Stellar Evolution, *edited by R. T. Rood and A. Renzini* (ISBN 0 521 59184 8)

Relativistic Gravitation and Gravitational Radiation, *edited by J.-A. Marck and J.-P. Lasota* (ISBN 0 521 59065 5)

Instrumentation for Large Telescopes, *edited by J. M. Rodríguez Espinosa, A. Herrero and F. Sánchez* (ISBN 0 521 58291 1)

Stellar Astrophysics for the Local Group, *edited by A. Aparicio, A. Herrero and F. Sánchez* (ISBN 0 521 63255 2)

Nuclear and Particle Astrophysics, *edited by J. G. Hirsch and D. Page* (ISBN 0 521 63010 X)

Theory of Black Hole Accretion Discs, *edited by M. A. Abramowicz, G. Björnsson and J. E. Pringle* (ISBN 0 521 62362 6)

Interstellar Turbulence, *edited by J. Franco and A. Carramiñana* (ISBN 0 521 65131 X)

Globular Clusters, *edited by C. Martínez Roger, I. Pérez Fournón and F. Sánchez* (ISBN 0 521 77058 0)

The Formation of Galactic Bulges, *edited by C. M. Carollo, H. C. Ferguson and R. F. G. Wyse* (ISBN 0 521 66334 2)

Very Low-Mass Stars and Brown Dwarfs, *edited by R. Rebolo and M. R. Zapatero-Osorio* (ISBN 0 521 66335 0)

Molecular Hydrogen in Space

Edited by
F. COMBES
Observatoire de Paris, DEMIRM

G. PINEAU DES FORÊTS
DAEC, Observatoire de Meudon
IAS, Université de Paris-Sud, France

PUBLISHED BY THE PRESS SYNDICATE OF THE UNIVERSITY OF CAMBRIDGE
The Pitt Building, Trumpington Street, Cambridge CB2 1RP, United Kingdom

CAMBRIDGE UNIVERSITY PRESS
The Edinburgh Building, Cambridge CB2 2RU, UK   http://www.cup.cam.ac.uk
40 West 20th Street, New York, NY 10011-4211, USA   http://www.cup.org
10 Stamford Road, Oakleigh, Melbourne 3166, Australia
Ruiz de Alarcón 13, 28014 Madrid, Spain

© Cambridge University Press 2000

This book is in copyright. Subject to statutory exception
and to the provisions of relevant collective licensing agreements,
no reproduction of any part may take place without
the written permission of Cambridge University Press.

First published 2000

Printed in the United States of America

10/12 pt. Typeset in $\text{L\!A\!T\!E\!X}$ by the authors

*A catalog record for this book is available from
the British Library.*

Library of Congress Cataloging-in-Publication Data is available

ISBN 0 521 78224 4 hardback

Contents

Preface xi

Conference participants xiv

Conference photograph / poster xviii

1. Physics of H_2 and HD

Astrophysical Importance of H_2 3
    A. Dalgarno

Radiative and Electronic Excitation of Lyman and Werner Transitions
in H_2 13
    E. Roueff, H. Abgrall, X. Liu and D. Shemansky

The Cooling of Astrophysical Media by H_2 and HD 23
    D. Flower, J. Le Bourlot, G. Pineau des Forêts and E. Roueff

Highly Excited Singlet Ungerade States of H_2 and their Theoretical
Description 31
    Ch. Jungen and S. C. Ross

Laboratory Studies of Long-range Excited States of H_2 39
    W. Ubachs

A Model of Interstellar Dark Matter 47
    J. Schaefer

Mass of H_2 Dark Matter in the Galactic Halo 57
    Y. Shchekinov, R. J. Dettmar and P. M. W. Kalberla

2. Formation - Destruction

Experiments with Trapped Ions and Nanoparticles 63
    D. Gerlich, J. Illemann, and S. Schlemmer

Laboratory Studies of Molecular Hydrogen Formation on Surfaces of
Astrophysical Interest 71
    V. Pirronello, O. Biham, G. Manicó, J. E. Roser and G. Vidali

The Formation of H_2 and Other Simple Molecules on Interstellar
Grains 85
    E. Herbst

The Interaction of H Atoms with Interstellar Dust Particles: Models ... 89
   V. Sidis, L. Jeloaica, A. G. Borisov and S. A. Deutscher

The Energetics and Efficiency of H_2 Formation on the Surface of Simulated Interstellar Grains 99
   D. A. Williams, D. E. Williams, D. Clary, A. Farebrother, A. Fisher, J. Gingell, R. Jackman, N. Mason, A. Meijer, J. Perry, S. Price and J. Rawlings

Probing the Connection between PAHs and Hydrogen (H, H_2) in the Laboratory and in the Interstellar Medium 107
   C. Joblin, J. P. Maillard, I. Vauglin, C. Pech and P. Boissel

3. Observations and Models

Non Stationary C-shocks: H_2 Emission in Molecular Outflows ... 117
   G. Pineau des Forêts and D. Flower

The Ortho/Para Ratio in C and J-type Shocks 123
   D. Wilgenbus, S. Cabrit, G. Pineau des Forêts and D. Flower

Theoretical Models of Photodissociation Fronts 131
   B. Draine and F. Bertoldi

ISO Spectroscopy of H_2 in Star Forming Regions 139
   M. Van den Ancker, P. R. Wesselius and A. G. G. M. Tielens

Observations of the H_2 Ortho-Para Ratio in Photodissociation Regions 143
   S. Ramsay, A. Chrysostomou, P. Brand, M. Burton and P. Puxley

H_2 Emission from CRL618 151
   F. Herpin, J. Cernicharo and A. Heras

Hydrogen in Photodissociation Regions: NGC2023 and NGC7023 . 155
   D. Field, J. L. Lemaire, J. P. Maillard, S. Leach, G. Pineau des Forêts, E. Falgarone, F. P. Pijpers, M. Gerin, F. Rostas, D. Rouan and L. Vannier

A Pre-FUSE View of H_2 161
   M. Jura

H_2 Absorption Line Measurements with ORFEUS 165
   P. Richter, H. Bluhm, O. Marggraf and K. S. de Boer

Ultraviolet Observations of Molecular Hydrogen in Interstellar Space 171
   T. Snow

FUSE and Deuterated Molecular Hydrogen 179
   R. Ferlet, M. André, G. Hébrard, A. Lecavelier, M. Lemoine, G. Pineau des Forêts, E. Roueff and A. Vidal-Madjar

ISO-SWS Observations of H_2 in Galactic Sources 189
   C. Wright

H_2 in Molecular Supernova Remnants 193
   W. Reach and J. Rho

3D Integral Field H_2 Spectroscopy in Outflows 197
   J. Tedds, P. Brand and M. Burton

Near-Infrared Imaging and [OI] Spectroscopy of IC443 using 2MASS
and ISO 201
   J. Rho, S. Van Dyk, T. Jarrett, R. M. Cutri, and W. T. Reach

ISOCAM Spectro-imaging of the Supernova Remnant IC443 205
   P. Cox, D. Cesarsky and G. Pineau des Forêts

Spatial Structure of a Photo-Dissociation Region in Ophiucus 211
   F. Boulanger, E. Habart, A. Abergel, E. Falgarone, G. Pineau des
   Forêts and L. Verstraete

Tracing H_2 Via Infrared Dust Extinction 217
   J. Alves, C. Lada and E. Lada

The Small Scale Structure of H_2 Clouds 221
   P. Boissé, S. Thoraval, J. C. Cuillandre, G. Duvert, and L. Pagani

Hot Chemistry in the Cold Diffuse Medium: Spectral Signature in the
H_2 Rotational Lines 225
   E. Falgarone, L. Verstraete, P. Hily-Blant and G. Pineau des Forêts

H_2 Observations of the OMC-1 Outflow with the ISO-SWS 231
   D. Rosenthal, F. Bertoldi and S. Drapatz

4. Extragalactic and Cosmology

The Role of H_2 Molecules in Cosmological Structure Formation . . 237
   T. Abel and Z. Haiman

The Role of H_2 Molecules in Primordial Star Formation 247
   F. Palla and D. Galli

Evolution of Primordial H_2 for Different Cosmological Models 259
   D. Puy

Dynamics of H_2 Cool Fronts in the Primordial Gas 263
   M. Ibanez, and M. Bessega

Is Reionization Regulated by H_2 in the Early Universe? 269
    A. Ferrara, B. Ciardi and P. Todini

H_2 in Galaxies 275
    F. Combes

Transformation of Galaxies within the Hubble Sequence 285
    D. Pfenniger

Extragalactic H_2 and its Variable Relation to CO 293
    F. Israel

The Galactic Dark Matter Halo: Is it H_2? 297
    P. Kalberla, J. Kerp and U. Haud

Observations of H_2 in Quasar Absorbers 301
    J. Bechtold

H_2 Emission as a Diagnostic of Physical Processes in Starforming Galaxies 307
    P. van der Werf

5. Outlook

H_2 in the Universe: Perspectives 317
    J. Black

Author index 325

Preface

This book gathers all contributions to the International Conference on *H₂ in Space* held in Paris, France, on September 28- October 1st, 1999. The attendance was 106 participants from 16 countries. The goal was to gather together representatives of three communities: experts in the physics of the molecule, including experimentalists, observers of the interstellar medium, in particular of the warm H_2 detected in infrared lines, and theoreticians studying H_2 formation and cooling in astrophysical objects, from the early universe to the present galaxies.

The electronic structure of the H_2 molecule has been well studied in the past, but it was shown that recent progress has been made on the theory of highly excited and long-range Rydberg states, both from calculations of line-strengths and comparison with experiments. New and more accurate data are now available for the rates of collisional excitation of rovibrational transitions by neutrals, as well as protons or electrons. It has been known for several decades that interstellar H_2 is formed on dust grains, but, until now, the efficiency of this process was poorly known. Recent experiments have reproduced this formation process on silicates and amorphous carbon and shown that the efficiency is strongly dependent on temperature. In particular, the mobility of H atoms on grains is much lower than previously thought. Laboratory experiments with trapped ions and nanoparticles have opened new avenues of investigation. The relationship of H_2 with PAHs is being addressed both in the laboratory and by means of astronomical observations.

H_2 infrared emission is produced in photodissociation regions (PDRs) and shocks. An impressive harvest of data has been gathered by ISO, and the comparisons with models have considerably increased our knowledge of the physics and chemistry of star-forming regions. The rovibrational lines can help to distinguish the nature of the shocks (C-type, J-type, dissociative or not) and their age. The ortho:para ratio is an indicator of the state in which H_2 is formed, the conditions of excitation in the shocks, or the history of molecular clouds. H_2 emission has been observed and interpreted in sources as diverse as Herbig-Haro objects, star-forming molecular clouds and the associated PDRs, planetary nebulae, supernovae remnants, and X-ray excited gas.

Most observations of the H_2 molecule are of warm gas and pertain to only

1% or less of the molecular mass. The bulk of the material is cold H_2 that will be traced by FUSE, through H_2 UV absorption, or has been traced by CO millimetric emission (but with a highly uncertain ratio for conversion to H_2), by infrared measurements of dust extinction, or by thermal dust emission. These tracers are strongly dependent on metallicity, and there exists the tantalizing possibility that we are missing large amounts of cold H_2 in the outer parts of galaxies. This is suggested by ISO mapping of pure rotational lines of H_2 in galaxies, by the excess gamma-ray emission, and by the detection of small-scale (10 AU) structure in the ISM. In the primordial gas, H_2 formed in the gas phase, in the absence of grains; H_2 and HD cooling then triggered the collapse of the first structures. The re-ionization of large regions of the Universe is then necessary to avoid a "cooling catastrophe". The observation of the HD/H_2 ratio at various redshifts is a crucial test of the baryonic fraction of the Universe, and our hopes rest with FUSE to resolve recent controversies.

The conference confirmed that progress is currently being made in a number of crucial areas, including laboratory simulations of H_2 formation on grains and cosmological scenarios. There is already a wealth of observational data, and their on-going interpretation will lead to further advances. A big step will be made in the near future, when results begin to flow from FUSE. Essential to the undoubted success of the meeting was the interaction between the three communities, which do not usually attend the same conferences: laboratory experts learned about the astrophysical applications, whilst astronomers were provided with more information on the basic physics of the molecule.

We are grateful to the members of the Scientific Organizing Committee for their help in the choice of speakers and participants: F. Bertoldi (MPIfR-Bonn, Germany), J. Black (Onsala Space Observatory, Sweden), F. Boulanger (I.A.S., France), F. Combes (Obs. de Paris, France), P. Cox (I.A.S., France), B. Draine (Princeton, USA), D. Field (Univ. Aarhus, Denmark), D. Flower (Durham, UK), J-L. Lemaire (Univ. Cergy-Pontoise, France), G. Pineau des Forêts (Obs. de Meudon, France), A. Vidal-Madjar (IAP, France).

The conference was sponsored by Programme National Chimie du Milieu Interstellaire (PCMI-CNRS), Collaborative Computational Project Number 7 (CCP7, UK), Observatoire de Paris (MEN), Ministere des Affaires Etrangeres (MAE), University Cergy-Pontoise, University Paris XI (IAS), and Institut d'Astrophysique de Paris (IAP).

This meeting could not have gone smoothly without the active collaboration of all LOC members and the graduate students Pierre Hily-Blant and David Wilgenbus. We want also to thank Lionel Provost, for his web registration software, and Jean Mouette, for his technical expertise and for taking photographs. The participants greatly appreciated being invited by Senator Loridant to the Palais du Luxembourg, where the conference dinner was held.

Paris, December, 1999

The Editors,

Françoise COMBES (Observatoire de Paris)
Guillaume PINEAU des FORETS (Observatoire de Meudon)

Participant List

Abel Tom, Center for Astrophysics Cambridge – USA *tabel@cfa.harvard.edu*
Abergel Alain, IAS Orsay – FRANCE *abergel@ias.fr*
Abgrall Hervé, Observatoire de Paris, Meudon – FRANCE *herve.abgrall@obspm.fr*
Allard Nicole, Observatoire de Meudon – FRANCE *allard@obspm.fr*
Alves Joao, ESO Garching – GERMANY *jalves@eso.org*
Bertoldi Frank, MPIfR Bonn, Bonn – GERMANY *bertoldi@mpifr-bonn.mpg.de*
Black John, Onsala Space Observatory Onsala –SWEDEN *jblack@oso.chalmers.se*
Boissé Patrick, Lab. de radioastronomie Paris – FRANCE *boisse@ensapa.ens.fr*
Boucard Stephane, DIAM, Univ. Paris 6, Paris – FRANCE *stephane@diam.jussieu.fr*
Boulanger François, IAS Orsay – FRANCE *boulanger@ias.fr*
Brechignac Philippe, LPPM, Univ. Paris-Sud, Orsay – France
  phb@ppm.u-psud.fr
Cabrit Sylvie, Observatoire de Paris – FRANCE *sylvie.cabrit@obspm.fr*
Cernicharo Jose, CSIC. IEM Madrid – SPAIN *cerni@astro.iem.csic.es*
Combes Françoise, Observatoire de Paris – FRANCE *francoise.combes@obspm.fr*
Cox Pierre, IAS Orsay –FRANCE *cox@ias.fr*
Dalgarno Alexander, Center for Astrophysics Cambridge – USA
  adalgarno@cfa.harvard.edu
Dowek Danielle, LCAM Orsay – FRANCE *dowek@lcam.u-psud.fr*
Draine Bruce, Princeton University Princeton – USA *draine@astro.princeton.edu*
Dubernet-Tuckey Marie-Lise, Obs. de Besancon, Besancon –FRANCE
  mld@obs-besancon.fr
Dulieu François, Université Cergy Pontoise, Cergy Pontoise – FRANCE
  francois.dulieu@lamap.u-cergy.fr
Ellinger Yves, ENS Paris – FRANCE *ellinger@letmex.ens.fr*
Falgarone Edith, ENS Paris – FRANCE *falgarone@ensapa.ens.fr*
Feautrier Nicole, Observatoire de Meudon – FRANCE *Nicole.Feautrier@obspm.fr*
Ferlet Roger, IAP Paris – FRANCE *ferlet@iap.fr*
Ferrara Andrea, Osservatorio Astrofisico Arcetri Firenze – ITALY
  ferrara@arcetri.astro.it
Field David, University of Aarhus Aarhus –DENMARK *dfield@ifa.au.dk*
Fillion Jean-Hugues, Observatoire De Meudon, Meudon –FRANCE
  jean-hughes.fillion@lamap.u-cergy.fr
Flower David, University of Durham, Durham –UK *david.flower@dur.ac.uk*
Gee Christelle, Universite de Cergy-Pontoise, Cergy-Pontoise – FRANCE
  gee@paris.u-cergy.fr

Participant List

Gerbaldi Michele, Institut d'Astrophysique PARIS – FRANCE *gerbaldi@iap.fr*
Gerlich Dieter, Institut fur Physik Chemnitz –GERMANY *gerlich@physik.tu-chemnitz.de*
Gingell Jon, University College London –UK *uccajmg@ucl.ac.uk*
Giraud Edmond, Observatoire, Marseille- FRANCE *Edmond.Giraud@observatoire.cnrs-mrs.fr*
Glownia James, IBM Research Division Yorktown Heights – USA *glownia@us.ibm.com*
Gry Cecile, ESA Vilspa Madrid – SPAIN *cgry@iso.vilspa.esa.es*
Habart Emilie, I.A.S Orsay – FRANCE *habart@ias.fr*
Hennebelle Patrick, Demirm, ENS Paris – FRANCE *hennebel@lra.ens.fr*
Herbst Eric, Ohio State University Columbus Ohio – USA *herbst.6@osu.edu*
Herpin Fabrice, CSIC Madrid –SPAIN *herpin@isis.iem.csic.es*
Hily-Blant Pierre, Demirm, ENS Paris – FRANCE *hilyblan@elbereth.obspm.fr*
Humbert Eric, DIAM, Univ. Paris 6, Paris – FRANCE *p6atom55@cicrp.jussieu.fr*
Ibanez Miguel, Universidad de los Andes Merida – VENEZUELA *ibanez@ciens.ula.ve*
Israel Frank, Leiden Observatory Leiden – NETHERLANDS *israel@strw.leidenuniv.nl*
Joblin Christine, CESR-CNRS Toulouse Cdx 04 – FRANCE *joblin@cesr.fr*
Jolicard George, Observatoire de Besancon, Besancon – FRANCE *george@obs-besancon.fr*
Jungen Christian, Univ. Orsay, Orsay – FRANCE *jung1@ss10.lac.u-psud.fr*
Jura Michael, UCLA Los Angeles – USA *jura@clotho.astro.ucla.edu*
Kalberla Peter, Radioastronomical Inst. Bonn, Bonn – GERMANY *pkalberla@astro.uni-bonn.de*
Kawamura Akiko, University of Tokyo, Tokyo – JAPAN *kawamura@ioa.s.u-toyko.ac*
Kulesa Craig, University of Arizona Tucson – USA *ckulesa@as.arizona.edu*
Lakhlifi Azzedine, Observatoire De Besancon, Besancon – FRANCE *lakhlifi@obs-besancon.fr*
Leach Sydney, Observatoire de Paris Meudon – FRANCE *leach@obspm.fr*
Le Bourlot Jacques, Observatoire de Meudon – FRANCE *Jacques.Lebourlot@obspm.fr*
Le Page Valéry, Paris – FRANCE *Vlepage@aol.com*
Lee Jung-Kyu, Univ. of New South Wales Sydney – AUSTRALIA *jklee@phys.unsw.edu.au*
Lemaire Jean Louis, Université de Cergy-Pontoise, Cergy – FRANCE *lemaire@obspm.fr*
Liszt Harvey, NRAO Charlottesville – USA *hliszt@nrao.edu*
Mccarroll Ronald, Univ. P et M Curie Paris – FRANCE *ron@diam.jussieu.fr*

Minh Young Chol, Korea Astron. Observatory Taejon – SOUTH-KOREA
  minh@hanul.issa.re.kr
Miville-Deschenes Marc-Antoine, I.A.S. Orsay – FRANCE *mamd@ias.fr*
Noriega-Crespo Alberto, IPAC Pasadena – USA *alberto@ipac.caltech.edu*
Oosato Takahiro, Ibaraki University Mito – JAPAN *oosato@atlas.riken.go.jp*
Omont Alain, IAP, Paris, FRANCE *omont@iap.fr*
Palla Francesco, Osservatorio Di Arcetri Firenze – ITALY *palla@arcetri.astro.it*
Parneix Pascal, LPPM, Univ. Paris-Sud, Orsay – France
  parneix@venus.ppm.u-psud.fr
Pauzat Francoise, ENS Paris – FRANCE *pauzat@letmex.ens.fr*
Perault Michel, ENS Paris – FRANCE *perault@lra.ens.fr*
Perry James, University College London London – UK *uccajsp@ucl.ac.uk*
Pfenniger Daniel, Univ. of Geneva Sauverny – SWITZERLAND
  daniel.pfenniger@obs.unige.ch
Philippe Laurent, DIAM, Univ. Paris 6, Paris – FRANCE *philippe@diam.jussieu.fr*
Pineau Des Forêts Guillaume, Observatoire de Meudon, FRANCE
  forets@obspm.fr
Pirronello Valerio, Universita' di Catania Catania – ITALY *vpirrone@ing.unict.it*
Puy Denis, Institute of Theoretical Physics Zurich – SWITZERLAND
  puy@physik.unizh.ch
Ramsay Howat Suzanne, Royal Observatory Edinburgh – UK
  skr@roe.ac.uk
Reach William, Caltech Pasadena – USA *reach@ipac.caltech.edu*
Rho Jeonghee, Caltech Pasadena – USA *rho@ipac.caltech.edu*
Richter Philipp, Sternwarte, Bonn – GERMANY *prichter@astro.uni-bonn.de*
Rosenthal Dirk, Max-Planck-Institut Garching – GERMANY *rosenthal@mpe.mpg.de*
Rostas François, Observatoire de Paris-Meudon, Meudon – FRANCE
  Rostas@obspm.fr
Roueff Evelyne, Observatoire de Paris Meudon – FRANCE *roueff@obspm.fr*
Schaefer Joachim, MPI Garching – GERMANY *jas@mpa-garching.mpg.de*
Schermann Catherine, DIAM, Paris 6, Paris – FRANCE *schermann@diam.jussieu.fr*
Shchekinov Yuri, Rostov State University Rostov on Don – RUSSIA
  yus@rsuss1.rnd.runnet.ru
Sidis Victor, LCAM Orsay – FRANCE *sidis@lcam.u-psud.fr*
Snow Theodore, University of Colorado Boulder, CO – USA *tsnow@casa.colorado.edu*
Sorokin Peter, IBM Research Division Yorktown Heights – USA
  sorokin@watson.ibm.com
Spielfiedel Annie, Observatoire de Meudon – FRANCE *Annie.Spielfiedel@obspm.fr*
Szczerba Ryszard, N. Copernicus Astr. Center Torun – POLAND

szczerba@ncac.torun.pl
Talbi Dahbia, ENS Paris – FRANCE *talbi@letmex.ens.fr*
Tappe Achim, Onsala Space Observatory Onsala –SWEDEN *achim@oso.chalmers.se*
Tchang-Brillet Lydia, Observatoire Meudon – FRANCE
 Lydia.Tchang-Brillet@obspm.fr
Tedds Jonathan, University of Leeds,Leeds – UK *jat@ast.leeds.ac.uk*
Tiné Stefano, Observatoire Meudon – FRANCE *stine@obspm.fr*
Ubachs Wim, Vrije Universiteit Amsterdam – NETHERLANDS *wimu@nat.vu.nl*
Usuda Tomonori, Subaru Telescope, NAOJ Hilo –USA *usuda@naoj.org*
Van Den Ancker Mario, Astronomical Inst. Amsterdam –
 NETHERLANDS *mario@astro.uva.nl*
Van Der Werf Paul, Leiden Observatory Leiden – NETHERLANDS
 pvdwerf@strw.leidenuniv.nl
Vannier Laurence, Observatoire de Meudon, Meudon – FRANCE
 laurence.Vannier@obspm.fr
Verstraete Laurent, I.A.S. Orsay – FRANCE *verstra@ias.fr*
Vidal-Madjar Alfred, IAP Paris – FRANCE *alfred@iap.fr*
Vidali Gianfranco, Syracuse University Syracuse – USA *gvidali@syr.edu*
Wesselius Paul, SRON, Groningen – NETHERLANDS *paul@sron.rug.nl*
Wilgenbus David, Observatoire de Paris – FRANCE *wilgenbu@mesioq.obspm.fr*
Williams David, University College London – UK *daw@star.ucl.ac.uk*
Wright Christopher, Univ. College, ADFA, UNSW Canberra –
 AUSTRALIA *wright@ph.adfa.edu.au*

Conference Photograph

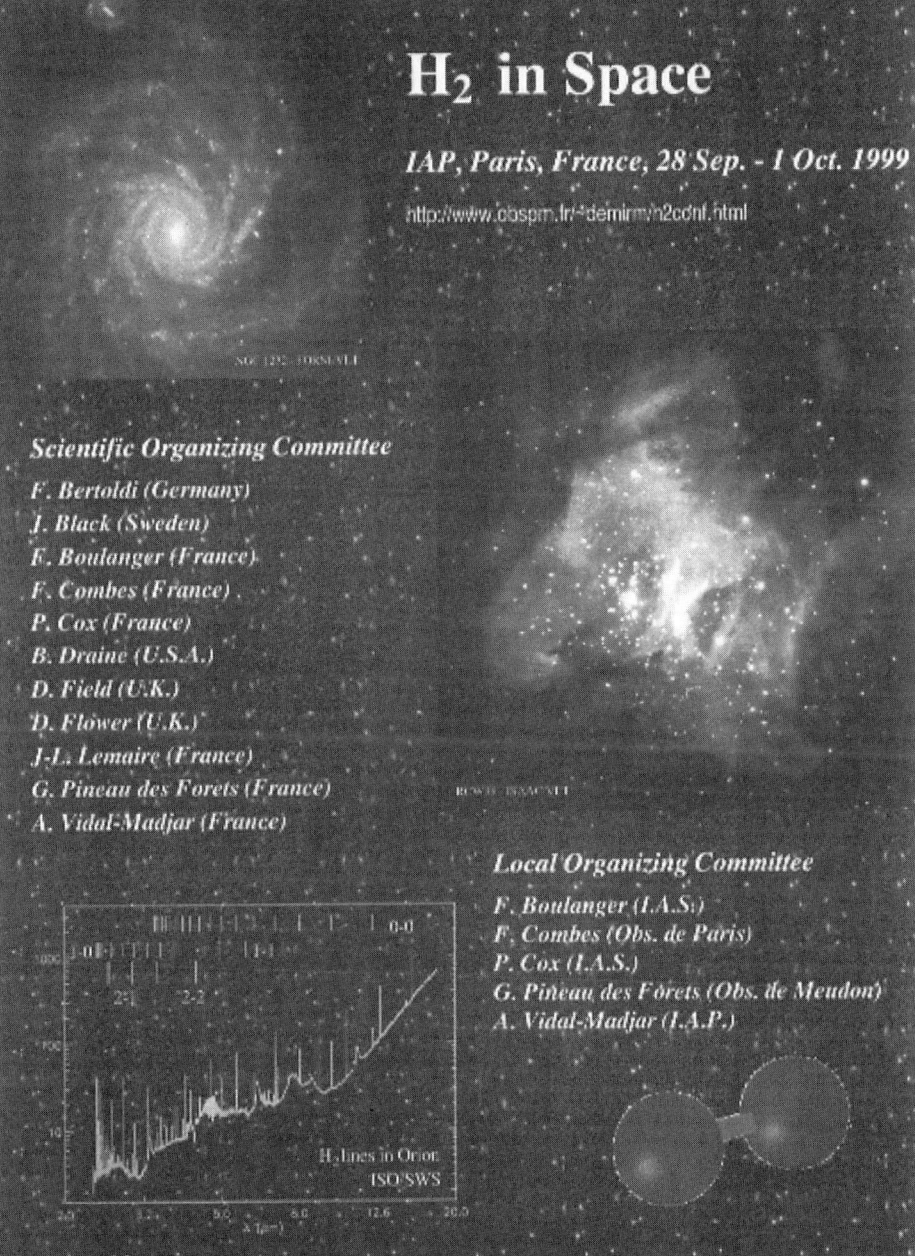

1. Physics of H$_2$ and HD

Michael Jura and Pierre Cox, talking about H_2 in supernovae remnants.

Astrophysical Importance of H_2

By A. Dalgarno

Harvard-Smithsonian Center for Astrophysics, Cambridge, MA 02138, USA

A brief history of observations of H_2 and HD molecules is presented. The properties of H_2 and HD that make observations of them a uniquely powerful diagnostic probe are pointed out. The interpretations of the observations in the ultraviolet, near infrared and infrared made of a diverse range of astrophysical objects are discussed and the influence of H_2 and HD on their evolution is described.

1. Introduction

The hydrogen molecule occupies a central place in astrophysics. There are many reasons. The hydrogen molecule was the first neutral molecule to be formed in the Universe and it played a crucial role in the collapse of the first cosmological objects. It is the most abundant molecular species in the Universe, able to survive in hostile environments and found to exist in diverse astronomical objects ranging from planets to quasars. Hydrogen molecules have been detected by the absorptions and emissions at ultraviolet and infrared wavelengths to which they give rise. Molecular hydrogen has specific radiative and collisional properties that make it a diagnostic probe of unique capability.

2. Ultraviolet absorption

The first detection of H_2 beyond the solar system was accomplished by Carruthers (1970) who employed a rocket-borne ultraviolet spectrometer and detected absorption bands of the Lyman system of H_2 looking towards the star ξ Persei. More extensive data at much higher spectral resolution were obtained with a spectrometer aboard the Copernicus satellite (Spitzer and Jenkins 1975). The resolution was good enough to separate out individual rotational transitions. The resulting rotational populations were characterized by two temperatures, one for the low J levels and one for the high J levels. The two temperatures arise because of the rapid change in the radiative lifetimes of the rotational levels with J and the slow change of the rate coefficients for rotational energy transfer in collisions with H atoms. It happens that at densities characteristic of diffuse clouds the collision times fall between the low and high J radiative lifetimes. Thus the low J population distribution measures the kinetic temperature of the ambient gas and the high J population distribution reflects directly the efficiency with which the levels are populated. The transition between the two populations occurs where the radiative and collision time scales are equal and depends directly on the density.

The levels are populated by ultraviolet pumping. Following the initial absorption into the excited $B^1\sum_u^+$ and $C^1\pi_u$ states, the levels spontaneously radiate back to a distribution of rovibrational levels in the ground electronic state. The corresponding oscillator strengths and spontaneous transition probabilities were calculated by Allison and Dalgarno (1970) and with improved precision by Abgrall et al.(1993). The rovibrational levels then decay by quadrupole radiation cascades into J states of the lowest vibrational level (Black and Dalgarno 1976). Thus the high J population provides a measure of the intensity of the interstellar radiation field that is driving the pumping mechanism. The measured high J populations were used by Black and Dalgarno (1973,1977) and Jura

(1975a,b) to infer the radiation field intensities, often finding intensities in excess of the average field. Ultraviolet pumping produces a transient population of excited vibrational states and an important confirmation of the mechanism was obtained with the successful detection with the Hubble Space Telescope of absorption by H_2 in vibrational level v = 3 (Federman et al.1995).

The radiative decays from the electronically excited $B^1\sum_u^+$ and $C^1\pi_u$ states do not all terminate in discrete vibrational levels. Some fraction end in the vibrational continuum leading to the destruction of the molecule, as pointed out by Solomon and Wickramasinghe (1969) and Stecher and Williams (1967). The specific fractions have been calculated quantum-mechanically by Dalgarno and Stephens (1970) and with improved molecular data by Abgrall et al. (1997).The astrophysical calculations led to the identification of the continuum emissions seen in laboratory spectra (Dalgarno, Herzberg and Stephens 1973).

Because the radiation field intensities are known, so are the photodissociation lifetimes. In equilibrium, formation and destruction proceed at the same rate . Formation in the interstellar gas takes place on the surface of grains (Hollenbach and Salpeter 1971) and Jura (1975a,b) has derived rate coefficients for the grain formation of H_2 that suggest recombination occurs on every collision of an H atom with a grain surface.

Another useful property of H_2 is its existence in ortho and para states corresponding to parallel and antiparallel nuclear spins. If the spins are parallel, J must be odd and if the spins are antiparallel J must be even. Radiative coupling of odd and even J levels is highly improbable. The observed odd and even J level populations for low J appear to be in thermal equilibrium. Exchange collisions of H atoms with H_2 have a substantial barrier and occur too slowly in cold gas to be effective in bringing about equilibrium.Thus the level population data indicate the presence of protons which in contrast can undergo exchange reactions that change the nuclear spins with high efficiency (Dalgarno, Black and Weisheit,1973) Hence the ratio of the $J=1$ and $J=0$ populations provides a reliable estimate of the kinetic temperature but the odd-even ratio tells us little about the formation temperature.

The conclusion that there are protons present in the gas does not establish but strongly indicates the clouds are subjected to ionizing cosmic rays. A more useful indication of cosmic rays is the observed abundance of the deuterated molecule HD. According to Dalgarno, Black and Weisheit (1973) and Watson (1973), HD is mostly produced by the reaction sequence

$$H^+ + D \rightarrow H + D^+$$

$$D^+ + H_2 \rightarrow HD + H^+.$$

The abundance of H^+ is proportional to the cosmic ray ionizing flux so that it is possible to derive from the HD/H_2 ratio the cosmic ray ionization rate (Black and Dalgarno 1973, Hartquist, Black and Dalgarno 1978) provided the D/H ratio is known. Alternatively if the cosmic ray rate is known from say the study of ionic or molecular abundances the ratio of total D to total H can be inferred.

A recent demonstration of the utility of ultraviolet absorption observations is given by the analysis of Orfeus II echelle spectra taken towards the star BD+393226 by Bluhm et al.(1999).

3. Ultraviolet Emission

The radiative transitions out of the $B^1\sum_u^+$ and $C^1\pi_u$ states give rise to spectral emission lines in the ultraviolet between 90 and 170 nm with prominent features around 110 nm and 160 nm (Sternberg 1988). They have been observed coming from the reflection nebula IC 63 (Witt et al. 1989, Hurwitz 1998). They also arise from solar fluorescence in the atmospheres of the Jovian planets though the Jovian spectra contain additional contributions at ultraviolet wavelengths from photoelectron impact excitation of H_2 (Yelle et al.1987, Liu and Dalgarno 1996a).

The ultraviolet photons may modify the chemistry of diffuse interstellar clouds and shocked gas by causing photodissociation and photoionization of other atoms and molecules such as O_2 and H_2O that may be present in the gas, as was pointed out by Prasad and Tarafdar (1983) for the photons generated in interstellar clouds by excitation of H_2 by impacts of the secondary electrons produced by cosmic ray ionization. The electron impact induced spectrum has been calculated by Sternberg, Dalgarno and Lepp (1987) and Gredel et al.(1989). The same spectrum has been observed in Jupiter aurorae (Feldman et al.1993, Trafton et al.1994, Clarke et al.1994, Kim et al.1995) and has led to a determination of the temperature of the emitting gas (Liu and Dalgarno 1996b).

Yet another astrophysically useful attribute of H_2 is its preferential absorption of lines emitted by other atoms,ions and molecules.The photodissociation of CO takes place through line absorption into pre-dissociating states,lines which are partly absorbed by H_2. Thus H_2 acts as a shield limiting the destruction of CO (van Dishoeck and Black 1987). Absorption of Lyman alpha radiation by H_2 from atomic hydrogen has been investigated in detail by Black and Van Dishoeck (1987a). It leads to emissions from specific rovibrational levels of the $B^1 \Sigma_u^+$ state of H_2 which have been detected in sunspots (Jordan et al.1977,1978), in T Tauri and Burnham's nebula (Brown et al.1981) and in Herbig-Haro objects (Schwartz 1983).Emissions stimulated by absorption of Lyman beta are apparent in the Jovian day-glow (Liu and Dalgarno1996a). Near coincidences between absorption wavelengths of H_2 and emission lines of some positive ions leading to fluorescence in the Lyman and Werner systems of H_2 also occur (Feldman and Fastie 1973,Jordan et al.1978). Liu and Dalgarno (1996a) have suggested that HD might be detected in Jupiter from lines pumped by the strong OVI solar line at 103.191 nm which coincides with the Q(3) line of the (1,1) band of the Werner system of HD at 103.186 nm.

4. Infrared Emission

The ground state rovibrational levels populated in the ultraviolet pumping process radiate by electric quadrupole transitions in the infrared and near infrared (Osterbrock 1962, Gould and Harwit 1963). Accurate quantum-mechanical calculations of the spontaneous transition probabilities have been carried out (Turner, Kirby-Docken and Dalgarno 1977, Wolniewicz, Simbotin and Dalgarno 1998). Complete solutions, taking account of the initial absorptions and the subsequent ultraviolet and infrared cascades have been presented by Black and Dalgarno (1976), Black and van Dishoeck (1987b), Sternberg (1988) and Neufeld and Spaans (1997) who predicted the resulting relative intensities of the strongest emission lines in the infrared and near infrared.

The intensity ratios of the lines offer a powerful diagnostic of the condition of the emitting gas. Particularly useful is the ratio of the S(1)(2-1) and S(1)(1-0) intensities, predicted to equal 0.54 (Black and Dalgarno 1976). A value of 0.54 was measured for the reflection nebula Parsamyan 18 by Sellgren (1986) and by Ryder et al. (1998) and for the planetary nebula Hubble 12 by Dinerstein et al.(1988) and Ramsay et al. (1993). The most persuasive example is the reflection nebula NGC 2023 where there is a close match for many lines between theoretical models (Black and van Dishoeck 1987, Sternberg 1989,

Draine and Bertoldi 1996) and observations (Sellgren et al.1992, Burton et al.1992, Field et al.1994,1998, Martini, Sellgren and DePoy 1999) though some discrepancies remain.

Individual rovibrational levels can decay by many radiative paths at an extended range of wavelengths. Because of the reliability of the quantum mechanical predictions any departure of the measured ratios from the theoretical values for any chosen initial rovibrational state is due to selective extinction. An example is the study of Bertoldi et al.(1999).

A characteristic feature of ultraviolet pumping is the relatively strong emission from high vibrational levels and low rotational levels. Extended emission in the v=6-4 transition at 1.601 microns from the Orion A molecular cloud has been used by Luhman et al.(1998) to map the outer boundary of the cloud.

The first detection of infrared emission from H_2 was made by Gauthier et al. (1976) looking towards the Becklin Neugebauer object and the Kleinmann-Low nebula. They attributed the emission lines to thermal excitation in a warm gas heated by shocks. Many other examples of shock-heated gas have been found. The S(1)((2-1) to S(1)(1-0) ratio for a gas at 2000K is 0.13, much less than the value of 0.54 appropriate to ultraviolet pumping.

The clear distinction between thermal gas and irradiated gas becomes lost as the gas density is increased and collisions tend to bring about thermal equilibrium,no matter the nature of the exciting source (Sternberg and Dalgarno 1989). Detailed models for various gas densities and radiation field intensities have been constructed by Sternberg and Dalgarno (1989) which show that the approach to thermal equilibrium occurs at different rates for different levels. Thus relative intensities can be used to infer densities and radiation fields and separate the contributions from fluorescence and collisions. Observations of emissions from external galaxies most often indicate a mixture of the two mechanisms (Mouri 1994, Sugai et al.1997). A diffuse weakly ionized gas subjected to X-rays yields a value of 0.06 (Tiné et al.1997). A value of 0.061 ± 0.029 has been found for the Seyfert galaxy NGC 1275.The same ratio occurs in a thermal gas at 1800K. Observations of additional lines should settle the question.

The models depend on the rate coefficients for rotational and vibrational energy transfer. Considerable progress has been made in recent years to improve the reliability of the cross sections but there remain uncertainties and the data are incomplete (Roueff 2000). Of particular interest is the conversion of vibrational energy into rotational energy, which may contribute to the population of high J states. Quasi-resonant scattering may occur (Stewart et al. 1988, Forrey et al.1999) by which transitions such that $\Delta j = -4\Delta v$ are preferred. Proton impacts will play a major role in ionized gases (Tiné et al. 1997).

A large uncertainty in predicting the infrared response to ultraviolet radiation is the contribution to the level populations by the formation mechanism. Because the rate of formation equals the rate of destruction and the destruction efficiency is 0.11 ultraviolet pumping is nine times more effective in populating the rovibrational levels. However the formation mechanism may be concentrated into specific levels. Burton et al.(1992) have suggested that the influence of formation pumping may be manifested in NGC 2023.

It may prove possible to infer the formation population rates from observations. In a dense cloud formation and pumping occur at similar rates (Le Bourlot et al.1995), as they do in clouds subjected to radiation by X-ray sources (Gredel and Dalgarno 1995,Tine et al. 1997). In dissociative shocks (Hollenbach and McKee 1989, Neufeld and Dalgarno 1989) H_2 is formed in the post-shock relaxation layer and recombination on the grain surfaces may dominate the population of the rovibrational levels. There will also be a contribution from associative detachment

$$H + H^- \to H_2 + e$$

which produces H_2 in an extended distribution of J and v levels (Bieniek and Dalgarno 1979, Bieniek 1980, Launay, Le Dourneuf and Zeippen 1991). The contribution of associative detachment to infrared emission in the planetary nebula NGC 7027 has been considered by Black, Porter and Dalgarno (1981).

Pure rotational transitions of H_2 have been observed in a variety of sources (Beck, Lacy and Geballe 1979, Parmar, Lacy and Achtermann 1994) and recently by Valentijn et al.(1996), Sturm et al.(1996), Kunze et al.(1996), Moorwood et al.(1996), Wesselius et al.(1996), Wright et al. (1996) and Timmerman et al.(1996) using the ISO satellite. Because of the long radiative lifetimes they are usually indicators of the gas kinetic temperature but care is needed in interpreting the data. Sternberg and Neufeld (1999) have pointed out that the apparent departure of the ortho-para ratio from its equilibrium value can occur because of optical depth effects on the pumping.

Infrared emission from the deuterated molecule HD has been investigated by Sternberg (1990) for ultraviolet pumping and by Timmermann (1996) for shocks and has now been detected by Wright et al. (1999) and Bertoldi et al.(1999). Because the radiative lifetimes are much reduced below those of H_2, collisions are less effective in modifying the populations. The observations are important sources of information on the cosmic [D]/[H] ratio and will be particularly instructive if the emissions can be detected in external galaxies.

5. Infrared absorption

Infrared absorption by quadrupole transitions by H_2 was observed in the interstellar gas by Lacy et al. (1994) and by Black and Kulesa (2000). In Jupiter, Saturn, Uranus and Neptune, H_2 was discovered by quadrupole absorption and by collision-induced absorption (Hanel et al. 1979). More recently dipole absorption measurements of HD on Jupiter and Saturn with ISO have been successful (Griffin et al.1996, Encrenaz et al.1996, Davis et al.1996).

6. Chemistry

The hydrogen molecule is of fundamental importance in the evolution of the Universe because of its high efficiency as a coolant compared to a hydrogen atom. It is formed in a warm dust free plasma by reactions initiated by He^+, H^+ and H^-. HD is formed by similar reactions and by reactions of D and D^+ with hydrogen atoms and molecules (Galli and Palla 1998, Stancil, Lepp and Dalgarno 1998) and cooling through collisional excitation of HD is effective at low temperatures (Dalgarno and Wright 1972, Puy et al.1993).In the collapse of the first cosmological objects, the abundance of H_2 was enhanced by three body recombination as the density increased (Palla, Salpeter and Stahler 1983). The response of H_2 to the ultraviolet radiation introduced into the Universe by the collapsed objects depended on the shape of the ultraviolet spectrum. Ionization initially enhances the abundance of H_2 whereas photodissociation diminishes it. Which occurred substantially influenced the further evolution of the Universe and the formation of galaxies. There have been many studies (Haiman, Thoul and Loeb 1996, Tegmark et al. 1997, Haiman, Rees and Loeb 1996, 1997, Parravano and Pech 1997, Ferrara 1998, Omukai and Nishi 1998, Abel et al. 1998, Brown, Coppi and Larson 1999, Nakamura and Umemura 1999).

The formation of H_2 on grains in interstellar regions initiated the chemistry of interstellar clouds. The H_2^+ ions produced by cosmic ray ionization of H_2 reacted with H_2 to create H_3^+, recently detected in the interstellar gas by infrared absorption (Geballe and Oka 1996, McCall et al. 1998, Geballe et al.1999), and the H_3^+ ions participated in proton transfer reactions with neutral atoms and molecules, leading to a rich chemistry and an array of complex molecules. Without the grain formation of H_2 the molecular composition would have been severely limited in abundance and variety. Molecular formation would have been initiated by radiative association of heavy element atoms and ions with hydrogen atoms. Some H_2 would emerge as products of sequences such as

$$C^+ + H \to CH^+ + \nu$$

$$CH^+ + H \to C^+ + H_2.$$

7. Conclusions

Molecular hydrogen is the simplest of molecules but it has complexity enough to serve as a probe of a wide range of physical environments. Because of its radiative and collision properties which we understand reasonably well we can construct realistic models of the response of H_2 to its surroundings. H_2 is found in nearly every corner of the Universe and it touches most of the important questions in astronomy. Because of its chemistry H_2 is central to the evolution of the early Universe and the formation of galaxies and in the interstellar gas the chemistry beginning with the formation of H_2 controls the ionization and thermal balance and the mechanisms of star formation.

Perhaps of equal potential importance to the discovery of H_2 in different environments is the possible existence of large amounts of undetected H_2. H_2 may go unseen because its usual signature CO is under-abundant or because the gas is too cold for excitation to occur. Pfenniger et al. (1994) and Combes and Pfenniger (1997) have proposed that a substantial fraction of dark matter is cold molecular hydrogen.

This work was supported by the National Science Foundation, Division of Astronomy, and by the National Aeronautics and Space Administration under grant NAG5-4986.

REFERENCES

ABEL, T., ANNINOS, P., NORMAN, M.L. & ZHANG, Y. 1998 *ApJ* **508**, 51.

ABGRALL, H., ROUEFF, E., LIU, X. & SHEMANSKY, D.E. 1997 *ApJ* **481**, 557.

ABGRALL, H., ROUEFF, E., LAUNAY, F., RONCIN, J.Y. & SUBBIL, J.L. 1993 *A&AS* **101**, 273 & **101**, 323.

ALLISON, A.C. & DALGARNO, A. 1970 *Atom. Data Tables* **1** 289.

BERTOLDI, F., TIMMERMAN, R., ROSENTHAL, D., DRAPATZ, S. & WRIGHT, C.M. 1999 *A&A* **346**, 267.

BECK, S.C., LACY, J.H. & GEBALLE, T.R.1979 *ApJ* **234**, L213.

BIENIEK, R.J. 1980 *J. Phys. B.* **13**, 4405.

BLACK, J.H. & DALGARNO, A. 1973 *ApJ* **184**, L101.

BLACK, J.H. & DALGARNO, A. 1976 *ApJ* **203**, 132.

BLACK, J.H. & DALGARNO, A. 1977 *ApJS* **34**, 405.

BLACK J.H. & KULESA, C. 2000 in preparation.

BLACK, J.H. & VAN DISHOECK, E.F. 1987 *ApJ* **322**, 412.

BLUHM, H., MARGGRAF, O., DE BOER, K.S. RICHTER, P. & HEBER, U. 1999, *A&A* **352, 287**.

BROWN, V., COPPI, P.S. & LARSON, R.B. 1999, it ApJLin press.
BROWN, A., JORDAN, C., MILLAR, T.J., GONDHALEKAR, P. & WILSON, R., 1981 *Nature* **290**, 341.
BURTON, M.G., HOLLENBACH, D.J. & TIELENS, A.G.G.M. 1992 *ApJ* **399**, 563.
CARRUTHERS, G.R. 1970 *ApJ* **161**, L81.
CLARKE, J.T., BEN JAFFEL, L., VIDAL-MADJAR, A., GLADSTONE, G.R., WAITE, J.H., PRANGÉ, R., GÉRARD, J-C. & AJELLO, J. 1994 *ApJ* **430**, L73.
COMBES, F. & PFENNIGER, D. 1997 *A&A* **327**, 453.
DALGARNO, A. & STEPHENS, T.L. 1970 *ApJ* **160**, L107.
DALGARNO, A. & WRIGHT, E.L. 1972 *ApJ* **174**, L49.
DAVIS, G.R. ET AL.1996 *A&A* **315**, L393.
DINERSTEIN, H.L., LESTER, D.F., CARR, J.S. & HARVEY, P.M. 1988 *ApJ* **327**, L27.
DRAINE, B.T. & BERTOLDI, F. 1996 *ApJ* **468**, 269.
ENCRENAZ, TH. ET AL. 1996 *A&A* **315**, L397.
FEDERMAN, S.R., CARDELLI, J.A., VAN DISHOECK, E.F., LAMBERT, D.L. & BLACK, J.H. 1995 *ApJ* **445**, 325.
FELDMAN, P. & FASTI, W.G. 1973 *ApJ* **185**, L101.
FELDMAN, P.D., MCGRATH, M.A., MOOS, H.W., DURRANCE, S.T., STROBEL, D.F. & DAVIDSEN, A.F. 1993 *ApJ* **406**, 279.
FERRARA, A. 1998 *ApJ* **499**, L17.
FIELD, D., GERIN, M., LEACH, S., LEMAITRE, J.L., PINEAU DES FORÊTS, G., ROSTAS, F. ROWAN, D. & SUNOWS, D. 1994 *A&A* **286**, 909.
FIELD, D., LEMAITRE, J.L., PINEAU DES FORÊTS, G., GERIN, M., LEACH, S., ROSTAS, F. & ROWAN, D. 1998 *A&A* **333**, 280.
FORREY, R.C., BALAKRISHNAN, N., DALGARNO, A., HEGGERT, M.R. & HELLER, E.J. 1999 *Phys.Rev.Lett* **82**, L2657.
GALLI, D. & PALLA, F. 1998 *A&A* **335**, 403.
GAUTHIER, T.N., FINK, V., TREFFERS, R.R. & LARSON, H.P. 1976 *ApJ* **207**, L129.
GEBALLE, T.R. & OKA, T. 1996 *Nature* **384**, 334.
GEBALLE, T.R., MCCALL, B.J., HINKLE, K.H. & OKA, T. 1999 *ApJ* **510**, 251.
GOULD, R.J. & HARWIT, M. 1963 *ApJ* **137**, L694.
GREDEL, R., S. LEPP, DALGARNO, A. & HERBST, E. 1989 *ApJ* **347**, 289.
GREDEL, R. & DALGARNO, A. 1995 *ApJ* **446**, 852.
GRIFFIN, M.J. ET AL. 1996 *A&A* **315**, L389.
HAIMAN, Z., REES, M.J. & LOEB, A. 1997,*ApJ* **476**, 458 & **484**, 985.
HAIMAN, Z., REES, M.J. & LOEB, A. 1996 *ApJ* **467**, 522.
HAIMAN, Z., THOUL, A. & LOEB, A. 1996, *ApJ* **464**, 323.
HANEL, R. ET AL. 1979 *Science* **204**, 972.
HARTQUIST, T.W., BLACK, J.H. & DALGARNO, A. 1978 *MNRAS* **186**, 643.
HOLLENBACH, D. & MCKEE, C.F. 1989 *ApJ* **342**, 306.
HURWITZ, M. 1998 *ApJ* **500**, L67.
JORDAN, C., BRUECKNER, G.E., BARTOE, J-D.F., SANDLIN, G.D. & VAN HOOSIER, M.E. 1977 *ApJ* **226**, 687.
JORDAN, C., BRUECKNER, G.E., BARTOE, J-D.F., SANDLIN, G.D. & VAN HOOSIER, M.E. 1977 *Nature* **270**, 326.
JURA, M. 1975a *ApJ* **197**, 375.
JURA, M. 1975b *ApJ* **197**, 575.
KIM, Y.H., CALDWELL, J.J. & FOX, J.L. 1995 *ApJ* **447**, 906.

KUNZE, D. ET AL. 1996 *A&A* **315**, L101.
LACY, J.H., KNACKE, R., GEBALLE, T.R. & TOKUNAGA, A.T. 1994 *ApJ* **428**, L69.
LAUNAY, J.M., LEDOURNEUF, M. & ZEIPPEN, C.J. 1991 *A&A* **252**, 842.
LE BOURLOT, J., PINEAU DES FORÊTS, G., ROUEFF, E., DALGARNO, A. & GREDEL, R. 1995 *ApJ* **449**, 178.
LIU, W. & DALGARNO, A. 1996a *ApJ* **462**, 502.
LIU, W. & DALGARNO, A. 1996b *ApJ* **467**, 446.
LUHMAN, K.L., ENGELBRACHT, C.W. & LUHMAN, M.L. 1998 *ApJ* **499**, 479.
MARTINI, P., SELLGREN, K. & DEPOY, D.L. 1999 *ApJ* **526**, 772.
MCCALL, B.J., GEBALLE, T.R., HINKLE, K.H. & OKA. T. 1998 *Science* **279**, 1910.
MOORWOOD, A.F.M. ET AL. 1996 *A&A* **315**, L109.
MOURI, H. 1994 *ApJ* **427**, 777.
NAKAMURA, F. & UMEMURA 1999 *ApJ* **515**, 239.
NEUFELD, D.A. & DALGARNO, A. 1989 ApJ **340**,869.
NEUFELD, D.A. & SPAANS, M. 1997 *ApJ* **473**, 894.
OMUKAI, K. & NISHI, R. 1998 *ApJ* **508**, 141.
OSTERBROCK, D 1962 *ApJ* **136**, 359.
PALLA, D., SALPETER, E.E. & STAHLER, S.W. 1983 *ApJ* **271**, 632.
PARRAVANO, A. & PECH, C. 1997 *A&A* **327**, 1262.
PARMAR, P.S., LACY, J.H. & ACHTERMANN, J.M. 1994 *ApJ* **430**, 786.
PFENNIGER, D., COMBES, F. & MARTINET, L. 1994 *A&A* **285**, 79.
PRASAD, S.S. & TARAFDAR, S.P. 1983 *ApJ* **267**, 603.
PUY, D., GRENACHER, L. & JETZER, P. 1999 *A&A* **345**, 723.
PUY, D., ALECIAN, G., LE BOURLOT, J., LÉORAT, J. & PINEAU DES FORÊTS, G. 1993 *A&A* **267**, 337.
RAMSAY, S.K., CHRYSOSTOMOU, A., GEBALLE, T.R., BRAND, F.W.J., MOUNTAIN, M. 1993 *MNRAS* **263**, 695.
ROUEFF, E. 2000 *this volume*.
RYDER, S.D., ALLEN, L.E., BURTON, M.G., ASHLEY, M.C.B. & STOREY, J.W.V. 1998 *MNRAS* **294**, 338.
SCHWARTZ, R.D. 1983 *ApJ* **268**, L37.
SELLGREN, K., WERNER, M.W. & DINERSTEIN, H.L 1992 *ApJ* **400**, 238.
SOLOMON, P.M. & WICKRAMASINGHE, N.C. 1969*ApJ* **158**, 449.
SPITZER, L. & JENKINS, E.B. 1975 *ARAA* **13**, 133.
STANCIL, P.C., LEPP, S. & DALGARNO, A. 1998 *ApJ* **409**, 1.
STECHER, T.P. & WILLIAMS, D.A. 1967 *ApJ* **149**, L29.
STERNBERG & A., DALGARNO 1989 *ApJ* **338**,197.
STERNBERG, A. & NEUFELD, D.A. 1999 *ApJ* **516**, 371.
STERNBERG, A., DALGARNO, A. & LEPP, S. 1987 *ApJ* **320**, 676.
STERNBERG, A. 1989 *ApJ* **347**, 863.
STERNBERG, A. 1990 *ApJ* **361**, 121.
STEWART, B., MAGILL, P.D., SCOTT, T.P., DEROUARD, J. & PRITCHARD, D.E. 1988 *Phys. Rev. Lett.* **60**, 282.
STURM, E., ET AL. 1996 *A&A* **315**, L133.
SUGAI, H., MALKAN, M., WARD, M.J., DAVIES, R.J. & MCLEAN, J.S. 1997 *ApJ* **481**, 186.
TEGMARK, M., SILK, J., REES, M.J., BLANCHARD, A., , T. & PALLA, F. 1997*ApJ* **474**,1.
TIMMERMAN, R. ET AL. 1996 *A&A* **315**, L281.

TINÉ, S., LEPP, S., GREDEL, R. & DALGARNO, A. 1997 *ApJ* **481**, 282.
TRAFTON, L.M., GÉRARD, J.C., MUNHOVEN, G. & WAITE, J.H. 1994 *ApJ* **421**, 816.
TURNER, J., KIRBY-DOCKEN, K. & DALGARNO, A. 1977 *ApJS* **35**, 281.
VALENTIJN, E.A. ET AL. 1996 *A&A* **315**, L145.
VAN DISHOECK, E.F. & BLACK, J.H. 1989 *ApJ* **334**, 771.
WESSELIUS, P.R. ET AL. 1996 *A&A* **315**, L197.
WITT, A.N., STECHER, T.P. BOROSON, T.A. & BOHLIN, R.C. 1989 *ApJ* **336**, L21.
WOLNIEWICZ, L., SIMBOTIN, I. & DALGARNO, A. 1998 *ApJS* **115**, 293.
WRIGHT, C.M., VAN DISHOECK, E.F., COX, P., SIDHER, S. & KESSLER, M.F. 1999 *ApJ* in press.
WRIGHT, C.M. ET AL. 1996 *A&A* **315**, L301.
YELLE, R.V., MCCONNELL, J.C., SANDEL, B.R. & BROADFOOT, A.L. 1987 *J. Geophys. Res.* **92**, 15110.

Bruce Draine illustrating the H_2 excitation for Alex Dalgarno.

Radiative and electronic excitation of Lyman and Werner transitions in H_2

By Evelyne Roueff[1], Hervé Abgrall[1], Xianming Liu[2], and Don Shemansky[2]

[1]Departement d'Astrophysique Extragalactique et de Cosmologie & UMR 8631 du CNRS, Observatoire de Paris-Meudon 92190 Meudon, France,

[2]Department of Aerospace and Mechanical Engineering, University of Southern California, Los Angeles, CA 90089, USA

We describe the present status of knowledge in the first excited electronic states of H_2 which are radiatively connected to the ground state. The astrophysical significance of radiative and electronic excitation of H_2 is emphasized. Future needs and perspectives are underlined.

1. Introduction

Molecular hydrogen, a two electrons system, could be considered as the simplest molecular system from the theoretical point of view and can thus serve as a pertinent test of the fundamental basic quantum mechanics. We stress out the different peculiarities of this molecule:

• the Coulomb degeneracy of the infinite separation system renders the electronic potential curves pattern very intricate,

• H_2 is composed by the lightest nuclei; then the non adiabatic effects, which are inversely proportional to the reduced mass of the system become important and lead to the coupling between nuclear and electronic motions,

• and finally, it is a symmetric molecule without permanent dipole moment. An additional symmetry i.e., the inversion of the electrons respective to the center of mass is generated; the electronic states become sorted according to their g and u symmetries. The protons have a nuclear spin of 1/2 so that the total (*nuclear* × *electronic*) wave function is antisymmetric relative to the permutation of nuclei which obey to Fermi statistics. Ortho nuclear states correspond to a symmetric nuclear spin wavefunction (total spin I = 1) and an antisymmetric spatial wavefunction involving odd values of J, the rotational quantum number when the electronic state is totally symmetric as in the $^1\Sigma_g^+$ ground state. Inversely, para states involve then an antisymmetric nuclear spin wavefunction (total spin I = 0) and a symmetric spatial wavefunctions with even values of J. The degeneracy of para and ortho levels is then 2J+1 and $3 \times (2J + 1)$ respectively. Radiative ortho - para transitions are not allowed unless very small relativistic terms in the matter-radiation interaction are taken into account. The rate is about 10^{-12} yr^{-1} and thus of no astrophysical significance (Raich and Good (1964), Dodelson (1986)).

Experimental studies of molecular hydrogen are equally difficult. The absorption electronic spectroscopy has to be performed in a vacuum ultraviolet region since the $^1\Sigma_g^+$ ground state is radiatively linked to $^1\Sigma_u^+$ and $^1\Pi_u$ states, i.e. B and C are the first levels attainable in absorption experiments.

Due to the large number of electronic states, electronic emission spectroscopy can occur from the infrared to the vacuum ultraviolet wavelength region. No obvious regularities arise in the corresponding spectra due to the large non adiabatic effects already mentioned. Finally, no electric dipole transitions are taking place within the ground state due to the symmetry of the molecule. Then rovibrational spectroscopy can only be performed via Raman studies or involves electric quadrupole transitions with $\Delta J = 0$, ± 2.

We also mention the deuterated substitutes of H_2, HD and D_2. As the inversion symmetry of the electrons respective to the center of charge is preserved at first approximation, the exchange of nuclei is not feasible in HD. So, as the center of mass and the center of charge are no longer coincident, a small permanent dipole moment is present in HD and rovibrational electric dipole transitions may occur within the ground electronic potential curve. These electric dipole transition probabilities are larger than the electric quadrupole transition probabilities (Abgrall, Roueff and Viala 1982). For D_2, the non adiabatic effects are slightly reduced compared to H_2 due to the larger reduced mass. Ortho and para states are also available with nuclear degeneracy respectively equal to 6 and 3 since deuteron is a boson particle of nuclear spin equal to 1. Now, even (odd) values of the rotational levels within the ground electronic states correspond to ortho (para) states.

After the development of some theoretical considerations in section 2, we describe radiative and electronic excitation studies in section 3 and 4. We finally give our conclusion and perspectives.

2. Theoretical considerations

2.1. The different point of views

Pure theoreticians are only concerned by the ab-initio approach: in this case, the goal is to calculate the electronic and nuclear molecular properties from basic quantum theory. This approach implies to build the molecular potentials by considering *all* contributions to the Hamiltonian. Then, the calculation of non adiabatic couplings between nuclear and electronic motions is performed and finally, the resolution of the coupled nuclear Schrödinger equation allows to obtain an accuracy of about 5 wavenumbers on the energy terms. The most prominent studies in this field are from Kolos, Wolniewicz and Dressler.

In astrophysical or geophysical observations, the spectral information involves both the wavelengths positions and the observed intensities. Then, one can determine the molecular column densities and physical conditions when the radiative transfer has been performed. Such calculations require the knowledge of the oscillator strengths, or equivalently the Einstein A coefficients via the matrix elements of the electronic transition moments and the collisional excitation rates by the various partners present in the considered environment.

As far as we are concerned, we want to improve the accuracy on the wavenumbers since high spectral resolution is achieved in recent experiments and observations. We allow us to use slightly adjusted molecular potential curves for the excited spectroscopic levels which are *not* affected by rotational couplings. Moreover, we use the experimental ground state level positions to derive the transition wavenumbers. Solving the coupled equations between excited states, we get both eigenvalues and eigenfunctions. The resulting absolute accuracy on the transition wavenumbers becomes less than 0.5 cm^{-1}. However, the precision on the rovibronic transition matrix elements, then on the oscillator strengths or Einstein A coefficients is more difficult to evaluate. When the emission transition probabilities are large, the relative accuracy is about a few per cent. However, the uncertainties are highly enhanced for low values of the emission transition probabilities ($A \leq 10^5$ s^{-1}) when destructive interference may take place in the calculations of the rovibronic transition matrix elements. Detailed comparison between spectroscopic studies including the measurement of intensities is then determinant to conclude.

	$B\ ^1\Sigma_u^+$, $v_1 J$	$C\ ^1\Pi_u^+$, $v_2 J$	$B'\ ^1\Sigma_u^+$, $v_3 J$	$D\ ^1\Pi_u^+$, $v_4 J$
$B\ ^1\Sigma_u^+$, $v_1 J$	adiabatic potential $+ \frac{\hbar^2[J(J+1)]}{2\mu\ R^2}$	rotational coupling	radial coupling	rotational coupling
$C\ ^1\Pi_u^+$, $v_2 J$	rotational coupling	adiabatic potential $+ \frac{\hbar^2[J(J+1)-1]}{2\mu\ R^2}$	rotational coupling	radial coupling
$B'\ ^1\Sigma_u^+$, $v_3 J$	radial coupling	rotational coupling	adiabatic potential $+ \frac{\hbar^2[J(J+1)]}{2\mu\ R^2}$	rotational coupling
$D\ ^1\Pi_u^+$, $v_4 J$	rotational coupling	radial coupling	rotational coupling	adiabatic potential $+ \frac{\hbar^2[J(J+1)-1]}{2\mu\ R^2}$

TABLE 1. Hamiltonian matrix between the four B, C^+, B' and D^+ excited states in H_2 (or D_2).

2.2. Chronology of theoretical treatments in the calculations of transition probabilities for the Lyman and Werner transitions in H_2

We do not describe the extensive theoretical studies devoted to H_2 and will mainly restrict to the papers which are most often referred in the astrophysical and geophysical literature. A decisive work has been performed in 1970 by Allison & Dalgarno (1970) who calculated all band transition probabilities of H_2, HD and D_2 with ab-initio values of potential curves and transition moments. In this paper, the authors solve the one dimension Schrödinger nuclear equation without rotation (i.e. no centrifugal barrier) for the excited and ground electronic states. Such a treatment assumes the separation of the electronic, vibrational and rotational motions. The transition probabilities involving specific rotational quantum numbers are then obtained by multiplying the J-independent transition probabilities by the Hönl London factors reported in all molecular textbooks (cf Herzberg 1989). Abgrall et al. (1987) considered the effects of centrifugal potential and rovibronic coupling between the B and C^+ electronic states which are radiatively connected to the ground state via $\Delta J = +1$ (R) and $\Delta J = -1$ (P) transitions. In this case, the C^- state becomes uncoupled and radiatively connected to the ground state via $\Delta J = 0$ (Q) transitions. The wavelengths and transition probabilities of the Lyman and Werner band systems of H_2 have been calculated up to $J = 10$ by Abgrall and Roueff (1989) and an application to the atomic to molecular transition in interstellar clouds has been performed (Abgrall et al. 1992). Since 1993, the same authors (Abgrall et al. 1993a,b,c, 1994) consider the rovibronic coupling between the four B, B', C and D electronic excited states. Rotational coupling splits the C and D $^1\Pi_u$ potentials into $^1\Pi_u^+$ and $^1\Pi_u^-$. Tables 1 and 2 display the Hamiltonian matrix for the two symmetries.

The diagonal terms include here the centrifugal barrier; μ is the reduced mass of the molecule and R is the internuclear distance. We note that there is no change in the rotational quantum number involved. There are no levels with $J = 0$ for C and D; then only radial coupling remains between B and B' when $J = 0$. This property is taken into account when fitting the electronic potential curves. There is no coupling with the g states. We assume here that spin orbit coupling is negligible. Figure 1 shows the mentioned electronic potential curves. The resolution of the corresponding Schrödinger equations for the bound states within the procedure already described for deriving the

	$C\ {}^1\Pi_u^-,\ v_1 J$	$D\ {}^1\Pi_u^-,\ v_2 J$
$C\ {}^1\Pi_u^-,\ v_1 J$	adiabatic potential $+\ \frac{\hbar^2[J(J+1)-1]}{2\mu R^2}$	radial coupling
$D\ {}^1\Pi_u^-,\ v_2 J$	radial coupling	adiabatic potential $+\ \frac{\hbar^2[J(J+1)-1]}{2\mu R^2}$

TABLE 2. Hamiltonian matrix between C^- and D^- excited states in H_2 (or D_2).

potential curves has allowed a detailed comparison with emission spectra recorded with the 10 meter VUV spectrograph at Meudon Observatory where both positions and intensities have been analyzed.

2.3. Expressions of the radiative properties

We recall now the various expressions involving the electronic transition matrix elements. The spontaneous emission probability is given by the expression:

$$A(v_j, v_i; J_j, J_i) = \frac{4}{3\hbar^4 c^3 (2J_j+1)} (E_{v_j J_j} - E_{v_i J_i})^3 |M_{S\alpha}|^2 \quad (2.1)$$

where E_{vJ} is the energy of the level (v,J), $M_{S\alpha}$ is the electric dipole matrix element between wavefunctions of excited $S(v_j, J_j)$ and ground electronic $X(v_i, J_i)$ states, and α indicates whether the spectroscopic branch label is P, Q or R. When the emission takes place into the X continuum states, the expression is modified as:

$$A(v_j, e_i; J_j, J_i) = \frac{4}{3\hbar^4 c^3 (2J_j+1)} (E_{v_j J_j} - E_{e_i J_i})^3 |M_{S\alpha}|^2 \quad (2.2)$$

where the kinetic energy of dissociating atoms e_i replaces the vibrational index v_i. The total part of the emission which produces dissociation is:

$$A_c(v_j; J_j) = \sum_{J_i} \int_0^\infty A(v_j, e_i; J_j, J_i) de_i \quad (2.3)$$

The mean kinetic energy of dissociating products is defined by:

$$\bar{E}_k(v_j; J_j) = \frac{\sum_{J_i} \int_0^\infty e_i A(v_j, e_i; J_j, J_i) de_i}{A_c(v_j; J_j)} \quad (2.4)$$

The total emission probability is:

$$A_t(v_j; J_j) = A_c(v_j; J_j) + \sum_{v_i, J_i} A(v_j, v_i; J_j, J_i) \quad (2.5)$$

When absorption spectra are concerned, oscillator strengths become the convenient quantities. Their expression is the following:

$$f(v_i, v_j; J_i, J_j) = \frac{2m}{3\hbar^2 e^2 (2J_i+1)} (E_{v_j J_j} - E_{v_i J_i}) |M_{S\alpha}|^2 \quad (2.6)$$

To go beyond the adiabatic approximation, the rovibronic wavefunction of excited states is expressed as an expansion over the electronic Born Oppenheimer (B.O.) wavefunctions. It is a good approximation to limit the expansion to the 4 B, C, B' and

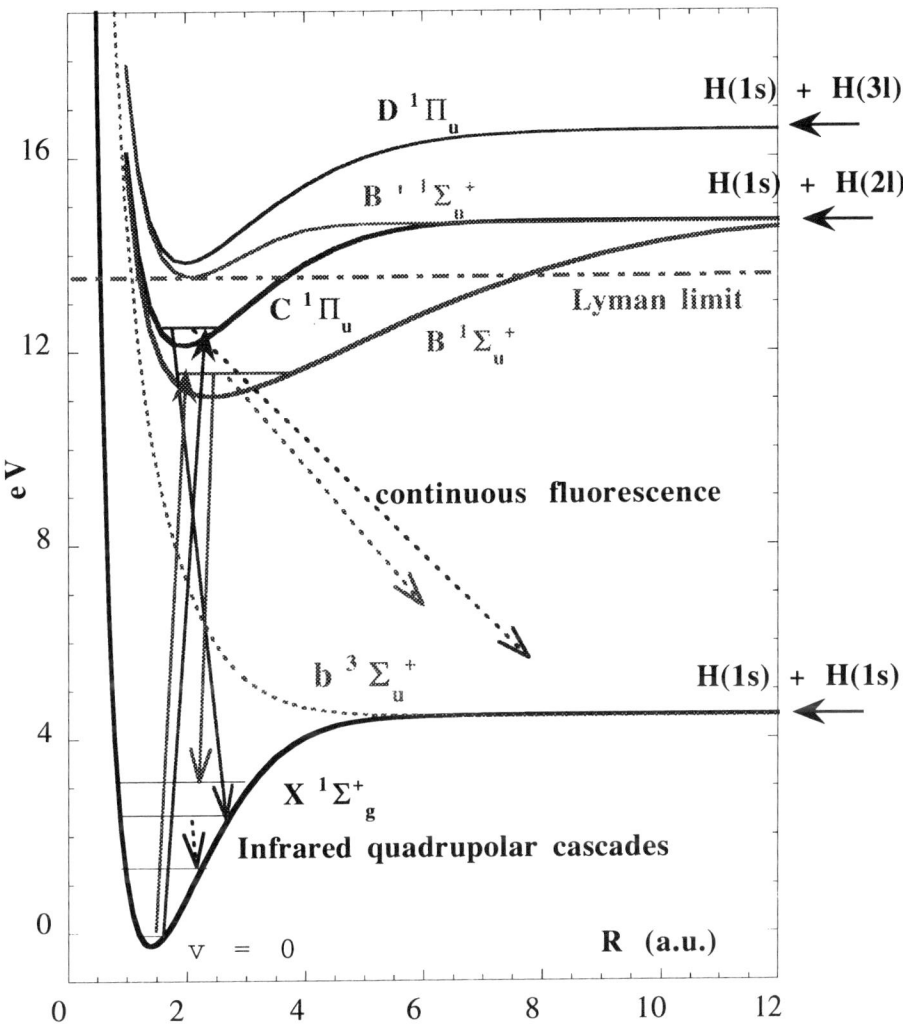

FIGURE 1. Potential curves involved in the Lyman and Werner systems in H_2.

D states and $\Phi_{SvJ} = \sum_T \Psi_{TJ} f_{STvJ}$ where S, T are dummy labels for B, C, B' and D. Each Ψ is the product of the electronic B.O. wavefunction and the pure rotational nuclear wavefunction. The vibrational functions f_{STvJ} and the energy levels $E_{v_j J_j}$ are obtained by searching the eigenvalues of the coupled equations whose matrix is displayed in Tables 1 and 2. The formalism is described in detail by Senn, Quadrelli and Dressler (1988).

One can define

$$\rho(T) = \int (f_{STvJ}(R))^2 dR \tag{2.7}$$

as the fraction of T in the state SvJ (the normalisation is such that $\rho(B)+\rho(C)+\rho(B')+\rho(D) = 1$). Except for very few cases, one value of T is preponderant and by convention, the label S stands for the state of greatest weight.

The dipole matrix elements $M_{S\alpha}$ involved in eq. (2) are given by the following expressions † :

$$M_{SP} = (J_j+1)^{1/2}\{\langle f_{Bv_jJ_j}|M_{BX}|f_{Xv_iJ_i}\rangle + \langle f_{B'v_jJ_j}|M_{B'X}|f_{Xv_iJ_i}\rangle\}$$
$$+ J_j^{1/2}\{\langle f_{C+v_jJ_j}|M_{CX}|f_{Xv_iJ_i}\rangle + \langle f_{D+v_jJ_j}|M_{DX}|f_{Xv_iJ_i}\rangle\}$$

$$M_{SQ} = (2J_j+1)^{1/2}\{\langle f_{C-v_jJ_j}|M_{CX}|f_{Xv_iJ_i}\rangle + \langle f_{D-v_jJ_j}|M_{DX}|f_{Xv_iJ_i}\rangle\}$$

$$M_{SR} =$$
$$J_j^{1/2}\{\langle f_{Bv_jJ_j}|M_{BX}|f_{Xv_iJ_i}\rangle + \langle f_{B'v_jJ_j}|M_{B'X}|f_{Xv_iJ_i}\rangle\} -$$
$$(J_j+1)^{1/2}\{\langle f_{C+v_jJ_j}|M_{CX}|f_{Xv_iJ_i}\rangle + \langle f_{D+v_jJ_j}|M_{DX}|f_{Xv_iJ_i}\rangle\}$$

were M_{BX}, M_{CX}, $M_{B'X}$ and M_{DX} are the real values of electronic transition moments calculated for each internuclear distance in the B.O. approximation. The signs of the electronic moments are chosen to be consistent with those of the matrix elements of Wolnewicz and Dressler (1988)

3. Radiative excitation

Figure 1 also displays the absorption and fluorescence processes occuring in interstellar photon dominated regions. The energy of the incident photons are less than 13.6 eV, so that since most hydrogen molecules are in their ground vibrational state, photodissociation can only take place via fluorescence towards the continuum of the ground electronic state. This mechanism has first been suggested by Solomon (1970). Such continuous fluorescent emission has indeed been detected towards IC63 by Witt et al. (1989) by the IUE satellite between 120 and 180 nm. Recent observations at shorter wavelengths in the discrete spectrum via the ORFEUS II mission (Hurwitz 1998) result in an absolute intensity fainter than an extrapolation from the previous observations by a factor of 10. The higher extinction at shorter wavelengths is probably the origin of the discrepancy but the question remains open. The hydrogen atoms are ejected with some kinetic energy in the dissociation process, a heating mechanism (Stephens and Dalgarno 1973) in the interstellar space. These properties have recently been reevaluated by Abgrall, Roueff and Drira (2000) in the 4 coupled electronic states approach. Table 3 gives a sample of results obtained in this study.

The first column of Table 3 displays a pure B level whereas the two levels with J = 1 show a strong interaction between B and C^+. Finally, the last column gives an example

† for M_{SQ} the $(2J_j + 1)^{1/2}$ factor was erroneously omitted in Abgrall et al. (1997) and in Abgrall and Roueff (1989). However the factor was taken into account in the calculations.

J	0	1	1	0
v	9	14	3	8
$\rho(B)$	1.00	.694	.307	.417
$\rho(C^+)$	0.	.306	.693	0.
$\rho(B')$	0.	0.	0.	0.583
$\rho(D^+)$	0.	0.	0.	0.
$E_{v,J}$ in cm^{-1}	100,842.91	105,689.42	105,661.12	118,336.05
A_t in s^{-1}	1.1(9)	9.7(8)	1.1(9)	6.1(8)
A_c in s^{-1}	4.6(8)	3.4(8)	1.5(8)	5.9(8)
E_k in eV	0.22	0.56	0.56	0.04

TABLE 3. Properties of some excited energy levels. numbers in parentheses indicate powers of 10. A_t refers to the inverse of the total radiative lifetime.

of radial coupling. These informations are available in Abgrall et al. (2000) up to values of J equal to 25 and can be obtained from the authors. It is now clear that such coupled results should preferably be used for quantitative purposes as shown in Abgrall et al. (1997). A similar study has been performed for D_2 (Abgrall et al. 1999).

4. Electronic excitation

Fluorescent emission also occurs after electronic excitation. A new experimental setup has been built at the Jet Propulsion Laboratory (Liu et al. 1995). High resolution experimental reference FUV spectra have been produced for wavelengths between 114 and 169 nm at a spectral resolution of 0.0136 nm. The operating mode is such that the electron energy is fixed at 100 eV while the emitted spectral intensity is recorded by wavelength scans. A careful study has allowed to perform quantitative comparison between the discrete and continuum fluorescence experimental spectra and theoretical simulations. The agreement between experimental records and models is greatly enhanced when coupled calculations are used as shown on Figure 2. A similar study has been performed on D_2 (Abgrall et al. 1999).

Such an analysis has allowed to determine electronic excitation cross sections of H_2 toward B and C electronic states and improve the previous results obtained in an older experimental setup (Shemansky, Ajello & Hall 1985). An new analytic formula is given in Liu et al. (1998) in function of the electron energy. The results are normalized at high energies with the Born formula. However, the experimentally derived cross sections of Liu et al. (1998) differ by approximately a factor of 2 from those calculated by Celibero and Rescigno (1993) and by Branchett, Tennyson and Morgan (1990) in the region around 20 eV. The experimental uncertainty arises primarily from instability of magnetically collimated electron gun and second electrons, and, to less extent, from the cascades of EF, GK, H $^1\Sigma_g^+$ upper states which can be electronically excited and fluoresce to the B or C states. Electronic excitation of H_2 occurs in the atmospheres of outer planets where the fluorescence is observed from the vacuum ultraviolet to the visible. High resolution spectra (0.56 Å) of the Jovian aurora have been obtained with the *Hubble Space Telescope* (HST) Goddard High-Resolution Spectrograph by Trafton et al. (1994), Clarke et al. (1994) and Kim, Caldwell and Fox (1995). Jupiter's auroral spectrum has also been obtained via Galileo spectrometer observations by Pryor et al. (1998). In this latter case, the fluorescence arises primarily from the triplet manifold. Ingersoll et al. (1998) observe with wide band filters between 300 and 1000 nm a possible signature

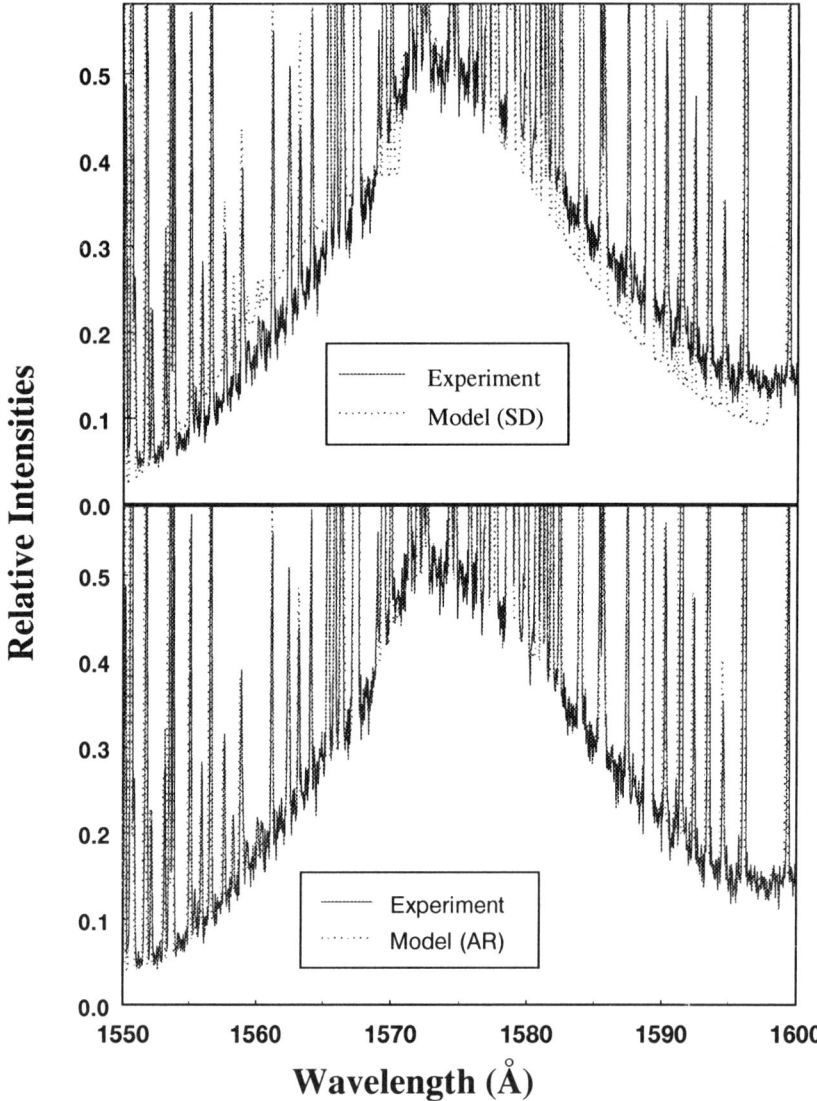

FIGURE 2. upper part: comparison between experiment and model calculations using the band continuum oscillator strengths of Stephens and Dalgarno (1972); lower part: the same with model calculations using the coupled treatment and including centrifugal barrier effects.

of the cascades from the electronic states of H_2 such as EF, GK, H $^1\Sigma_g^+$. The proper theoretical approach in these levels has still to be done and calculations are in progress.

5. Conclusions

Qualitative improvement in the knowledge and prediction of Lyman and Werner systems of H_2 and D_2 has been achieved in the last ten years thanks to a close interplay between theoretical and experimental studies. The states dissociating above the $H(1s) + H(2s)$ dissociation limit are presently not considered. The Lyman and Werner systems of HD remain to be studied experimentally in order to compare with ab initio results. Finally, the role of the cascades from E, F, G, K, H states to B and C has to be elucidated. The FUSE mission offers an exceptional opportunity to take advantage of these new data. The discrepancy between theoretical and experimental electronic excitation cross sections has yet to be resolved. Finally, astrophysical models of Photon or X rays dominated regions should also benefit from these improvements

REFERENCES

ABGRALL H., ROUEFF E., VIALA Y. 1982 *A&AS* **50**, *505-22*

ABGRALL H., LAUNAY F., ROUEFF E., RONCIN J.-Y. 1987 *J. Chem. Phys.* **87**, *2036-44*

ABGRALL H. AND ROUEFF E. 1989 *A&AS* **79**, *313-28*

ABGRALL H., LE BOURLOT J., PINEAU DES FORÊTS, ROUEFF E., FLOWER D.R., HECK L. 1992 *A&A* **253**, *525-36*

ABGRALL H., ROUEFF E., LAUNAY F., RONCIN J.-Y., SUBTIL J.L. 1993a *J. Mol. Spectr.* **157**, *512-23*

ABGRALL H., ROUEFF E., LAUNAY F., RONCIN J.-Y., SUBTIL J.L. 1993b *A&AS* **101**, *273-322*

ABGRALL H., ROUEFF E., LAUNAY F., RONCIN J.-Y., SUBTIL J.L. 1993c *A&AS* **101**, *323-62*

ABGRALL H., ROUEFF E., LAUNAY F., RONCIN J.-Y. 1994 *Can. J. Phys.* **72**, *856-65*

ABGRALL H., ROUEFF E., LIU X., SHEMANSKY D.E. 1997 *ApJ* **481**, *557-66*

ABGRALL H., ROUEFF E., LIU X., SHEMANSKY D.E. AND JAMES G.K. *1999 J. Phys. B: At. Mol. Phys* **32**, *3813-38*

ABGRALL H., ROUEFF E., DRIRA I. *2000 A&A , in press*

ALLISON, A. & DALGARNO, A. *1970 At. Nucl. Data Tables* **1**, *289-342.*

BRANCHETT J., TENNYSON J. , MORGAN G.K. *1990 J. Phys. B: At. Mol. Phys* **23**, *4625-39*

CELIBERO R., RESCIGNO T.N. *1993 Phys. Rev. A* **47**, *1939-45*

CLARKE J.T., BEN JAFFEL L., VIDAL-MADJAR A., GLADSTONE G.R., WAITE J.H., PRANGÉ R., GÉRARD J.-C., AJELLO J. *1994 ApJ* **430**, *L73-76*

DODELSON S. *1986 J. Phys. B: At. Mol. Phys* **19**, *2871-2879*

HERZBERG G. *1989 Molecular Spectra and Molecular Structure. I. Spectra of Diatomic Molecules Krieger Publishing Company, Malabar, Florida*

HURWITZ M. *1998 ApJ* **500**, *L67-9*

INGERSOLL A., VASADA A., BOLTON S.J., KLAASEN K.P., ANGER C.D., LITTLE B. *1998 Icarus* **135**, *251-64*

KIM Y.H., CALDWELL J.J., FOX J.L. *1995 ApJ* **447**, *906-14*

LIU X., AHMED S.M., MULTARI R.A., JAMES G.K. AJELLO J.M. *1995 ApJ Supp.* **101**, *375-99*

LIU X., AHMED S.M., MULTARI R.A., JAMES G.K. AJELLO J.M. *1998 J.Geophys. Res.* **103**, *26,739-58*

PRYOR W.R., AJELLO J.M., TOBISKA W.K., SHEMANSKY D.E., JAMES G.K., HORD C.W., STEPHENS S., WEST R.A., STEWART I.F., MCCLINTOCK W.E. BARTH C.A., SIMMONS K.E. *1998 J.Geophys. Res.* **103**, *20,149-58*

RAICH J.C., GOOD R.H. *1964 ApJ* **139**, *1004-13*

SENN P., QUADRELLI P., DRESSLER K. *1988 J. Chem. Phys.* **89**, *7401-27*

SOLOMON P.M. *1970 IAU Symposium* **36**, *320.*
STEPHENS T.L. AND DALGARNO A. *1973 J. Quant. Spectrosc. Radiat. Transfer* **12**, *569-86*
STEPHENS T.L. AND DALGARNO A. *1973 ApJ* **186**, *165-7*
TRAFTON L.M., GÉRARD J.C., MUNHOVEN G., WAITE J.H. *1994 ApJ* **421**, *816-27*
WITT A.N., STECHER T., BOROSON T.A., BOHLIN R.C. *1989 ApJ* **336**, *L21-24.*
WOLNIEWICZ L. AND DRESSLER K. *1988 J. Chem. Phys.* **88**, *3861-70.*

The cooling of astrophysical media by H_2 and HD

By David FLOWER[1], Jacques LE BOURLOT[2], Guillaume PINEAU DES FORÊTS[2] AND Evelyne ROUEFF[2]

[1] Physics Department, The University, Durham DH1 3LE, UK
[2] Observatoire de Paris, DAEC, UMR 8631 du CNRS, F-92195 Meudon, France

We summarize the results of recent quantum mechanical calculations of cross sections and rate coefficients for the rovibrational excitation of H_2 and HD by the principal perturbers, H, He, and H_2. These results have been used to evaluate the rate of cooling of astrophysical media by H_2 and HD molecules; these calculations are also described. The cooling of the primordial gas by rotational transitions of H_2 is considered as a special case.

All the numerical results and related software are available from http://ccp7.dur.ac.uk/.

1. Introduction

Molecular hydrogen is recognized as a major contributor to the cooling of astrophysical media. Its role is all the more significant under conditions, such as those which prevailed in the primordial gas, where few other coolants were present; but H_2 is also an important, sometimes the dominant coolant of low density interstellar gas, for kinetic temperatures $T > 100$ K. Interstellar gas can be heated to such temperatures by shock waves, by the dissipation of turbulence, or by absorbing energy from the local ultraviolet radiation field, as in photon-dominated regions.

Although the elemental abundance of deuterium is approximately 5 orders of magnitude less than that of hydrogen, it turns out that cooling by HD must often be taken into account, essentially for two reasons. First, chemical fractionation can, in media which are only partially molecular, enhance the abundance of HD, relative to that of H_2. In the primordial gas, for example, this enhancement is believed to be about 2 orders of magnitude. Second, the smaller rotational constant of HD, and the fact that the $J = 0 \to 1$ rotational transition may be induced by non-reactive collisions, ensure that the contribution of HD to the rate of cooling increases, relative to that of H_2, as T decreases.

We have recently undertaken a systematic study of the excitation of H_2 and HD by the most abundant perturbers, H, He, and H_2. These quantum mechanical calculations incorporated not only rotational but also vibrational transitions, which become important at high temperatures $(T > 1000$ K). This work is summarized below, where reference to related studies is made. Our objective was to provide a consistent and reliable set of data from which the Òcooling functionÕ (rate of cooling per molecule) could be calculated for both H_2 and HD. The calculations of the cooling functions are outlined below. All the collisional data and the associated cooling functions are available electronically from the CCP7 website (http://ccp7.dur.ac.uk/).

2. Collisional rate coefficients

2.1. H_2

2.1.1. H_2 - He

Schaefer and Köhler (1985) computed cross sections for rotationally elastic and inelastic scattering of He on H_2, using the quantum mechanical method. Earlier studies had included additionally the vibrational degree of freedom but employed less accurate H_2 - He interaction potentials and dynamical approximations.

Muchnick and Russek (1994) have provided a parametric fit to the results of their ab initio calculations of the H_2 - He potential, incorporating the dependence on the intramolecular (vibrational) coordinate. This potential is the most complete and accurate available to date. It has been used by Flower, Roueff and Zeippen (1998) to compute cross sections and hence rate coefficients for transitions, induced by He, between the rovibrational levels of H_2 up to approximately 20 000 K above the ground state, i.e. up to $(v, J) = (3, 7)$ of ortho-H_2 and $(v, J) = (3, 8)$ of para-H_2. The quantum mechanical coupled channels method was used, without further dynamical approximations, apart from representing the vibrational eigenstates by harmonic oscillator functions (Eastes and Secrest 1972). The spectroscopically measured values of the rovibrational energy levels (Dabrowski 1984) were used. The rate coefficients for rotational transitions within the $v = 0$ vibrational ground state were compared with the calculations of Schaefer and found to agree to within a few per cent. The rate coefficient for vibrational relaxation, $v = 1 \to 0$, summed over the final rotational states and averaged over the initial rotational states, assuming a Boltzmann distribution within the $v = 1$ manifold, was also calculated in the kinetic temperature range $100 < T < 3000$ K. The results were found to agree with laboratory measurements (Audibert et al. 1976; Dove and Teitelbaum 1974) to within a factor of 2 over this entire range of temperatures, which corresponds to 6 decades in the value of the relaxation coefficient. We conclude that the rate coefficients for pure rotational transitions in H_2, induced by He, are known to high accuracy, and those for rovibrational transitions to an accuracy which is acceptable for many astrophysical applications, notably for the purpose of determining the contribution to the H_2 cooling function.

Independent quantum mechanical calculations by Balakrishnan et al. (1999a, b), using numerically exact vibrational eigenfunctions and extending up to $v = 6$, were found to be generally in good agreement with the results of Flower et al. (1998).

2.1.2. H_2 - H_2

Schaefer and Meyer (1979), Monchick and Schaefer (1980), and Danby et al. (1987) studied rotational relaxation in H_2 - H_2 collisions, building on earlier work by Sheldon Green and his collaborators but taking advantage of more accurate determinations of the H_2 - H_2 interaction potential. Subsequently, Schwenke (1988) provided a fit to the results of his own ab initio determination of the potential; this fit has recently been used by Flower (1998) and by Flower and Roueff (1998a, 1999a) to compute cross sections and rate coefficients for rotational and rovibrational transitions in H_2, induced by other H_2 molecules.

Transitions between the two forms of molecular hydrogen – ortho (J odd) and para (J even) – involve a change in the total nuclear spin ($I = 1$ in ortho, $I = 0$ in para) and are induced by proton-exchanging reactions of H_2 molecules with proton-transferring ions (H^+, H_3^+) or with H atoms. In H_2 - H_2 collisions, only non-reactive scattering needs to be considered, and the ortho and para forms remain distinct. It follows that ortho - ortho, ortho - para, and para - para scattering need to be considered.

In practice, it is found that simultaneous excitation of both molecules in a collision is much less likely than single excitation. Accordingly, calculations were performed for excitation of ortho-H_2 by ortho-H_2 and para-H_2, constrained to their ground states, $J = 1$ and $J = 0$, respectively. Similarly, results were obtained for excitation of para-H_2 by ground state ortho- and para-H_2. The cross sections were found to be insensitive to the rotational state ($J = 1$ or $J = 0$) of the perturber (Flower and Roueff 1998a, 1999a).

For pure rotational transitions, the rate coefficients agree with the earlier calculations by Monchick & Schaefer (1980) and by Danby et al. (1987) to better than 30 per cent. Regarding vibrational relaxation, the level of agreement with the measurements of Audibert et al. (1974, 1975) and of Dove and Teitelbaum (1974) is satisfactory at low temperatures ($T \approx 100$ K) and at high temperatures ($T > 1000$ K), but, at $T = 500$ K, the calculated value is almost an order of magnitude smaller than measured. This discrepancy might be attributable to inaccuracies remaining in the H_2 - H_2 interaction potential or to errors in the now 25 years old measurements. On the basis of their semi-classical calculations, Zenevich and Billing (1999) have suggested that constraining the perturber molecule to its rotational ground state may result in the rate coefficient for vibrational relaxation being underestimated. However, their calculations involved a number of approximations, notably the use of classical mechanics to treat the relative motion of the two molecules, and the validity of their results at temperatures as low as 500 K remains to be demonstrated.

2.1.3. H_2 - H

Although ostensibly the simplest of the three systems, H_2 - He, H_2 - H_2, and H_2 - H, the scattering of H on H_2 proves to be the most difficult to treat theoretically. At low energies, the cross sections for rotational transitions are small, owing to the weak anisotropy (angle dependence) of the interaction potential (Flower 1997a). These cross sections are found also to be sensitive to the representation of the $v = 0$ vibrational eigenfunction (Forrey et al. 1997; Flower and Roueff 1999b).

A further complication, at high collision energies, is that reactive scattering can occur. The barrier is approximately 5000 K, and reactive scattering assumes importance for temperatures $T > 1000$ K. Calculations which have incorporated this process have been performed by Sun and Dalgarno (1994), using a quantum mechanical approach but including only the rotational levels $J \leq 3$ of $v = 0$, and by Mandy and Martin (1993), Martin and Mandy (1995), and Lepp et al. (1995), who employed the quasi- classical trajectory (QCT) Monte-Carlo method. However, all these studies predate the most recent and possibly the most accurate determination of the H_2 - H interaction potential (Boothroyd et al. 1996).

At temperatures $T < 3000$ K, rate coefficients for vibrational relaxation in non- reactive scattering, determined by means of the quantum mechanical method, fall increasingly below the values which derive from the QCT method, indicating the failure of the quasi-classical approach (Flower 1997b). In our calculations of the H_2 cooling function (see Section 3 below), the rate coefficients for non- reactive scattering calculated by Flower and Roueff (1998b), using the potential of Boothroyd et al. (1996) and a quantum mechanical treatment of the collision, have been used, together with rate coefficients for reactive scattering determined semi-empirically (Le Bourlot et al. 1999).

2.2. HD

The interaction potentials H_2 - X (X = H, He, H_2) may also be used to study HD - X scattering. Allowance must be made for the displacement of the centre of mass of HD from the mid-point of the internuclear axis, and for the effect of the higher value of the reduced molecular mass on the vibrational eigenfunctions.

The shift of the centre of mass has the important consequence that $\Delta J = 1$ transitions are allowed in non-reactive HD - X scattering. Indeed, $\Delta J = 1$ transitions dominate rotational population transfer in HD. However, the associated absence of the ortho/para dichotomy in HD implies that all the rotational states have to be included simultaneously in the collision calculations, which increases considerably the computing time for a given set of molecular eigenstates.

Schaefer (1990) calculated rate coefficients for the excitation of HD by He and H_2, but only for the rotational levels $J \leq 4$ and $J \leq 2$, respectively. We have recomputed these data, using the interaction potentials referenced in Section 2.1 above, including the levels $v = 0$, $J \leq 9$ for HD - He scattering (Roueff and Zeippen 1999a) and $v = 0$, $J \leq 8$ for HD - H_2 (Flower 1999). In addition, we have computed the rate coefficients for HD - H scattering, with the basis $v = 0$, $J \leq 9$ (Roueff and Flower 1999). Where comparison is possible, results for HD - He and HD - H_2 are found to agree well with the previous calculations of Schaefer (1990). For HD - H, where no previous calculations exist, the rate coefficients for transitions $\Delta J = 1$ and $\Delta J = 2$ are found to be similar in magnitude to those for HD - He scattering.

We have already mentioned that transitions $\Delta J = 1$ dominate population transfer in HD. At $T = 100$ K, the rate coefficients for de-excitation by H of the first excited state of HD, $J = 1$, to the ground state, $J = 0$, is of the order of 10^{-11} cm^3 s^{-1}; this may be compared with values in the range 10^{-13} - 10^{-14} cm^3 s^{-1} for the transitions, induced by H, from the first excited states of ortho- and para-H_2 to their respective ground states, i.e. $J = 3 \rightarrow 1$ and $J = 2 \rightarrow 0$. The much larger values of the rate coefficients for HD imply less sensitivity to the remaining uncertainties in the interaction potential and to the form of the vibrational eigenfunctions.

Flower and Roueff (1999c) have extended the collision calculations for the systems HD - H and HD - H_2 to excited vibrational states of HD, including levels $v \leq 2$, $J \leq 9$. Roueff and Zeippen (1999b) have studied HD - He, including levels of HD up to $(v, J) = (3, 3)$. Thus, the data relating to the collisional excitation of HD are approaching a level of completeness similar to those of H_2.

3. Cooling functions

For convenience in applications, we have fitted the dependence of the rate coefficients, $q(T)$, on the kinetic temperature, T, to the functional form

$$\log [q(T) \text{ cm}^3 \text{ s}^{-1}] = a + b/t + c/t^2$$

where $t = 10^{-3} T$ (K) $+ \delta t$ and a, b, c are transition-dependent constants; δt is also a constant, which prevents divergence of the fit at low temperatures. δt depends on the collision system (e.g. HD - H) but is independent of the transition.

Knowing the rate coefficients, spontaneous radiative transition probabilities, and the spectroscopic values of the energy levels, the level populations (of H_2 and HD) may be computed, in steady state, for given values of the total density, $n_H = n(H) + 2n(H_2)$, kinetic temperature, T (K), atomic to molecular hydrogen density ratio, $n(H)/n(H_2)$, and ortho:para density ratio, $n(\text{ortho-}H_2)/n(\text{para-}H_2)$. The rate of cooling per H_2 or HD molecule, the cooling function, is then given by

$$W(X) = \frac{1}{n(X)} \sum_{i>j} (E_i - E_j) n_i A(i \rightarrow j)$$

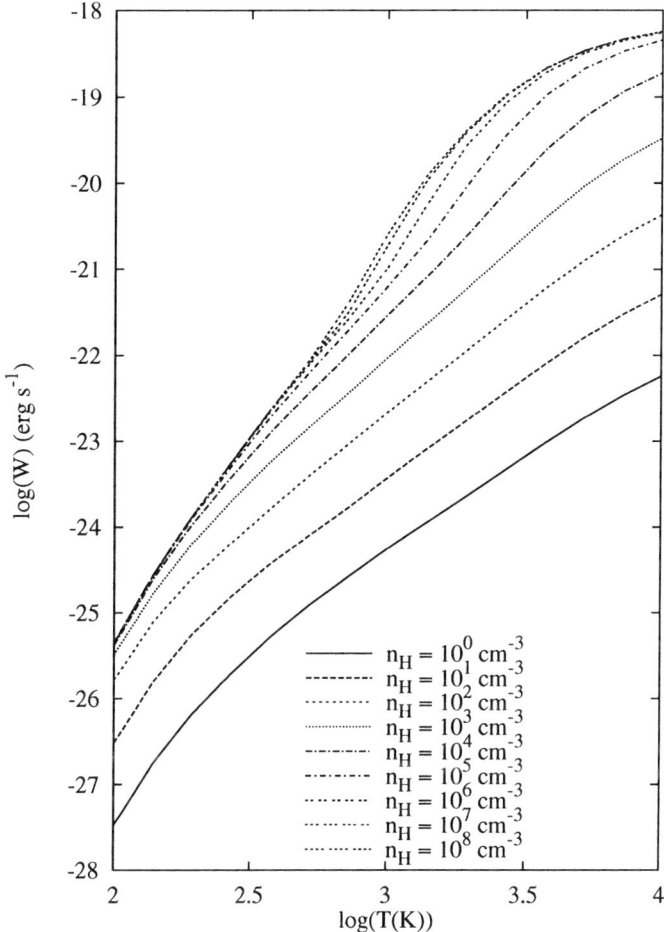

FIGURE 1. The cooling function, $W(H_2)$ (in units of 10^{-7} W), calculated for $1 \leq n_H \leq 10^8$ cm^{-3}, an ortho:para-H_2 density ratio of 1, and a H/H_2 abundance ratio of 1.

where X = H_2 or HD, E_i is the energy of level i, relative to the ground state, n_i is the density of population in this level, and $A(i \rightarrow j)$ is a spontaneous radiative transition probability; $W(X)$ is expressed in units of erg s^{-1} (10^{-7} W).

The H_2 cooling function, $W(H_2)$, has been calculated (Le Bourlot et al. 1999) for values of n_H, T, $n(H)/n(H_2)$, and $n(\text{ortho-}H_2)/n(\text{para-}H_2)$, within the ranges

$$1 \leq n_H \leq 10^8 \text{ cm}^{-3}$$
$$100 \leq T \leq 10^4 \text{ K}$$
$$10^{-8} \leq n(H)/n(H_2) \leq 10^6$$
$$0.1 \leq n(\text{ortho})/n(\text{para}) \leq 3$$

and taking $n(He)/n_H = 0.10$. $W(H_2)$ is obtained by numerical interpolation within a data set, le-cube; together with this data set, the interpolation program, interp.f, is provided on the website (http://ccp7.dur.ac.uk).

In the case of HD, the cooling function, $W(HD)$, has been computed (Flower et al. 2000) for the following ranges of the parameters

$$1 \leq n_H \leq 10^8 \text{ cm}^{-3}$$

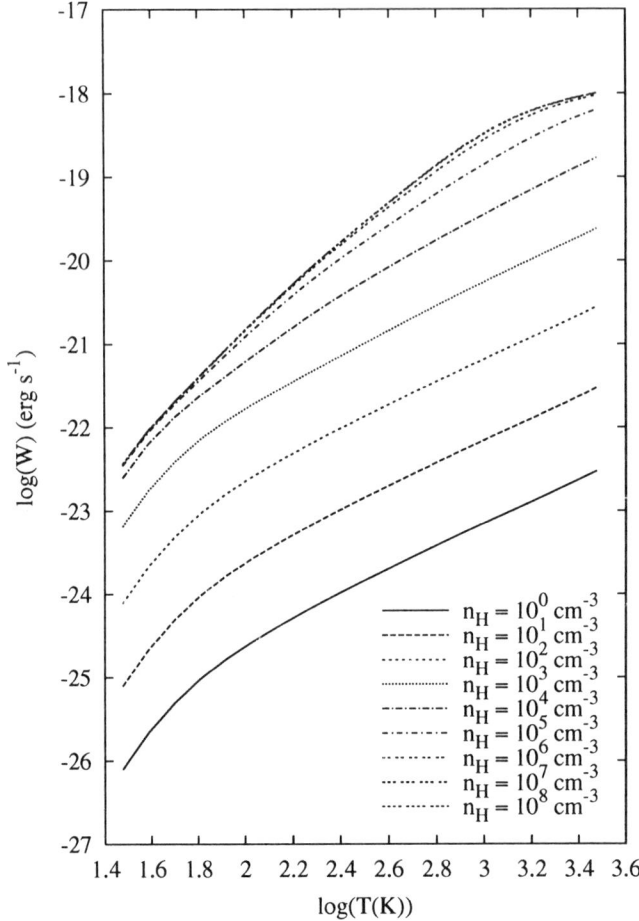

FIGURE 2. As Fig. 1, but for the cooling function $W(\mathrm{HD})$ (in units of 10^{-7} W).

$$30 \leq T \leq 3 \times 10^3 \text{ K}$$
$$10^{-4} \leq n(\mathrm{H})/n(\mathrm{H_2}) \leq 10^4$$
$$0.1 \leq n(\mathrm{ortho})/n(\mathrm{para}) \leq 3$$

and is available from the same website. Once again, we assumed $n(\mathrm{He})/n_H = 0.10$. $W(\mathrm{HD})$ was found to be insensitive to both $n(\mathrm{H})/n(\mathrm{H_2})$ and $n(\mathrm{ortho\text{-}H_2})/n(\mathrm{para\text{-}H_2})$. Its dependence on n_H and T may then be fitted by an appropriate and compact function, which is provided as a Fortran program, W-HD.f.

In Figs. 1 and 2, we plot the cooling functions, $W(\mathrm{H_2})$ and $W(\mathrm{HD})$ respectively, as functions of T, for $1 \leq n_H \leq 10^8$ cm^{-3}, $n(\mathrm{ortho\text{-}H_2})/n(\mathrm{para\text{-}H_2}) = 1$, and $n(\mathrm{H})/n(\mathrm{H_2}) = 1$. Particularly at low temperatures and high densities, the cooling rate per HD molecule is much larger than the cooling rate per H$_2$ molecule. The smaller rotational constant (and hence closer rotational level spacing) and larger radiative transition probabilities ensure that, as the Boltzmann limit is approached at low temperatures, a molecule of HD is a more efficient coolant than a molecule of H$_2$. Of course, the rate of cooling per unit volume is proportional to the molecular number density, and, as the elemental abundance ratio $n_D/n_H = 10^{-5}$, $n(\mathrm{HD}) \ll n(\mathrm{H_2})$. On the other hand, in media in which the transformation of H into H_2, and D into HD, is incomplete, chemical fractionation can, at low temperatures, enhance the ratio $n(HD)/n(H_2)$. The reaction $D^+(H_2, HD)H^+$, which is endothermic by 464 K in the reverse direction, is responsible for fractionation. In the primordial gas, for example, chemical fractionation may have increased the ratio $n(HD)/n(H_2)$ by about 2 orders of magnitude, compared with the D:H elemental abundance ratio (cf. Galli and Palla 1998).

4. The primordial gas

It is recognized that cooling, principally by H_2 and HD, must have been a controlling factor in the formation of the first condensations in the primordial gas, through gravitational contraction. As the fractional abundance of H_2 is small, H and He are responsible for its excitation, together with H^+, which we have not considered so far. It transpires that the fractional abundance of H^+, subsequent to freeze-out, is sufficient, when combined with the large values of the rate coefficients for transitions $J \to J'$, in which $\Delta J = |J - J'|$ may be both even and odd (Gerlich 1990), to ensure that H_2 - H^+ collisions are important contributors to the rate of cooling of the primordial gas (Flower and Pineau des Forêts 1999).

Under conditions where $n(H_2) \ll n(H)$ and at low temperatures ($T < 150$ K) and densities ($n_H \approx n(H) < 10$ cm^{-3}), the rate of cooling by H_2 is determined by transitions amongst the levels $J = 0, 1,$ and 2, i.e. by collisions $J = 0 \to 2$, induced by H^+, H, and He, and $J = 0 \to 1$ and $1 \to 2$, induced by H^+. The cooling function is then independent of $n(H)/n(H_2)$ and may be calculated by solving the rate equations, in steady state, for the level populations; only $n(H)$, T, and $n(H^+)/n(H)$ need to be specified. A Fortran program, W-H_2.f, which performs this calculation is available from the CCP7 website (http://ccp7.dur.ac.uk/).

We are grateful to the Royal Society and the CNRS for financial support under the European Science Exchange Programme. One of the authors (DRF) gratefully acknowledges the award of a research fellowship by the University of Durham.

REFERENCES

AUDIBERT M.-M., JOFFRIN C., DUCUING J., 1974 Chem. Phys. Lett. **25** 158

AUDIBERT M.-M., VILASECA R., LUKASIK J., DUCUING J., 1975 Chem. Phys. Lett. **31** 232

AUDIBERT M.-M., VILASECA R., LUKASIK J., DUCUING J., 1976 Chem. Phys. Lett. **37** 408

BALAKRISHNAN N., FORREY R.C., DALGARNO A., 1999a Astrophys. J. **514** 520

BALAKRISHNAN N., VIEIRA M., BABB J.F., DALGARNO A., FORREY R.C., LEPP S., 1999b Astrophys. J. **524** 1122

BOOTHROYD A.I., KEOGH W.J., MARTIN P.G., PETERSON M.R., 1996 J. Chem. Phys. **104** 7139

DABROWSKI I., 1984 Can. J. Phys. **62** 1639

DANBY G., FLOWER D.R., MONTEIRO T.S., 1987 MNRAS **226** 739

DOVE J.E., TEITELBAUM H., 1974 Chem. Phys. **6** 431

EASTES W., SECREST D., 1972 J. Chem. Phys. **56** 640

FLOWER D.R., 1997a MNRAS **288** 627

FLOWER D.R., 1997b J. Phys. B **30** 3009

FLOWER D.R., 1998 MNRAS **297** 334

FLOWER D.R., 1999 J. Phys. B **32** 1755

FLOWER D.R., LE BOURLOT J., PINEAU DES FORÊTS G., ROUEFF E. 2000 MNRAS in press
FLOWER D.R., PINEAU DES FORÊTS G., 1999 MNRAS submitted
FLOWER D.R., ROUEFF E., 1998a J. Phys. B **31** 2935
FLOWER D.R., ROUEFF E., 1998b J. Phys. B **31** L955
FLOWER D.R., ROUEFF E., 1999a J. Phys. B **32** 3399
FLOWER D.R., ROUEFF E., 1999b J. Phys. B **32** L171
FLOWER D.R., ROUEFF E., 1999c MNRAS **309** 833
FLOWER D.R., ROUEFF E., ZEIPPEN C.J., 1998 J. Phys. B **31** 1105
FORREY R.C., BALAKRISHNAN N., DALGARNO A., 1997 Astrophys. J. **489** 1000
GALLI D., PALLA F., 1998 A&A **335** 403
GERLICH D., 1990 J. Chem. Phys. **92** 2377
LE BOURLOT J., PINEAU DES FORÊTS G., FLOWER D.R., 1999 MNRAS **305** 802
LEPP S., BUCH V., DALGARNO A., 1995 Astrophys. J. Suppl. **98** 345
MANDY M.E., MARTIN P.G., 1993 Astrophys. J. Suppl. **86** 199
MARTIN P.G., MANDY M.E., 1995 Astrophys. J. **455** L89
MONCHICK L., SCHAEFER J., 1980 J. Chem. Phys. **73** 6153
MUCHNICK P., RUSSEK A., 1994 J. Chem. Phys. **100** 4336
ROUEFF E., FLOWER D.R., 1999 MNRAS **305** 353
ROUEFF E., ZEIPPEN C.J., 1999a A&A **343** 1005
ROUEFF E., ZEIPPEN C.J., 1999b A&A to be submitted
SCHAEFER J., 1990 A&AS **85** 1101
SCHAEFER J., KÖHLER W.E., 1985 Physica **129**A 469
SCHAEFER J., MEYER W., 1979 J. Chem. Phys. **70** 344
SCHWENKE D.W., 1988 J. Chem. Phys. **89** 2076
SUN Y., DALGARNO A., 1994 Astrophys. J. **427** 1053
ZENEVICH V.A., BILLING G.D., 1999 J. Chem. Phys. **111** 2401

Highly excited singlet ungerade states of H$_2$ and their theoretical description

By Ch. JUNGEN[1] AND S. C. ROSS[2]

[1] Laboratoire Aimé Cotton du CNRS, Université de Paris-Sud, 91405 Orsay, France
[2] Department of Physics, University of New Brunswick, Fredericton E3B 5A3, Canada

1. Introduction

Dipole absorption to excited states of diatomic hydrogen lying above 13.6 eV is not usually considered in the discussion of interstellar photophysical processes. The purpose of this contribution is to provide a brief survey of these states, their structure and decay dynamics, and in particular of the theoretical methods used to describe them.

Above about 14.6 eV excitation energy the density of electronic states of H$_2$ increases dramatically so that above 14.8 eV the spacing of successive electronic states becomes smaller than a vibrational quantum, and at an energy about 0.04 eV below the ionization potential (I.P. = 15.4254 eV) it becomes even smaller than a rotational quantum of energy. This means that the usual Born-Oppenheimer description of molecular structure becomes inadequate: rather than considering the rotational/vibrational motion of the nuclei as being slow and determined by the average field of the rapidly moving electrons, one must also take account of the opposite limit, corresponding to a rapidly rotating and vibrating ion core interacting with a highly excited, distant, and slowly orbiting electron. In terms of the level structure this means that for given electronic inversion symmetry (g/u) and electron spin (0/1) the electronic states $n, (l), \Lambda$ with associated vibrational structure v, N and parity $(-1)^p$ ($p = 0, 1$) are progressively reordered and eventually form Rydberg series. These series are appropriately labelled n, v^+, N^+ for each $(l), N$ and parity $(-1)^p$. l is the electron orbital quantum number which is is put into brackets because (albeit useful for book-keeping purposes) it is not always a good quantum number.

2. Quantum Defect Theory

Quantum defect theory[1-3] represents a whole Rydberg series by the single quantization condition

$$\tan \pi\nu + \tan \pi\mu = 0. \quad (1a)$$

ν is the effective principal quantum number,

$$\nu = (E^+ - E)^{-1/2}, \quad (1b)$$

E is the total energy, E^+ the ionization energy, and μ is the quantum defect which accounts for the fact that the core is not simply a point charge. μ is energy-dependent because the response of the core depends on the energy of the impinging electron, but in practice this variation is often weak on the energy scale of interest. Therefore an unperturbed series of states can be represented in essence by a single dynamical quantity, the number μ which embodies the deviations of the core field from that of a point charge. It is straightforward to see that Eqs. (1a) and (1b) together yield the familiar Rydberg equation,

$$E_n = E^+ - \frac{1}{(n-\mu)^2}, \quad (n = \text{integer}). \quad (1c)$$

The units of energy here are Rydberg. If wavenumber units are used the numerator on the r.h.s. of Eq. (1c) must be multiplied by 109737.316, (for infinite nuclear mass), if atomic units are used it must be multiplied by 1/2, and if eV are used it must be multiplied by 13.6054.

The molecular states n associated with each vibrational-rotational limit v^+, N^+ in H_2 interact, and therefore a quantization condition is required representing all series n, v^+, N^+ at once including their couplings. This is given by the generalised form of Eq.(1),

$$det|\tan \pi \nu_i \, \delta_{ij} + \tan \pi \mu_{ij}| = 0, \qquad (2a)$$

with

$$\nu_i = (E_i^+ - E)^{-1/2}. \qquad (2b)$$

Note that for a given total energy E Eq. (2b) defines an effective principal quantum number ν_i with respect to *each* vibrational-rotational ionization limit $E_i = E_{v^+, N^+}$. The quantum defects form a matrix, with the off-diagonal elements $i \neq j$ accounting for the interchannel couplings. If the off-diagonal elements vanish, Eq.(2) reduces to a set of independent series of the form of Eq.(1).

Since many core vibration-rotation levels are to be taken into account in H_2, and for each ionization limit v^+, N^+ several l partial waves are sometimes relevant, the number of quantum defect matrix elements μ_{ij} may become quite large in practice. The question then arises how these elements can be determined. The so-called frame transformation eigenchannel method[4] provides a solution to this problem, since it links Eq.(2) to the Born-Oppenheimer approximation in a way which we now discuss in some detail.

When ν is small (say, $\nu \leq 3$ or 4), all ν_i of Eq.(2b) are nearly equal ($\nu_i \approx \nu$) because the ion vibrational-rotational structure $E_i^+ - E_{i=1}^+$ is small compared to the total electronic binding energy $E - E_{i=1}^+$ ($E_{i=1}^+$ denotes the lowest ion level $v^+ = 0, N^+ = 0$). In this limit the quantization condition Eq.(2) then becomes equivalent to the eigenvalue equation for the quantum defect matrix μ. The frame transformation method exploits the fact that at the same time these low-ν levels correspond to the Born-Oppenheimer limit. Letting **U** be the eigenvector matrix of μ we have

$$\tan \pi \mu_\alpha \, \delta_{\alpha\alpha'} = \sum_{i,j} U_{\alpha i}^{tr} \tan \pi \mu_{ij} U_{j\alpha'}, \qquad (3)$$

and Eq.(2a) yields

$$\tan \pi \nu + \tan \pi \mu_\alpha = 0 \qquad (4)$$

for each eigenvalue $\tan \pi \mu_\alpha$. Eq. (4) represents the low ν limit of the more general Eq. (2). The usefulness of Eqs. (3) and (4) lies in the fact that the eigenvector matrix elements $U_{i\alpha}$ are often known *a priori* in a very good approximation. The transition from Eq.(4) to Eq.(2) with increasing effective principal quantum number ν corresponds physically to the un-coupling of the excited electron from the core and is reflected quantum-mechanically as a change of the most appropriate basis set. Thus, for a rotating-vibrating molecular core the matrix **U** connects coupled states $(l), \Lambda, R$ to uncoupled states $(l), v^+, N^+$, and its elements are products of a rotational factor which is essentially a vector coupling coefficient describing the un-coupling of l from the molecular axis, and a vibrational factor which describes the decoupling of the electron from the core vibrational motion. This latter factor is given by the set of vibrational wavefunctions of the ion core $\chi_{v^+}^{(N^+)}(R)$ which project R onto v^+. With the eigenvalue index α represented in the case considered here by $(l), \Lambda, R$, Eq.(4) now reads

$$\tan \pi \nu + \tan \pi \mu_{(l)\Lambda}(R) = 0 \qquad (5)$$

for each (l), Λ and R, and yields as solutions the R-dependent Rydberg equation

$$U_{n(l)\Lambda}(R) = U^+(R) - \frac{1}{[n - \mu_{(l)\Lambda}(R)]^2}, \quad (6)$$

where U^+ is the potential energy curve of the molecular ion and U for various values n is a set of molecular Rydberg potential curves, generated from a single quantum defect curve $\mu(R)$ for given $(l)\Lambda$. The significance of Eq. (6) is that if some potential energy curves $U_{n(l)\Lambda}(R)$ are known accurately, for example from quantum-chemical structure calculations, one can use Eq.(6) to extract $\mu_\alpha = \mu_{(l)\Lambda}(R)$ (and possibly its energy dependence). If electronically excited core states or several partial waves l are relevant in the problem (as in the singlet *gerade* manifold of states of H_2), the $R-$ dependent equation (5) itself has to be generalized to the multichannel form Eq. (2a).[5]

The body-fixed quantum defect functions can then be used together with Eq.(3) (or rather its inverse) to construct the matrix elements μ_{ij} based on the knowledge of the transformation matrix **U**. The important fact to note is that the resulting quantum defect matrix is able to describe *both* limiting situations, the Born-Oppenheimer limit at low energy as well as its opposite at higher energies, along with the intermediate region lying between. It thus accounts also for the non-adiabatic couplings which cause the transition from one limiting regime to the other. In the case that a single electronic channel forms an appropriate description (as for the p-type excitation channel primarily populated in photo-absorption of H_2) the rovibronic quantum defect matrix takes the specific form of a vibrational integral:

$$\tan \pi \mu^{(N,p)}_{v+N+,v+'N+'} =$$

$$\sum_\Lambda \int dR \left[\chi^{(N+)}_{v+}(R) \langle N^+|\Lambda\rangle^{(N,l,p)} \tan \pi \mu_{(l)\Lambda}(R) \langle \Lambda|N^{+'}\rangle^{(N,l,p)} \chi^{(N^{+'})}_{v+'}(R) \right], \quad (7)$$

where $\chi_{v+}(R)$ are the vibrational wavefunctions of the ion core and $\langle N^+|\Lambda\rangle$ is proportional to a vector coupling coefficient.[6]

A further important fact is that the thus constructed quantum defect matrix retains its meaning when the energy increases above the ionization potential: indeed, each quantum defect matrix element goes over smoothly into a corresponding electron phase shift (in units of π radians). The matrix formed by the elements $\tan \pi \mu_{ij}$ is thus equivalent to the reaction matrix **K** familiar from scattering theory. The quantization conditions Eq. (1a) and (2a) remain in fact virtually unchanged: in essence one has to replace the effective principal quantum number $\pi \nu_i$ for each open channel by the negative of the continuum electron eigenphase shift $-\pi\tau$. How this is done in detail has been spelled out in many papers (see Ref. 2).

A further generalization concerns *molecular dissociation*. The set of core vibrational wavefunctions $\chi_{v+}(R)$ is infinite and includes the bound state vibrational spectrum of the H_2^+ ion as well as representing the corresponding *vibrational* continuum. The quantum defect matrix $\mu^{(N,p)}_{v+N+,v+'N+'}$ of Eq. (7) therefore implicitly contains dynamical information on processes involving ionization continua as well as dissociation continua, such as dissociative recombination:

$$H_2^+ \;+\; e^- \rightarrow \; H \;+\; H.$$

This process has indeed been treated successfully by multichannel quantum defect theory a number of years ago.[7,8] A treatment of the competition, or interference, of ionization and dissociation processes (and in particular of autoionization and pre-dissociation) which starts out directly from the rovibronic quantum defect matrix elements of Eq. (7) has been introduced more recently.[9]

In summary we can say that multichannel quantum defect theory combined with the frame transformation concept provides the crucial link between the spectroscopy of low Rydberg states which is based on the standard Born-Oppenheimer approach, and the non-adiabatic inelastic scattering processes and chemical transformations occurring in the continuum. Put another way, this means that structure and dynamics are unified here in a single theoretical approach.

Dipole absorption can be calculated in terms of of energy-normalized channel dipole amplitudes $D_g^i(E)$ (where g denotes the initial ground state rovibrational level). The total dipole transition amplitude is then given as a coherent superposition of the channel transition amplitudes, of the form

$$D = \sum_i D_g^i(E) Z_i(E), \quad (8)$$

where $Z_i(E)$ are the channel mixing coefficients obtained by solving Eq. (2) for a given total energy E. The channel transition amplitudes $D_g^i(E)$ in turn can be related by the appropriate frame transformations to the clamped-nuclei dipole transition moments $d_{X\Lambda''}^{n(l)\Lambda'}(R,E)$ calculated by *ab initio* theory for transitions between the ground state $X\Lambda''$ and excited electronic states $n(l)\Lambda'$ (details can be found, e.g., in Refs. 6 and 10).

3. Quantum defects and potential energy curves

Eq. (6) may be used in two ways. If the quantum defect curve $\mu_{(l)\Lambda}(R)$ can be determined directly, for example from experiment, or by *ab initio* methods, it provides the knowledge of the fixed-nuclei electronic energies $U_{n(l)\Lambda}(R)$. Alternatively, if highly accurate quantum-chemical potential energy curves are available - and this has indeed become true in recent years for the H_2 molecule - the quantum defects $\mu_{(l)\Lambda}(R)$ maybe extracted from them. As an example, Fig. 1 displays the quantum defects $\mu_{(l=1)\Lambda}(R)$ for $\Lambda = 0$ and 1 which have been extracted from the best available *ab initio* calculations which extend from $n = 2$ to 5.[11,12] In order to represent all available potential energy curves to within $\approx 1 cm^{-1}$ by Eq.(6), it has been necessary to assume that the quantum defect curves depend somewhat on the energy: Fig. 1 also displays the curves $\partial \mu(R)/\partial E$ and $\partial^2 \mu(R)/\partial E^2$. The $n = 2$ members of this series correspond to the $B^1\Sigma_u^+$ ($\Lambda = 0$) and $C^1\Pi_u$ ($\Lambda = 1$) states which are the upper states of the Lyman and Werner bands of H_2, respectively. These bands contribute a major portion to the total oscillator strength in absorption of H_2. The $n = 3$ members correspond to the $B'^1\Sigma_u^+$ ($\Lambda = 0$) and $D^1\Pi_u$ ($\Lambda = 1$) pair of states, while the $n = 4$ members correspond to the $B''^1\Sigma_u^+$ ($\Lambda = 0$) and $D'^1\Pi_u$ ($\Lambda = 1$) pair of states. The advantage of the quantum defect representation Eq. (6) is that it may be used to predict quite reliably numerous higher states which have never been calculated by *ab initio* methods.

4. Rovibronic bound levels

Rovibronic level calculations based on Eqs. (2) and (17) start out from the quantum defects shown in Fig. 1. The H_2^+ adiabatic potential energy curve defines the vibration/rotation levels[13,14] of the ion core which are the ionization limits of the problem. In the calculations reported in Table 1 a total of 30 p wave ionization channels ($v^+ = 0 - 14$, $N^+ = J - 1, J + 1$) were included for upper levels with parity $(-1)^N$, while in the calculations for levels with parity $-(-1)^N$ 15 p wave ionization channels ($v^+ = 0 - 14$,

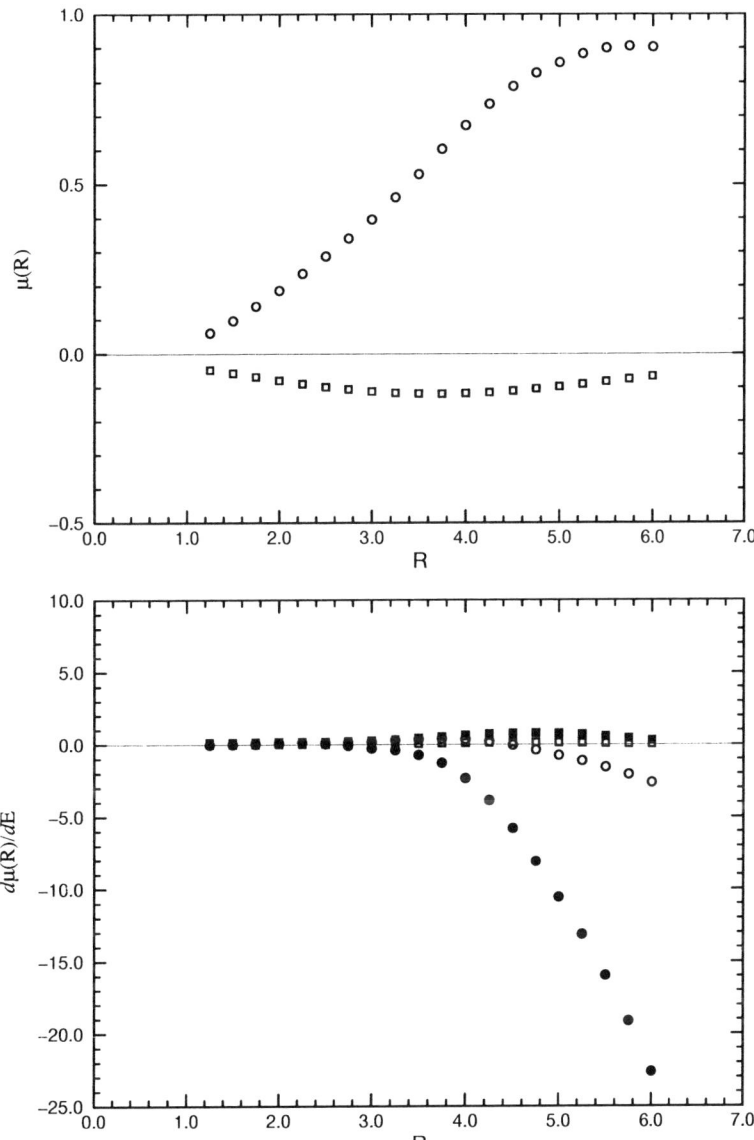

FIGURE 1. Quantum defects as functions of internuclear distance (in a.u.) for singlet ungerade H$_2$ (p channel): circles: $\Lambda = 0$, squares: $\Lambda = 1$. (a) quantum defects at threshold. (b) first (white symbols) and second (black symbols) derivatives with respect to the energy $E/a.u.$.

$N^+ = J$) were included. The energy-normalized electronic dipole transition moment for p wave excitation from the ground state was taken equal to $1.86 a.u.$ independent of the energy and internuclear distance (cf. Ref. 15 for details).

Table 1 presents an illustrative selection of levels of *singlet, ungerade* symmetry calculated in this way for $J = 0$ and 1 and $E > 112000 cm^{-1}$. They are compared with observed level energies available in the literature. Although the levels belong to various electronic states and coupling schemes and all input quantities have been taken from *ab initio* calculations without any adjustment, it can be seen that the deviations observed-

TABLE 1. Selected levels in singlet ungerade H_2.

N, p^b	approximate description^c	calculated^a energy/cm^{-1}	observed energy/cm^{-1}	obs. -calc.
approximate Born-Oppenheimer regime				
1, 0	$D^1\Pi_u^-, v = 0$	112930.9	112931.8^d	+0.9
1, 1	$D^1\Pi_u^+, v = 0$	112934.5	112935.5^d	+1.1
0, 0	$B'^1\Sigma_u^+, v = 2$	114076.1	114074.98^e	−1.1
1, 1	$B'^1\Sigma_u^+, v = 2$	114119.4	114118.25^e	−1.2
1, 0	$C^1\Pi_u^-, v = 11$	117151.2	117150.1^e	−1.1
1, 1	$D^1\Pi_u^+, v = 2$	117250.6	117251.6^d	+1.0
1, 1	$B'^1\Sigma_u^+, v = 3$	115648.4	115646.78^e	−1.6
0, 0	$B'^1\Sigma_u^+, v = 5$	117863.3	117861.41^e	−1.9
1, 0	$D^1\Pi_u^-, v = 3$	119214.7	119216.0^f	+1.3
transition region				
1, 1	$5p\sigma, v = 0$	119779.6	119777.7^f	−1.9
1, 1	$5p\pi, v = 0$	120240.9	120241.5^f	+0.6
1, 1	$6p\sigma, v = 0$	121249.5	121248.8^f	−1.7
1, 1	$6p\pi, v = 0$	121526.6	121527.5^f	+0.9
0, 0	$B''^1\Sigma_u^+, v = 3$	122658.9	122656.4^d	−2.5
1, 0	$D^1\Pi_u^-, v = 5$	122785.0	122786.2^d	+1.2
multichannel interactions				
1, 1	$14p0, v^+ = 0$	123814.7	123816.4^f	+1.7
1, 1	$14p0, v^+ = 0$	123860.1	123860.0^f	−0.1
1, 1	$15p0, v^+ = 0$	123918.1	123916.6^f	−1.5
1, 1	$5p\sigma, v = 2$	123923.8	123922.6^f	−1.2
1, 1	$13p2, v^+ = 0$	123940.6	123940.4^f	−0.2
1, 1	$16p0, v^+ = 0$	123988.0	123987.8^f	−0.2
1, 1	$14p2, v^+ = 0$	124020.1	124019.8^f	−0.3
1, 1	$17p0, v^+ = 0$	124040.9.0	124040.7^f	−0.2
1, 1	$18p0, v^+ = 0$	124077.0.0	124076.9^f	−0.1
1, 1	$15p2, v^+ = 0$	124096.2.0	124096.0^f	−0.2
1, 1	$4p\pi, v = 3$	124112.6	124113.4^f	+0.8

^{a}this work.
^{b}excited state: $N = J$, total parity $(-1)^p$.
^{c}in the multichannel regime: Hund's case d notation: $\nu l N^+, v^+$.
^{d}Ref. 16. ^{e} Ref. 17. ^{f} Ref. 18.

calculated are never larger than about 1 or 2 wavenumber units throughout the full energy range.

5. Pre-dissociation and autoionization resonances

Table 2 lists a few levels of H_2 which lie above the dissociation limit $H(1s) + H(n = 2)$ at $118376 cm^{-1}$ or the ionization limit at $124417.5 cm^{-1}$, and for which pre-dissociation and/or pre-ionization widths have been measured as well as calculated by quantum defect theory. Note that these calculations require no input in addition to that already used for the bound level calculations. Table 2 lists two types of resonances. The rovibronic levels

TABLE 2. Selected resonances in singlet ungerade H_2.

N, p^a	approximate description^b	calculated energy/cm^{-1}	observed energy/cm^{-1}	c calculated width/cm^{-1}	observed width/cm^{-1}
approximate Born-Oppenheimer regime, electronic pre-dissociation $\Delta n = -1$					
2,0	$D^1\Pi_u^+, v = 3$	119318.5^c	119320.8^d	14.1^c	14.5^e
2,0	$D^1\Pi_u^+, v = 5$	122876.5^c	122879.2^d	14.1^c	13.8^e
2,0	$D^1\Pi_u^+, v = 7$	125956.5^c	125959.3^d	11.3^c	11.0^e
2,0	$D^1\Pi_u^+, v = 9$	128561.0^c	128563.4^d	10.0^c	9.0^e
multichannel interactions, vibrational autoionization $\Delta v^+ = -1$					
1,1	$8p\sigma, v = 1$	124867.2^f	124866.6^g	4.0^h	5.1^i
1,1	$8p\pi, v = 1$	125017.4^f	125016.4^g	2.2^h	2.8^i
1,1	$8p\sigma, v = 2$	126917.0^f	126916.6^g	9.7^h	10.1^i
1,1	$8p\pi, v = 2$	127066.9^f	127065.7^g	2.9^h	4.2^i
1,1	$8p\sigma, v = 3$	128831.2^f	128828.8^g	11.7^h	16.2^i
1,1	$8p\pi, v = 3$	129009.9^f	129009.0^g	5.4^h	5.8^i

^{a} excited state: $N = J$, total parity $(-1)^p$
^{b} in the multichannel regime: Hund's case d notation: $\nu l N^+, v^+$.
^{c} Ref. 19. ^{d} Ref. 16. ^{e} Ref. 20. ^{f} this work.
^{g} Ref. 18. ^{h} Ref. 21. ^{i} Ref. 22.

of the $D^1\Pi_u^+$ state (corresponding to $n = 3$) with $v > 2$ state lie higher in energy than the $H + H(n = 2)$ dissociation limit. They are pre-dissociated through non-adiabatic Coriolis interaction with the $B'^1\Sigma_u^+$ state which for small R corresponds to $n = 3$, but dissociates at the $n = 2$ level into $H + H(n = 2)$. We may call this process a $\Delta n = -1$ *electronic pre-dissociation*. Table 2 shows that the agreement between observed and calculated widths is very good although no specific Coriolis coupling term has been built into the calculations. Instead, the decay dynamics is accounted for entirely by the frame-transformed quantum defect matrix elements [Eq. (7)] and a generalized variant of Eq. (2).

The levels $8p\sigma, v$ and $8p\pi, v$ for $v > 0$ lie above the ionization thresholds $E_{v^+=v-1,N^+}$. They can decay through non-adiabatic interaction between the bound $8p, v$ states and the ionization continuum built on the $v^+ = v - 1$ core. In the course of this $\Delta v^+ = -1$ *vibrational autoionization* process the core loses one vibrational quantum. The vibrational energy is transferred to the electron which thus manages to escape into the ionization continuum. The data collected in Table 2 again show quite good agreement between calculation and experiment.

6. Future work

The levels and resonances discussed in this contribution may be excited by photons with energies $< 13.6 eV$ if H_2 is vibrationally excited in the initial state. In order to make quantitative predictions of oscillator strengths the constant energy-normalized electronic transition moments used so far in the MQDT calculations should be replaced by more accurate energy and coordinate-dependent transition moments $d_{X\Lambda''}^{n(l)\Lambda'}(R)$. Inclusion of higher partial waves l and possibly core excited electronic channels should permit the extension of the MQDT treatment to larger $R-$ values, and hence to levels with higher vibrational excitation. Work along these lines is under way.

REFERENCES

1. M. J. Seaton, Rep. Prog. Phys. **46**, 167 (1983) (reprinted in Ref. 2).
2. *Molecular Applications of Quantum Defect Theory*, ed. Ch. Jungen (Institute of Physics Publishing, Bristol, 1996).
3. C. H. Greene and Ch. Jungen, Adv. At. Mol. Phys. **21**, 51 (1985).
4. U. Fano, Phys. Rev. **A 2**, 353 (1970) (reprinted in Ref. 2).
5. S. C. Ross and Ch. Jungen, Phys. Rev. **A 49**, 4353 (1994).
6. Ch. Jungen and G. Raseev, Phys. Rev. **A 57**, 2407 (1998).
7. A. Giusti, J. Phys. B: At. Mol. Opt. Phys. **13**, 3867 (1980) (reprinted in Ref. 2).
8. I. F. Schneider, O. Dulieu and A. Giusti-Suzor, J. Phys. B: At. Mol. Opt. Phys. **24**, L289 (1991) (reprinted in Ref. 2).
9. Ch. Jungen and S. C. Ross, Phys. Rev. **A 55**, R2503 (1997).
10. A. Matzkin, Ch. Jungen and S. C. Ross, Phys. Rev. **A 58**, 4462 (1998).
11. K. Dressler and L. Wolniewicz, Ber. Bunsenges. Phys. Chem. **99**, 246 (1995).
12. L. Wolniewicz (private communication).
13. H. Wind, J. Chem. Phys. **43**, 2956 (1965).
14. L. Wolniewicz and T. Orlikowski, Molec. Phys. **74**, 103 (1991).
15. Ch. Jungen and D. Dill, J. Chem. Phys. **73**, 3338 (1980) (reprinted in Ref. 2).
16. S. Takezawa, J. Chem. Phys. **52**, 2575, 5793 (1970).
17. T. Namioka, J. Chem. Phys. **40**, 3154 (1964), **41** 2141 (1964).
18. G. Herzberg and Ch. Jungen, J. Mol. Spectrosc. **41**, 425 (1972).
19. Hong Gao, Ch. Jungen and C. H. Greene, Phys. Rev. **A 47**, 4877 (1993).
20. M. Glass-Maujean, J. Breton and P. M. Guyon, Z. Phys. D - Atoms, Molecules and Clusters **5**, 189 (1987).
21. M. Raoult and Ch. Jungen, J. Chem. Phys. **74**, 3388 (1981).
22. P. M. Dehmer and W. A. Chupka, J. Chem. Phys. **65**, 2243 (1976).

Laboratory studies of long-range excited states of H$_2$

By W. Ubachs

Laser Centre, Vrije Universiteit, Department of Physics and Astronomy,
Amsterdam, The Netherlands

Present day laser technology has advanced such that multiple resonance excitation can be performed using several lasers of various wavelengths. Also narrowband tunable extreme ultraviolet laser radiation is readily available, to bridge the gap between the low-lying electronic ground state and the excited singlet states in molecular hydrogen. These methods have been employed to investigate a new class of excited states of H$_2$ that are confined to large internuclear separation.

1. Introduction

Molecular hydrogen, the smallest neutral chemical entity, is often considered to be the simplest molecule. For a spectroscopist, however, H$_2$ brings about a number of complications which make it a difficult object to study. First of all, from an experimental perspective, the electronic ground state is separated from the excited states by a large energy gap, which can be bridged only by photons in the domain of the extreme ultraviolet (XUV). Furthermore hydrogen is a light molecule with a very open rotational structure; the rotational lines are often so widely spaced that it is not obvious that they form a progression. Also, as a consequence of the small mass, deviations from the Born-Oppenheimer are most prominent and strongest in H$_2$. Non-adiabatic interactions shift the energy levels over several tens of cm^{-1}, so that the rovibronic structure becomes confused. As a result assignment of observed spectra, even with rotational quantum numbers only, is not straightforward. This point is illustrated by the *Dieke atlas* (Crosswhite 1972), a compilation of spectra pertaining to transitions between excited states, recorded in the visible domain with a classical spectrometer. Of the many thousands of spectral lines the majority have not yet been identified. Another complication is related to the energetic region to which the doubly-excited states of hydrogen descend. Opposed to the case of atomic helium, the alternate two-electron system where the doubly excited states lie well above the ionization potential, the $(2p\sigma_u)^2$ doubly excited state in hydrogen is iso-energetic with the first singly-excited state of $^1\Sigma_g^+$ symmetry. Due to a strong interaction between these states a double-well potential energy curve is formed, with two progressions of mutually perturbing rovibrational levels. Double-well structures are the rule in molecular hydrogen, rather than the exception. Many electronic states interact strongly with the doubly-excited states at intermediate internuclear separation thereby forming an entire class of double-well structures. In cases, where the barrier is sufficiently high, the vibrational motion is confined to large internuclear separations. These outer well states form the subject of our studies.

2. Single photon excitation with an XUV-laser source

The techniques of non-linear optics and four-wave mixing, using pulsed lasers with high peak powers have provided tunable sources of narrowband radiation in the domain of the extreme ultraviolet (XUV). In our laboratory we have available a source with a

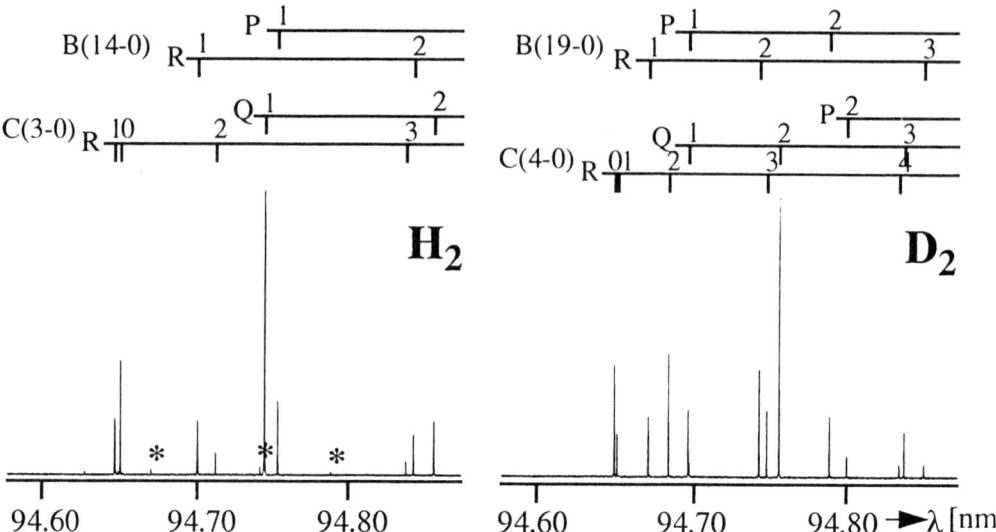

FIGURE 1. Recorded absorption spectra of H_2 and D_2 in the range 94.6 to 94.85 nm by the technique of 1 XUV + 1 UV resonance enhanced multi-photon excitation. Both spectra are recorded simultaneously by monitoring signal after a time-of-flight mass spectrometer gated at masses 2 and 4. Lines marked with an asterisk indicate D_2 resonances measured through the D^+ ions produced.

generic bandwidth of 0.25 cm^{-1} at 90-100 nm. When such a source of coherent light is used in combination with a collimated molecular beam to reduce Doppler broadening effects an absorption spectrum of hydrogen can be recorded at a line width of 0.25 cm^{-1}. An advantage of an XUV-source based on nonlinear up-conversion of visible laser light entails that the wavelength calibration can be performed in the visible domain, where wavelengths can be compared to the iodine reference standard. This allows for calibration of the frequencies of lines in the Lyman and Werner bands of H_2 and D_2 with an absolute accuracy of 0.03 cm^{-1}, beyond the capabilites of classical absorption or emission spectroscopy in the XUV. A typical spectrum in the region 94.60-94.85 nm is given in figure 1. Here the point on the complication of the hydrogen spectrum is illustrated. The order of the R(0) and R(1) lines in the open structured $C - X(3,0)$ Werner band is reversed by strong interactions between the $B^1\Sigma_u^+$ and $C^1\Pi_u$ systems. It is also illustrated in this figure that the spectrum of the D_2 isotopic species is not at all helpful in unravelling the structure of H_2. Note that the intensity distribution in the spectra is strongly affected by the mode of measurement: in 1 + 1 REMPI the ionization cross section partly determines the line intensities.

For details on the one-photon work we refer to earlier publications (Hinnen & al. 1994). Some of the Lyman and Werner bands of HD were also investigated with our narrowband laser system (Hinnen & al. 1995). A technical difficulty was overcome to run the laser system in the range 86-90 nm as well. In this range the spectroscopy of the $D^1\Pi_u, v = 1$ and $B'^1\Sigma_u^+, v = 0 - 2$ states were investigated, along with some higher vibrational levels in the Lyman and Werner systems (Reinhold & al. 1996).

The laser system can be improved by replacing the oscillator in the dye-laser by a pulsed-dye-amplification system using the ultra-narrow-band seed-input of a continuous wave ring-dye laser. This was demonstrated in a study of the resonance line of the helium atom at 58 nm, that was recorded with a linewidth of 0.02 cm^{-1} and a determination of the absolute accuracy to 0.0015 cm^{-1} (Eikema & al. 1997). In future we plan to perform an experimental reanalysis of the Lyman and Werner bands that should result in improved frequency positions by at least an order of magnitude. These measurements could be used to verify the coupled equations of the lowest excited states by the group of Roueff et al. (this conference).

3. Formation of double-well potential energy curves

A peculiarity of the electronic structure of the hydrogen molecule is that the $(2p\sigma_u)^2$ doubly excited state is so low in energy. The diabatic curves are drawn in figure 2a. In the diabatic picture in which the potential energy curves are allowed to cross it comes down in energy as low as the first excited configuration $(1s\sigma_g)(2s\sigma_g)$. In the adiabatic picture, when calculating Born-Oppenheimer potentials the Wigner-von Neumann non-crossing rules plays an important role. Due to the interaction between $(2p\sigma_u)^2$ and $(1s\sigma_g)(2s\sigma_g)$ two double-well structures are formed of $^1\Sigma_g^+$ symmetry: the $EF^1\Sigma_g^+$ and the $GK^1\Sigma_g^+$ states. The outer well of the lowest of these coincides with the ion-pair potential. Similarly the fourth state of $^1\Sigma_g^+$ symmetry is formed due to an interaction with a repulsive state converging to the (n=2) limit. As a result of an interaction with the ion-pair potential the outer well of the $H\bar{H}\,^1\Sigma_g^+$ state is formed. Similarly as a result of various interactions between pure electronic configurations a double-well structure of $^1\Pi_g$ symmetry is formed (Reinhold & al. 1998a). This state is drawn with a dashed line in figure 2b.

4. Double-resonance excitation of the $\bar{H}\,^1\Sigma_g^+$ outer well state

Since the XUV-laser source is based on up-conversion of lasers in the visible domain that are pumped by Q-switched lasers, the timing of the laser pulses (all of 3-5 ns duration) can be electronically controlled. Hence it is relatively simple to use multiple pulsed lasers in conjunction to perform double and multiple resonance studies. In figure 3 the excitation scheme is shown to study the hitherto unobserved fourth adiabatic state of $^1\Sigma_g^+$ symmetry. A first tunable XUV-photon is tuned into resonance with a singly-resolved rovibrational line in one of the Lyman bands, in this case either the $v = 18$ or $v = 19$ level. For these levels there is appreciable Franck-Condon overlap with the ground state at the classical inner turning point of the $B^1\Sigma_u^+$ state. A second photon then further excites the H$_2$ molecule into the $\bar{H}\,^1\Sigma_g^+$ outer well state using Franck-Condon overlap from the classical outer turning point on the $B^1\Sigma_u^+$ state potential as is graphically indicated in figure 3. For the generation of a spectrum signal must be produced. This is induced by a third laser at 355 nm, that promotes the population in the $\bar{H}\,^1\Sigma_g^+$ state to the energy region above the dissociation limit of the H$_2^+$-ion thus producing H$^+$ particles that can be detected after appropriate mass-selection in a time-of-flight mass-spectrometer. In some cases direct H$_2^+$-ion signal is observed as well, either from parasitic excitation routes (for this purpose the 3rd laser pulses is delayed in time by 30 ns, so that the population in the $B^1\Sigma_u^+$ state is decayed away) or from autoionization of the $\bar{H}\,^1\Sigma_g^+$ state.

In the spectrum of the $\bar{H}\,^1\Sigma_g^+$–$B^1\Sigma_u^+$ system a progression of P and R doublets is observable; the double resonance scheme has the marked advantage that the quantum numbers of the initially prepared state are known, hence from selection rules the excited $\bar{H}\,^1\Sigma_g^+, v, J$ states can be easily assigned. By accurately calibrating the wavelength of

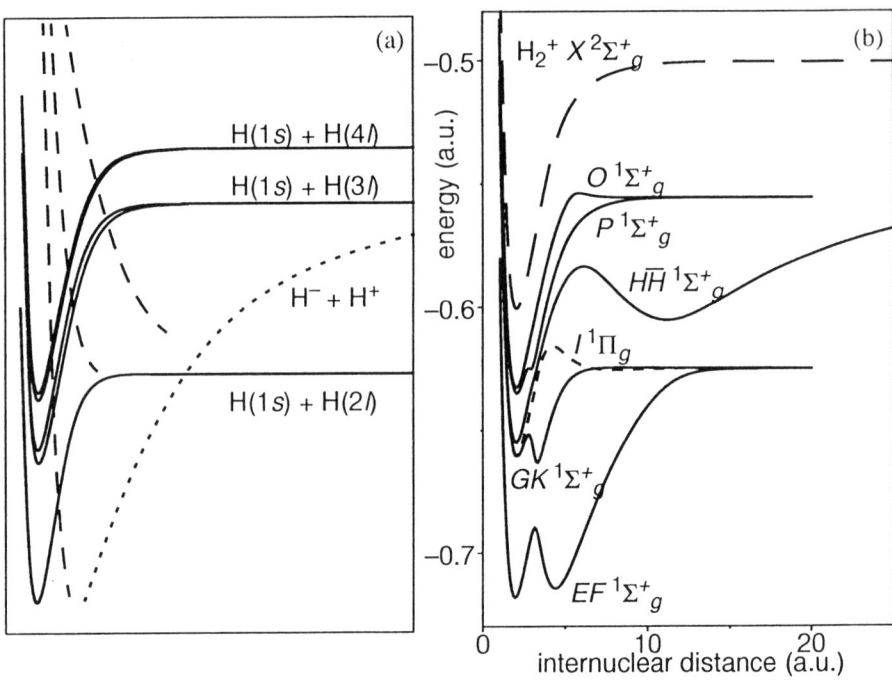

FIGURE 2. a: Simplified picture of diabatic potential curves. Full lines: singly excited states. Dotted line: the ion-pair potential. Dashed lines: the double-excited states. b: The calculated Born-Oppenheimer potentials for the $^1\Sigma_g^+$ states (full lines) and the $II'^1\Pi_g$ state (dashed).

the second laser the level positions of the $\bar{H}, v, J$ levels were determined to within 0.03 cm^{-1}. For an analysis of the rotational structure yielding molecular constants we refer to (Reinhold & al. 1997). The interesting decay dynamics of the higher lying vibrational levels in the $\bar{H}\,^1\Sigma_g^+$ outer well state in currently under investigation. Although in H$_2$ the $v = 5$ level already lies above the ionization limit the tunneling rate of these levels up to $v = 10$ is so low that autoionization is not observed. This is due to the potential energy barrier that dynamically inhibits autoionization; autoionization in this energy region can only occur at small internuclear separation, within the boundaries of the ground state potential of the H$_2^+$-ion. For the highest levels $v = 12 - 15$ tunneling is observed leading to autoionization and dissociation. There occurs a delicate interplay between these processes, which is dependent on the specific (v, J) level excited. Also coupling via resonant tunneling to levels at short internuclear distance, either in the $^1\Sigma_g^+$ inner well state or to other states in this energetic region, plays a role. However, the spectroscopy and dynamics of such states at short internuclear distance is not yet completely understood.

A remarkable effect of complete breaking of the $u - g$ inversion symmetry was observed in the $H\bar{H}$ state of the HD-isotopomer. The interaction between the $\bar{H}$ outer well state and another double-well state, referred to as $B''\bar{B}\,^1\Sigma_u^+$, is so strong that an additional vibrational progression of equally intense doublets could be observed, as shown in figure 4. For the homonuclear molecules H$_2$ and D$_2$ this $B''\bar{B}\,^1\Sigma_u^+$ has never been observed. Based on an accurate spectroscopic analysis of the data for H$_2$ and D$_2$, including mass-scaled Dunham coefficients a prediction was made for the level energies $\bar{H}(v, J)$ for HD. By comparing the experimentally observed values a strong perturbation in the form of an

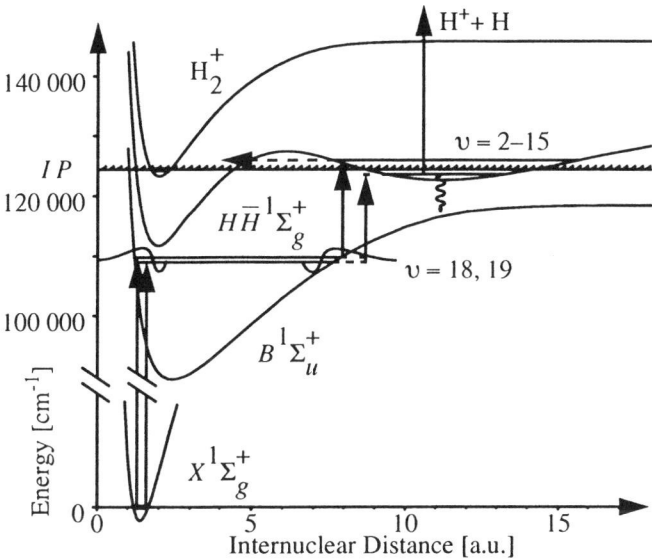

FIGURE 3. Scheme showing the two-photon excitation toward the outer well part of the $H\bar{H}\,^1\Sigma_g^+$ state. The tunable XUV and red lasers are spatially and temporally overlapped in the interaction region. A third laser at 355 nm is used to probe the $\bar{H}$-state population; if this laser is delayed in time exponential decay corresponding to the lifetime can be observed.

anti-crossing was found near $v = 14$ in the $\bar{H}$ state of HD. The presently observed perturbation is a textbook example of the symmetry-breaking mass-dependent term in the molecular Hamiltonian:

$$H_{gu} = \hbar^2 \frac{M_1 - M_2}{2M_1 M_2} \vec{\nabla}_{R\theta\phi} \cdot \sum_j \vec{\nabla}_j, \qquad (4.1)$$

where R, θ and ϕ refer to the nuclei and the index j runs over all electronic coordinates. For more details on this effect we refer to (Reinhold & al. 1998b).

The new and large set of level energies on the $H\bar{H}\,^1\Sigma_g^+$ state (82 for H_2 and 107 for D_2) allows for a comparison with renewed state-of-the-art *ab initio* calculations for higher lying states. With inclusion of adiabatic and relativistic effects the agreement is on the order of 1 cm^{-1} for H_2 and 0.5 cm^{-1} for D_2 (Reinhold & al. 1999). For HD calculations were performed with inclusion of the $u - g$ symmetry-breaking non-adiabatic term in the Hamiltonian, coupling the $H\bar{H}\,^1\Sigma_g^+$ and $B"\bar{B}\,^1\Sigma_u^+$ states, resulting in an agreement within 1.5 cm^{-1}, except for some levels close to the tunneling barrier.

5. Double-resonance excitation of the $II'\,^1\Pi_g$ outer well state

A second example of a long-range state hidden behind a barrier is that of the $II'\,^1\Pi_g$ state, where I' refers to the outer well part. This state has a maximum binding energy of only 200 cm^{-1} at $R \sim 8$ a.u. An XUV-IR double resonance scheme, with a third laser at 355 nm for probing the excited state population, was again employed to investigate the bound energy levels in this shallow well for H_2 and D_2. An overview spectrum, shown in figure 5 shows the $v = 0 - 2$ vibrational progression with some additional features and the onset of the (n=2) dissociation continuum. In excitation to a state of $^1\Pi_g$ symmetry

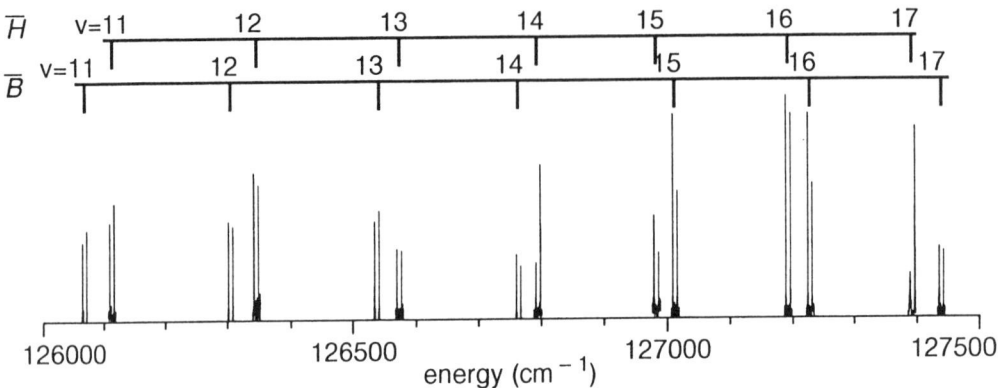

FIGURE 4. Recorded XUV-vis double resonance spectra of the $H\bar{H}\,^1\Sigma_g^+$ state in HD. The phenomenon of $u - g$ symmetry breaking is directly observable in the spectrum; forbidden transitions to states of u-symmetry are equally strong as the allowed ones.

a triplet of P, Q and R lines are easily identified. In D_2 a regular progression $v = 0 - 5$ was observed. The present experiment yields an onset of dissociation (in H_2) at 118377.2 $\pm$ 0.1 cm^{-1} in agreement with previous studies.

The data were analyzed to determine rotational constants but also following the method of (LeRoy & Bernstein 1970). At long range the I' state correlates with $(1s)$ + $(2p)$ atomic states and approaches the dissociation limit as $V(R) = -C_3 R^{-3}$, with C_3= 0.554929 a.u. (Stephens & Dalgarno 1974). Knowing the behaviour of the potential at large R the analysis of LeRoy and Bernstein provides a prediction on the number of vibrational levels confined in the potential. For H_2 up to $v = 6$ and for D_2 up to $v = 9$ should exist. These higher lying vibrational levels are not observed due to decreasing Franck-Condon overlap and a limited resolution of the laser system. A laser of narrower bandwidth could reveal the structure just below the (n=2) dissociation threshold. This energy region in the H_2 molecule is particularly interesting because the 12 possible molecular states (6 triplets and 6 singlets) converge to 12 possible combinations of atomic states with one electron in a $(1s)$ orbital and one in $(2p)$ or $(2s)$. Fine and hyperfine interactions make up for 12 possible combinations. In the last 0.5 cm^{-1} below the (n=2) dissociation limit a very perturbed structure is expected that remains a challenge for spectroscopists to unravel.

In the HD isotopomer a qualitatively different behaviour was found near the (n=2) dissociation limit. The adiabatic approximation breaks down at large internuclear separation and there is a splitting into two different limits, one associated with H(n=1) + D(n=2) and one with H(n=2) + D(n=1). The latter is 20 cm^{-1} lower in energy, consistent with the atomic isotope shift. This splitting can also be interpreted as a non-adiabatic interaction between the $C^1\Pi_u$ state, that converges repulsively to the upper limit and the $I'^1\Pi_g$ state that converges attractively to the lower limit. Again it is the mass-dependent $u - g$ symmetry breaking term that causes the interaction. Based on such an analysis it can be proven that there are only 4 bound vibrational levels $v = 0 - 3$ in the $I'^1\Pi_g$ state of HD. Interestingly, form the analysis it follows that the potential at large converges as a R^{-6} function; indeed a resonant dipole-dipole interaction, giving rise to a R^{-3} dependence, is only allowed in a fully inversion symmetric system. For more details we refer to (de Lange & al. 2000).

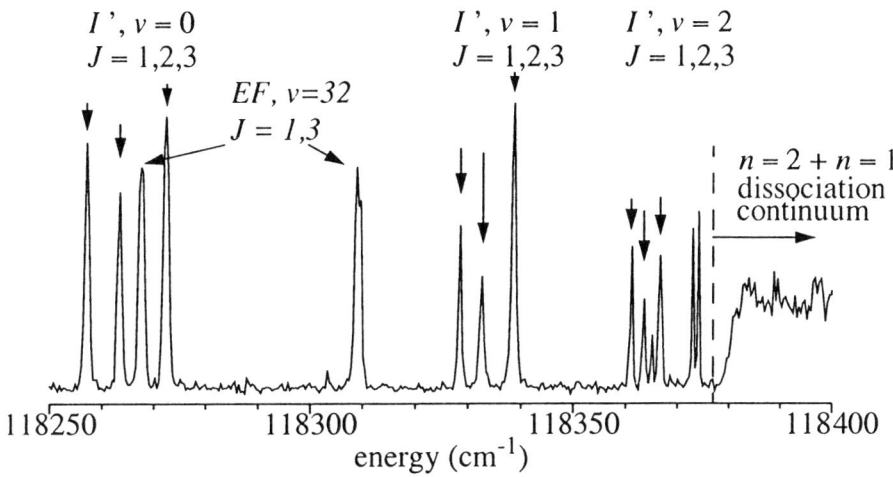

FIGURE 5. XUV+IR double resonance spectrum with the XUV-laser tuned to the $B\,^1\Sigma_u^+, v = 16, J = 2$ intermediate state.

6. Conclusions

Using present day advanced laser technology it has become possible to selectively excite singly-resolved rovibrational state in the $B^1\Sigma_u^+$ and $C^1\Pi_u$ states. Through double-resonance excitation with lasers tuned at various wavelengths two and three-photon excited states can be selectively probed. One advantage of the double-resonance excitation schemes is that unambiguous assignments can be easily made. Furthermore the level energies can be determined with unprecedented accuracy by using the usual frequency calibration techniques in modern laser spectroscopy. These experimental capabilities have opened a new line of research to systematically investigate the structure and dynamics of a class of long-range states in the hydrogen molecule.

The work presented here was performed in close collaboration with A. de Lange, E. Reinhold and W. Hogervorst at the Laser Centre Vrije Universiteit. Fruitful discussion and collaboration with L. Wolniewicz (Torun), Ch. Jungen (Orsay), P. Sorokin and J. Glownia (IBM, New York) and C.A. de Lange (Amsterdam) is gratefully acknowledged. The Netherlands Foundation for Fundamental Research of Matter (FOM) provided financial support.

REFERENCES

CROSSWHITE, H.M., *The Hydrogen Molecule Wavelength Tables of Gerhard Heinrich Dieke* (Wiley-Interscience, New York, 1972).

EIKEMA, K.S.E., UBACHS, W., VASSEN, W., & HOGERVORST, W., 1997 *Phys. Rev.* **A55**, 1866–1884.

HINNEN, P.C., HOGERVORST, W., STOLTE, S., & UBACHS, W. 1994 *Can. J. Phys.* **72**, 1032–1042.

HINNEN, P.C., WERNERS, S., HOGERVORST, W., STOLTE, S., & UBACHS, W. 1995 *Phys. Rev.* **A52**, 4425–4433.

DE LANGE, A., REINHOLD, E., HOGERVORST, W. & UBACHS, W. submitted to *Can. J. Phys.*.

LEROY, R.J., & BERNSTEIN, R.B. *J. Chem. Phys.* **52**, 3869.

REINHOLD, E., HOGERVORST, W. & UBACHS, W. 1996 *J. Mol. Spectr.* **180**, 156–163.

REINHOLD, E., HOGERVORST, W. & UBACHS, W. 1997 *Phys. Rev. Let.* **78**, 2543–2546.

REINHOLD, E., DE LANGE, A., HOGERVORST, W. & UBACHS, W. 1998 *J. Chem. Phys.* **109**, 9772–9782.

REINHOLD, E., HOGERVORST, W. & UBACHS, W. 1998 *Chem. Phys. Let.* **296**, 411–416.

REINHOLD, E., HOGERVORST, W., UBACHS, W. & WOLNIEWICZ, L. 1999 *Phys. Rev.* **A60**, 1258–1270.

STEPHENS, T.L., & DALGARNO, A. 1974 *Mol. Phys.* **28**, 1049.

A Model for Interstellar Dark Matter

By J. Schaefer

Max-Planck-Institut für Astrophysik, 85740 Garching, Germany

Two observable quantities have been calculated by using the data blocks which provide all details of interstellar UV absorption in H_2 gas from the electronic ground state of H_2 into 6 electronically excited states and fluorescence emission back into bound and continuum states of the electronic ground state. Both quantities describe details happening in the edges of interstellar H_2 abundances where UV radiation is very efficiently shielded and HI gas is produced by fluorescent radiative dissociation. The first is the fluorescence spectrum of H_2 in the wavelength range between 1350 and 1700 Å. Comparisons with published spectra show very nice agreement of the resolved features. The second is the velocity distribution of HI produced in fluorescent radiative dissociation. Subject of this comparison are the observed 21 cm *spin-flip* line profiles of the CNM and WNM (Cold and Warm Neutral Medium) HI species. The profile of the HI velocity dispersion, compared with the "narrow" WNM component, has been obtained at ≈ 9 km/s FWHM which is slightly below the statistical average, and this profile was found to be a molecular property of the H_2 gas rather than a function of intensity of the incident UV field or the temperature. The required long tail ("wide" Gaussian component) is provided by all LTE models.

The rest of the paper is devoted to an outline of a dark matter model which fulfills the condition of statistical equilibrium in an active gaseous interphase between the interstellar UV field and the constituent dark matter mass. I call the solid bodies "Jupiters", their masses are uncertain but large enough to keep the main part of the gas captured by gravitational attraction. In this picture, dark interstellar matter is discussed as an observable species due to interaction with the interstellar UV field.

---

1. Introduction

Two main galactic players are linked in a curious yet unexplained coincidence: HI and dark matter. Bosma found out in 1981 that for a sample of galaxies the surface density ratio of dark matter and HI is constant outside the optical disks. The ratio is about 10 (cf. Combes, this conference). This means: at large scales, dark matter and HI are concomitant radially, probably also vertically. I recall very briefly what we know about HI and continue to outline a model of dark matter at solar Galactic radius.

Kulkarni and Heiles (1988) summarized the findings of HI observations, i.e. of the *spin-flip* 21 cm line, in the diffuse interstellar medium. It is reasonable to distinguish between two species of neutral HI: the dense cold neutral HI of about 80 degrees Kelvin and the warm neutral HI of some thousand degrees Kelvin, I am citing "either surrounding the cold clouds in an envelope or pervading much of the space as an intercloud medium....Roughly half of all HI is warm neutral HI....The heating agent is unidentified." I may mention also a few well known numbers: the Einstein A coefficient for the 21 cm line is $2.85 \, 10^{-15} \, s^{-1}$. The spin-excited HI atoms are traditionally assumed to fly freely for about 11 Million years before they emit the *spin-flip* photon. Therefore, the mass of $3.5 \, 10^{14}$ spin-excited HI atoms counts for one observed *spin-flip* photon per second. The so determined HI mass has been used for the Bosma ratio and the whole Galaxy has been estimated to contain about $4.8 \, 10^9 M_\odot$ of HI (Henderson *et al.* (1982)).

The discussion of dark matter attracts enormous interest due to the well known fact that about nine times the known mass of the Milky Way has to be additionally located in the Halo. This is the requirement of dynamical models of the Galaxy. There are many dark matter candidates, most of them seem to be dark forever. I am in favor of those candidates which have at least a minimum chance of being observed in the future. They are composed of hydrogen, maybe of primordial gas in condensed gas clumps and located in the Halo as described in detail by Pfenniger et al. (1994) and Pfenniger and Combes (1994). One particular species of these candidates seems to be not dark at all: so-called dark hydrogen masses located in the optical Galactic disk. They are the main subjects of my talk. Their estimated abundance obtained by Combes and Pfenniger (1996) is quite impressive: with a Bosma ratio of 10, a decrease of HI $\sim R_G^{-1}$ and flaring of dark matter and HI starting at solar Galactic radius with a height $h_z = 0.03$ R_G, the dark matter fraction at solar Galactic radius was 50%, the surface density of dark matter 25 - 50 $M_\odot/pc^2$. If these numbers are confirmed, I think, models of dark hydrogen masses should become an important subject of ISM Physics. But these models cannot be simply added to the observed parts of the ISM without complications. A few of these parts might be explained in a different way and this could open up a chance for solving problems as e.g. the above mentioned missing heating agent of the warm neutral HI gas.

The model of dark interstellar matter is outlined in the next section. Changes of traditional convictions are discussed which seem to be unavoidable. This model has not yet been developed to the state where numerical simulations could be shown. But the crucial set of data is available for handling the interaction with the interstellar radiation field. These data, describing very accurately the absorption of the interstellar radiation in transitions from the electronic ground state of H_2 into 6 electronically excited states and the fluorescent emission back into the discrete ground states and into the ground state continuum, have been calculated and of course compared with the results published by Abgrall et al. (1997) in a series of papers. The agreement is very satisfying.

A simple application of this data set will be shown: the fluorescence spectrum of H_2 emitted from the edge of interstellar H_2 abundances, to be compared with UV spectra observed in the diffuse interstellar medium.

The same data set has been used in calculations of the velocity distributions of dissociating H Atoms. Some consequences will be discussed when dissociation is understood to be responsible for the velocity dispersion of the warm neutral HI 21 cm radiation.

2. The dark matter model

Dark H_2 gas can probably survive in condensed gas clumps in the Halo, if these clumps are very cold ($\approx$ 3 K), there is practically no metallicity, and the fractal hierarchical structure takes care for shorter collision intervals compared to the collapse time. These are the clouds of the model of Pfenniger and Combes. This model may be applied, according to the Bosma ratio, also to observed HI abundances in the Milky Way and in typical spiral galaxies which extend the optical disk radii by a factor of 2 to 4. I understand this finding as being due to absorption of intergalactic UV photons and, closer in, to Galactic UV radiation absorbed in the edges of the basic self-gravitating gas clumps (cf. Combes, this conference), followed by fluorescent radiation (partly) into the H_2 ground state continuum and *escape* of roughly half of the produced H atoms, 3/4 of which can therefore emit a *spin-flip* photon after 11 Million years time. It is a plausible explanation of the great patchiness of the observed HI sources.

The alternative action of the dissociated and spin-excited H atoms, escape and emission or recombination inside the "clumpuscules", allows to explain the observed velocity dispersion of HI in emission by the velocity distribution of HI produced in fluorescent dissociation. Quantitative results are shown below. As we will see, there is a warm component of HI evaporating off the surfaces, and a cold component, if LTE is applicable. An additional effect in these surface regions modifies the quantitative findings: ionization of the dissociated HI by the Lyman continuum ($\lambda < 912$ Å) accounts for "extreme scattering effects" (ESE), if we accept the interpretation by Walker and Wardle (1998) who attributed ESE to radio wave refraction by a cloud skin of free electrons crossing the line of sight from a Quasar.

The gaseous surfaces activated by UV radiation should be numerically simulated (which needs more programming work) starting with the weak interaction phenomena in the extended HI disk and then proceeding to more developed phases of interaction selected towards the Galactic center. This should provide not only details about the gas clumps but also about clumpiness and filling factors of observed sources and about diffuse gas pervading the surrounding space. Observed data can then be discussed in contrast to the traditional cloud models.

Collisions between the basic gas clumps of the Pfenniger-Combes model gain a new quality inside the optical disk where the metallicity increases exponentially radially, at large scales but locally even more so, and the radiation intensity of the UV field can vary over orders of magnitude. If the former is sufficiently admixed to the collapsing gas cloud, it promotes freezing and formation of solid H_2, while the latter raises the temperature and makes collisions inefficient to stop collapse. (Pfenniger and Combes, 1994). These conditions combined with the density of the gas clumps are found in star formation regions. Therefore, I assume that gas clumps can collapse under these conditions and, in following the model of Pfenniger and Combes, I call the collapsed bodies "Jupiters", but I may note that smaller and larger masses cannot be excluded simply because a combined setup of collapsing and colliding gas clouds yields a wide complexity of results.

I am hopeful that observed radiation from star formation regions can be preferentially attached to Jupiter atmospheres, as e.g., when high density and temperature gradients are required to describe phenomena which are usually attached to shock waves or turbulence. Conditions like this seem to be available in the gaseous interphase bound by the Jupiters. Of course, there are also gas sources of lower density populating the ISM which contain the well known particles and dust. The question is, what kind of source model gives the best interpretation of an observed signal. Interpretations could be different because at least one basic convention of ISM observation procedures is definitely violated by the dark matter Jupiter model. The sources in ISM, i.e. atoms, molecules and dust, observed in a beam along the line of sight, are generally distributed homogeneously over astronomical length scales which unavoidably restricts the objects of observation to low density clouds. The results are presented in surface brightness maps or integrated line emission maps or other maps which show "observed" clumpiness and structures like sheets and filaments. Better spatial resolution resolves more details and new structures in the maps, but there remains an open question: what is unresolved ?

The existence of a certain amount of Jupiters adds some new facts to the traditional understanding of the ISM. Really new is the high density of the gravitationally bound gas. As we know from our heavy planets, the density can reach parts of the standard amagat unit at lower temperatures and even more at higher temperatures. There is permanent energy injection provided by the radiation field, but evaporation of HI and HII is only possible off the rather light species. We have normally perfectly achieved statistical equilibrium between ionisation and recombination of HI and between dissociation and recombination of H_2 because the dissociated H atoms are either decelerated in the gas where they can recombine in 3-body collisions, thereby refreshing the ortho/para ratio of 3/1, or they are kept available for recombination by the gravitational field. But due to this efficient recombination, the HI gas in these atmospheres can only emit *spin-flip*

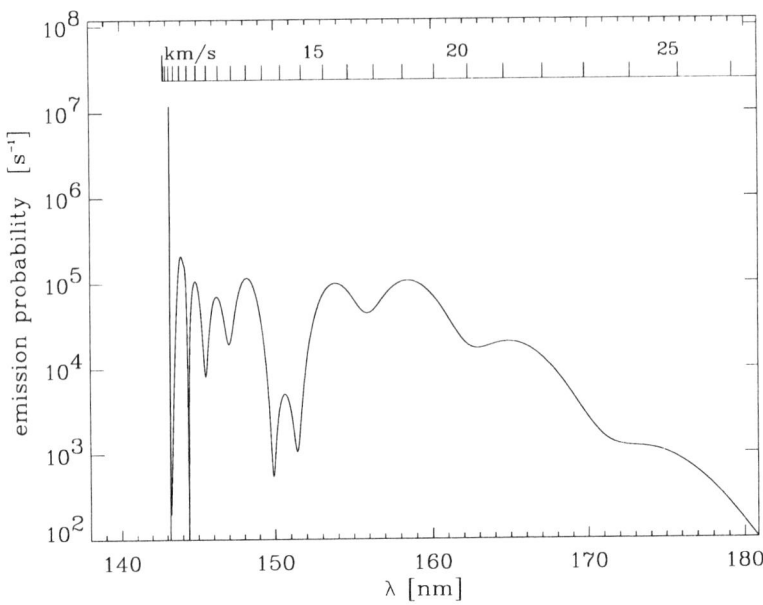

FIGURE 1. Emission probability from a B state, (J,v) = (14,12), of H_2 into the ground state continuum versus wave length and relative velocity of the dissociated H atoms.

photons rather immediately after fluorescent dissociation. One may think of collision induced magnetic dipole transitions due to collisions with electrons, ions (HII, H_2^+) and rotating H_2 molecules, but these effects seem to be inefficient. Hence, HI gas in the Jupiter atmospheres turns out to be no *spin-flip* emission source. This could contribute to the understanding of the radial HI deficiency often observed in the optical disks of spiral galaxies, as shown by Tacconi and Young (1986) for the NGC 6946 galaxy.

Another abundance decrease applies to a considerable amount of dust grains and other condensation sites which are crucial ingredients of the collapse. They become constituent part of the Jupiters.

The effective contribution of the Jupiters to the entire dark matter mass cannot be significant, since the local anti-correlation of HI and dark matter is only a small and not observed correction of the Bosma ratio, valid at large scales.

Up to now the microlensing (ML) collaborations MACHO (Massive Compact Halo Objects) and EROS (Expérience de Recherche d'Objects Sombres) have not seen the proposed Jupiters in front of the Magellanic Clouds. The estimated upper limit for the planetary "MACHOS" is now about 10% of the total dark matter mass, with high uncertainty. Searching for Jupiters in an ML setup requires observable moving stars behind star formation regions, knowledge of the distance to these stars and to the lensing Jupiters and of the relative angular velocities between the stars and the Jupiters. This necessary setup seems to be rather rare, probably due to the blending of the sources with the surrounding medium of the Machos in the optical light. There is still hope to find microlensing events of Jupiters in a special attempt, and even lensing of the condensed gas clumps is possible, as proposed by Draine (1998).

3. H_2 fluorescence emission

Here is my first application of the data set which provides the input for UV absorption of H_2 and subsequent fluorescence emission. Observed UV spectra of the diffuse interstellar medium show apparently common specific features. Witt et al. detected these features in 1989 and Martin et al. found them again in 1990. A selection of four spectra is shown from the latter paper and used for a comparison below, of spectra measured at different targets in the diffuse interstellar medium (Taurus, Ursa Major and Lindblad clouds). Sternberg failed in 1989 in reproducing the features by trying models with input data extended to all Werner and Lyman bands of H_2. He concluded that the features should be due to emission into the continuum. He was perfectly right as I will show.

A simple model is assumed: H_2 gas at the edge of the gaseous interphase of dark matter interacting with a Draine field . Self-shielding has been applied for an X state partitioning function of 2000 K, an ortho-para ratio of 3:1 and a rather small H_2 column density of 10^3 cm^{-2}. Obviously, this column density does not mean something like a complete surface model, but a first step of a simulation starting at the far outside of an H_2 abundance. Even ionization of HI is neglected. There is no doubt that quite some self-shielding steps are necessary to obtain absorption in the wings of the lines, outside of their saturated cores. E.g., Abgrall et al.(1992) did 50 steps to reach the optical depth $\tau_\nu = 1$. Subsequent steps have not yet been done in this project.

It is reasonable to neglect re-absorption of the fluorescence radiation at wavelengths $\lambda \geq 140$ nm because the populations of upper X state levels are small. This is the wavelength range where fluorescence emission into the continuum contributes significantly to the spectrum. The shape of this contribution is rather crude which is due to the fact that each of the emitting states has its own continuum spectrum starting on the red side of its line spectrum with peaks (resonances) and followed by a well understood undulatory shape. I show an example in Fig.1 , where the emission rate is plotted versus increasing wavelength and relative velocity of the dissociated atoms.

The comparison with the observed fluorescence spectra is shown after convolution with a Hanning function of apodized 150 cm^{-1} FWHM. Fig.2 shows the accurately placed maxima and minima. It means a very successful comparison. Identification of the continuum bumps is impossible and, as test calculations have shown, there is little variation as long as the vibrational-rotational ground states dominate the state population. I may note: if fluorescence into the continuum contributes significantly to the observed features, this indicates statistical equilibrium between H_2 dissociation and recombination.

I am waiting for more observed spectra to do more comparisons of frequency resolved details.

4. The heating agent

UV absorption, fluorescence and heating due to fluorescent dissociation happen in the same layer of H_2 gas. It will be shown in this section that this arrangement provides the yet missing heating agent of the warm neutral HI, in and outside of clouds. By using simple theoretical models we will find both species of HI, the "cold" and the "warm", produced in fluorescent dissociation. I am discussing two aspects of the heating: the statistics of decomposed spectra of neutral HI at solar Galactic radius and the extraordinary line profiles of tracer molecules like CO.

Statistics of HI observations at solar Galactic radius ("nearby regions" at b $\sim 30^o$, $0 < l < 360^o$) have been provided by Mebold (1972) who decomposed about 1200 spectra into a "wide" Gaussian component of $5 < \sigma < 17$ km/s representing 40% of the total HI

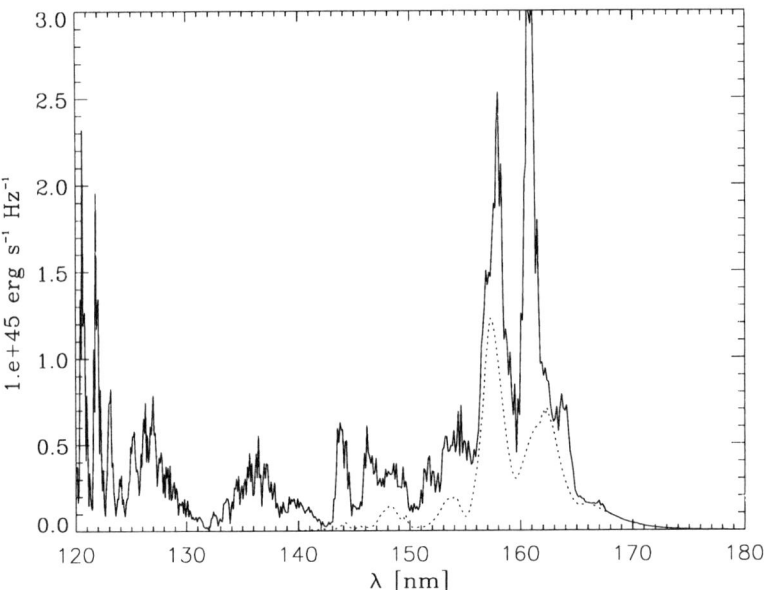

FIGURE 2. Fluorescence spectrum for T = 2000K, $\chi = 1$, $N_{H2} = 10^3 cm^{-2}$ and ortho-para ratio 3, convolved with a Hanning function of apodized 150 cm^{-1} FWHM. The dotted curve shows the emission into continuum.

population and a "narrow" Gaussian component of $\sigma < 5$ km/s which has always been observed together with a "wide" one. Heiles (1980) extended the statistics for $|b| > 10°$ and discussed the pervasiveness of the "wide" component which he defined by a velocity dispersion of $\sigma > 5$ km/s. Clouds show narrower components. He concluded that this "not strongly absorbing" (NSA) HI gas in the nearby region is distributed pervasively below $|b| = 40°$ and at an averaged height (above the Galactic plane) of 186 pc although the observed profiles show only factor-of-two agreement with the expected dispersion, and about 10 - 20% of the volume contains large holes with sizes up to 400 pc.

Line shapes of ^{12}CO and ^{13}CO in high latitude clouds (HLCs, $|b| > 25°$) need also a heating source as discussed by Blitz (1987). These clouds are observed at an average distance of 100 pc, with diameters of 2 pc and masses ≈ 40 M$_\odot$. They show considerable clumpiness, probably below the size scales which have been resolved yet.

Now I show a perspective of how the interstellar UV light acts as the heating agent of HI. The same simple model as described above has been applied to calculate velocity distributions of the dissociated H atoms as shown in Fig.3 for a set of temperatures. The peak at very low velocities is partly due to the fact that continuum radiation from low J's of the BΣ state normally starts up with a peak at the short wavelength limit, and a major contribution is caused by the first resonance above the dissociation limit of the ground $X^1\Sigma_g^+$ state, quasi-bound in the $l = 4$ centrifugal well at 0.86 cm^{-1} †. Therefore, the strength of this peak depends critically upon the population of (v = 0), J = 3, 4, 5 states in the H$_2$ gas. It has an equivalent kinetic temperature of ≈ 5 K. I may note: convolution with an instrument function of a few km/s resolution - as normally reported

† This is obtained from a $X^1\Sigma_g^+$ potential which gives $D_{00} = 36117.573$ cm^{-1}.

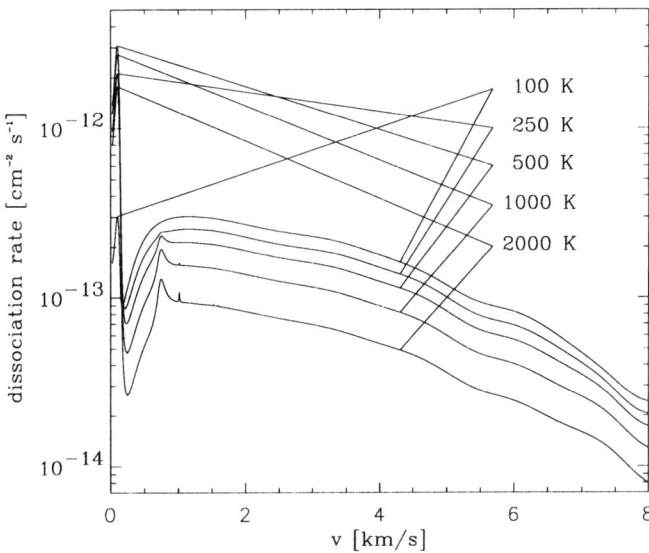

FIGURE 3. Velocity distributions of the dissociating H atoms obtained with model parameters mentioned in Fig.2, but at varying temperature.

for HI observations - could certainly simulate a cold HI gas of 10 - 50 K temperature. We see here the "cold" and the "warm" Doppler component of HI emitted by the same source although they are attached to two different species and separately spaced in the reviewed HI statistics. This has been explained above: the "warm" component evaporates off the clouds and emits after a free-flight time of 11 Million years. Generally, each single continuum spectrum of an excited B state contributes one or a few resonance peaks and continues with an undulating structure as shown in Fig.1. The latter is washed out in the composed velocity spectrum. The FWHM of all the rest curves in Fig.3 is reached at about 4.5 km/s ($\sigma \approx 3.8$ km/s). So, the σ of the "narrow" component appears to be clearly below 5 km/s, but one could use the long tail to introduce another Gaussian component with a σ close to and above 5 km/s.

I show also a few velocity profiles of the 2000 K distribution obtained with different Draine fields. Again, changes with field factors $\chi = 10$, 100 and 1000 do not change the profile significantly (Fig.4). The integrated dissociation rate increases with the χs due to the increased transition rates from X states to electronically excited states, in competition with quadrupole emission. Therefore, a few more resonance features are included at the higher χs. Similar plots could be shown for the lower temperatures.

This can be summarized by saying that LTE in the absorption, fluorescence and dissociation layer of the H_2 gas gives the same "narrow" $\sigma \approx 3.8$ km/s or 9 km/s FWHM component, independent of temperature and χ factor. Therefore, the obtained profiles are kind of a specific property of the H_2 gas. There is a remarkable increase of the integrated dissociation rate with decreasing temperature and the important long tail ("wide" component) makes an increasing relative contribution at higher temperatures. The "cold" HI component is larger at the higher temperatures and reaches a maximum between 250 K and 1000 K (Fig.3) due to emission into the $l = 4$ resonance. Thus the numbers so far obtained give a hint to an explanation of observed velocity dispersions

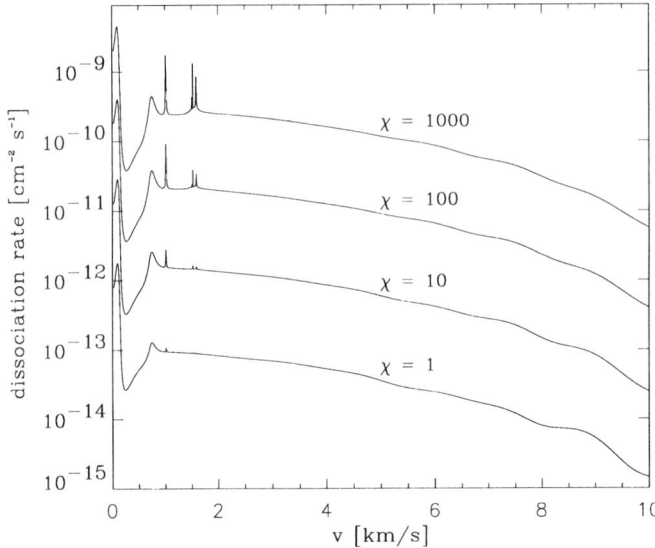

FIGURE 4. Selected velocity distributions of the dissociating H atoms obtained with model parameters mentioned in Fig.2, but with varying field factors.

of HI. By the way, there is far and wide no competing candidate. The sources showing only the "wide" Gaussian component are not covered by the simple LTE models used in the calculations. They need a different treatment.

A simple example may be added to illustrate how the quantitative findings fit to the observed abundances of H_2 (indirect from CO) and HI. I apply these findings to the above mentioned high latitude clouds, where CO maps have diameters of 2 pc and HI emission in the overlaid maps extend the diameter to the double and threefold size. An H atom with 4 km/s speed flies freely over a distance of roughly 4.3 pc in 11 Million years time. It is hard to believe that the supersonic H atoms recombine mainly on the available dust grain surfaces. These cloud models exhibit rapid expansion.

5. More problems

There are quite a few problems raised with the introduced dark matter Jupiter models and their gaseous interphase. Molecular dynamics problems are raised by the unusual high densities of the gas and the interactions with the highly supersonic HI atoms. 1) The dynamical setup makes possible line shape effects on tracer molecules which could not appear in traditional clouds because of their low densities. 2) Little is known about the dissociation of H_2 in collisions with high speed HI and also the opposite, recombination of slow HI with catalytic assistance of H_2 and He in 3-body collisions. These two processes start from or populate highly excited vibrational-rotational X states of a realistic state population, as we know, since these states have been observed in the ISO/SWS mission. For the latter I have found a few experimental low temperature results discussed in a paper by Schwenke (1988) and a first theoretical attempt to solve this problem. 3) High enough densities of both H_2 and HI could also cause collision induced dipole radiation of H_2 in the far infrared which is as yet entirely explained by a 2-component dust radiation.

In summary: the H_2 – HI and the H_2 – H_2 systems play an important role in the dark matter Jupiter models. The most important molecular processes besides UV radiation are collisions between HI and H_2.

New problems are also raised at large scales when the Jupiters are added to the known dynamical masses: think of maybe ten thousand Jupiters or the equivalent clumps of condensed gas added to each high latitude cloud. They give rise to "geometry effects", complex models of the radiating sources, a new understanding of clumpiness and probably interaction with each other.

REFERENCES

ABGRALL, H., ROUEFF, E., XIANMING LIU & SHEMANSKY, D. E. 1997 *Astrophys. J.* **481**, 557–566 and references therein.

BLITZ, L. 1987, in *Physical Processes in Interstellar Clouds*, (ed. G. E. Morfill & M. Scholer), pp. 35–58. D. Reidel Publishing Company.

BOSMA, A. 1981 *Astron. J.* **86**, 1791–1824.

COMBES, F., & PFENNIGER, D. 1996, in *New Extragalactic Perspectives in the New South Africa*, (ed. D. L. Block & J. M. Greenberg), pp. 451–466 Kluwer Academic.

DRAINE, B. T. 1978 *Astrophys. J. Suppl.* **36**, 595–619.

DRAINE, B. T. 1998 *Astrophys. J. Lett.* **509**, L41–L44.

HEILES, C. 1980 *Astrophys. J.* **235**, 833–839.

HENDERSON, A. P., JACKSON, P. D. & KERR, F. J. 1982 *Astrophys. J.* **263**, 116–122.

KULKARNI, S. R. & HEILES, C. 1988, in *Galactic and Extragalactic Radio Astronomy*, 2nd ed., (ed. G. L. Verschuur & K. I. Kellermann), A & A Library, pp. 95–153. Springer.

LANDOLT BÖRNSTEIN II/4 1961, p.180.

MARTIN, C., HURWITZ, M. & BOWYER, S. 1990 *Astrophys. J.* **354**, 220–228.

MEBOLD, U. 1972 *Astron. & Astrophys.* **19**, 13–26.

PFENNIGER, D., COMBES, F. & MARTINET, L. 1994 *Astron. & Astrophys.* **285**, 79–93.

PFENNIGER, D. & COMBES, F. 1994 *Astron. & Astrophys.* **285**, 94–118.

SCHWENKE, D. W. 1988 *J. Chem. Phys.* **89**, 2076–2091.

STERNBERG, A. 1989 *Astrophys. J.* **347**, 863–874.

TACCONI, L. J. & YOUNG, J. S. 1986 *Astrophys. J.* **308**, 600–610.

WALKER, M. & WARDLE, M. 1998 *Astrophys. J. Lett.* **498**, L125–L128.

WITT, A. N., STECHER, T. P., BOROSON, T. A. & BOHLIN, R. C. 1989 *Astrophys. J. Lett.* **336**, L21–L24.

Harvey Liszt and Peter Kalberla deeply meditating in the sun shine.

Mass of H_2 dark matter in the galactic halo

By Yu. A. SHCHEKINOV[1,3], R.- J. DETTMAR [1] AND P. M. W. KALBERLA[2]

[1]Astronomisches Institut, Ruhr Universität Bochum, Universität str. 150, D-44780 Bochum, GERMANY

[2]Radioastronomisches Institut, Universität Bonn, Auf dem Hügel 71, D-51121 Bonn, GERMANY

[3]Department of Physics, Rostov State University, Sorge 5, 344090 Rostov on Don, RUSSIA

We present a model for the interpretation of the residual gamma-ray background emission, which we attribute to interaction of high energy cosmic rays (CR) with dense molecular clumps in a flattened halo distribution. In a wide range of clump parameters we calculate the expected gamma-photon flux and put constraints on masses and sizes of clumps and their contribution to the galactic surface density: the best fit model implies dense clumps to be optically thick to gamma-photons with only about ~ 10 % contributing to the gamma-ray background. This circumstance increases substantially their contribution to the galactic surface density, up to 140 $M_\odot$ pc^{-2}. We discuss how this model relates to the extreme-scattering events.

---

1. Introduction

It is quite well established now that about 85 % of the baryon mass is still undetected: the mass of luminous baryons $\Omega_b^L \simeq 0.007$ is only about 15 % of the baryons inferred from cosmological nucleosynthesis, $\Omega_b \simeq 0.05$. The question of where these missing baryons are is of primary importance for galactic dynamics and star formation history. In the recent past it has been proposed that a considerable fraction of baryons can be contained in very dense and cold H_2 molecular clouds (MC) Pfenniger & Combes (1994), De Paolis et al. (1995), Gerhard & Silk (1996), Kalberla & Kerp (1998).

Recently, Walker & Wardle (1998) have reported arguments in favor of this understanding. They have suggested that the observed strong variations of fluxes of compact radio quasars, the extreme-scattering events (ESEs) discovered by Fiedler et al. (1987), are connected with refraction in HII atmospheres of compact MC ($M \sim 10^{-3} M_\odot$, $R \sim 1$ AU) and shown that the estimated covering factor of the ESE clouds and the dual-frequency light curves of radio quasars are compatible with the assumption that H_2 MC contribute to the galactic surface density as much as $\Sigma \geq 10^2 M_\odot$ pc^{-2}. Draine (1998) has demonstrated that optical lensing in such MC can improve the statistics of the microlensing events toward the LMC, and concluded that gravitational microlensing searches allow MC with $M \simeq 10^{-3} M_\odot$ and $R \simeq 10$ AU to be present in the halo.

Of many possible observational manifestations the diffuse γ-emission from interaction of CR protons with protons of MC and subsequent π^0 decay: $\pi^0 \to 2\gamma$, is the most robust. Several groups have recently studied this possibility, and concluded that EGRET data are consistent with the presence of dark matter (DM) in halo in the form of cold molecular gas, De Paolis et al. (1999), Dixon et al. (1998), Sciama (1999). In this paper we reanalyse EGRET data from the point of view that MC can be optically thick to both, high energy CR, and $\pi^0 \gamma$ photons – this issue was mentioned recently by Sciama (1999), while all other authors implicitly assume DM gas to be optically thin. We show that optically thick DM halo gas fits EGRET data best and that it contains much more mass than is usually thought.

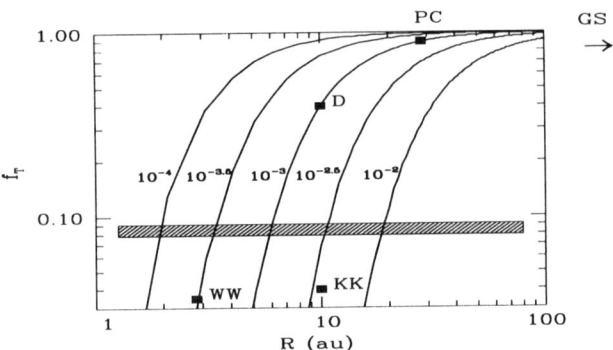

FIGURE 1. f_T versus clump radius for a set of clump masses; the curves are marked with mass in $M_\odot$; squares correspond to MC described by Draine (1998) – D, Kalberla & Kerp (1998) – KK, Pfenniger & Combes (1994) – PC, Walker & Wardle (1998) – WW, the arrow GS shows limits for MC of Gerhard & Silk (1996), dashed area shows the interval of f_T derived in this paper from fitting EGRET data.

2. Clump opacity to γ photons and CR

The fact that MC can be opaque to both CR and γ photons is readily seen from a comparison of the attenuation lengths $\Sigma_{CR} \simeq 40$ g cm^{-2} for high energy CR, and $\Sigma_p \simeq 80$ g cm^{-2} for γ photons, with a typical mass column density of a clump, $N_c \simeq 4 \times 10^6 M/R^2$ g cm^{-2}, (M and R are in solar mass and AU, respectively). A simple estimate of the fraction of baryons which can be seen in γ- emission has been given by Kalberla et al. (1999):

$$f_T = \frac{3L_{CR}^2 L_p^2}{4R^5} \int_0^R dr F_{CR}(r) F_p(r), \qquad (2.1)$$

where $L_i = \Sigma_i/\rho$, $i =$ CR, p,

$$F_i(r) = e^{-(R-r)/L_i}[1 + (R-r)/L_i] - e^{-(R+r)/L_i}[1 + (R+r)/L_i], \qquad (2.2)$$

Figure 1 shows the dependence of f_T versus cloud radius for different masses. It is readily seen that in the range of masses assumed for MC there is a significant absorption of CR and/or γ photons for $R \leq 20$ AU. Clouds with similar parameters have been described or restricted by Draine (1998), Kalberla & Kerp (1998), Walker & Wardle (1998). The total mass they contain is $M_{tot} = f_T^{-1} M_\gamma$, where M_γ is the mass observed in γ-emission.

3. Models

We calculated the distribution of γ-emission (with contributions from inverse Compton processes, bremsstrahlung and π^0 photons) over the sky including *observed* components and baryonic DM halo. To make more definite conclusions and to determine the best

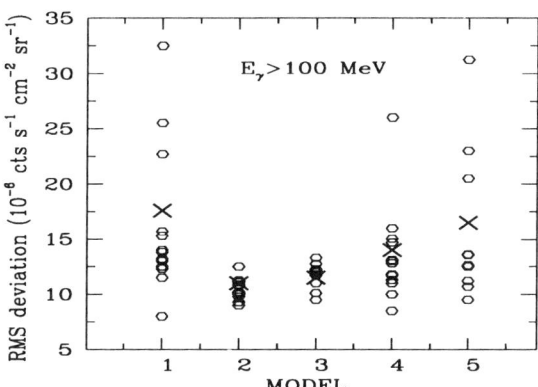

FIGURE 2. RMS deviations between predicted and observed fluxes of photons with $E_\gamma > 100$ MeV; circles show the RMS scatter derived for individual scans for constant longitudes, crosses show the mean RMS deviation for models.

model, we calculated the RMS deviations of predicted fluxes from the observed ones. The result is shown in figure 2.

Obviously the best fit is given by Model 2. The following components are parts of the models: the disk, extraplanar diffuse ionised gas, and the halo with *observed* parameters. For all gas components we have used the distribution in the form, Kalberla & Kerp (1998):

$$\rho_i(R,z) = \rho_{0i} g(R) \exp\left[\frac{-\Phi(z)}{\sigma_i^2(1+\alpha_i+\beta_i)}\right], \quad R = \sqrt{x^2+y^2} \qquad (3.3)$$

where i corresponds to the HI disk, the extraplanar DIG, and the gaseous halo with HI and hot plasma, $g(R)$ is a universal function describing radial distributions of gas components. In Model 2 we also described DM halo baryons by eq. (3.3). In Models 3–5 the DM halo was assumed to have the distribution proposed by De Paolis et al. (1999):

$$\rho_{H_2}(x,y,z) = \rho_0(q) \frac{a^2 + R_\odot^2}{a^2 + x^2 + y^2 + (z/q)^2} \qquad (3.4)$$

where $\rho_0(q)$ is the local DM density, q, the flattening parameter, $a = 5.6$ kpc.

In Models 1–2 we assumed CR in pressure equilibrium with gas components. As shown by Kalberla & Kerp (1998) this assumption ensures the ISM (including gas components, CR and magnetic field) to be stable against large scale perturbations, and explains the all sky survey in radio continuum, 21 cm and X-ray emission. From this point of view the best fit model (Model 2) can be regarded as self-consistent. Instead, in Models 3–5 CR were taken with constant energy density 0.12 eV cm^{-3}, as in De Paolis et al. (1999).

From a comparison of gas density which has been found to reproduce the background γ-emission with the value required to explain the rotation velocity ($n_0^d = 0.7$ cm^{-3} in Model 2), we can find the fraction of baryons missing in the EGRET observations. Model 1 does not produce significant γ- emission from the halo, and thus f_T is indefinite; Model 2 corresponds to only about 10 % DM baryons contributing to γ-emission from the halo; in Model 3 this fraction is close to one, while in Model 4 is about 0.75; Model 5 does not reproduce the observed rotation curve.

model	Eqn.	flattening[1]	cosmic rays[2]	DM density[3] EGRET	DM density[4] rot. curve
1	(3.3)		PE	0.0025	–
2	(3.3)	$q \sim 0.3$	PE	0.0625	0.7
3	(3.4)	$q = 0.3$	C	0.55	0.5
4	(3.4)	$q = 1$	C	0.18	0.24
5	(3.4)	$q = 1$, H	C	0.35	undefined

TABLE 1. [1]H – a central hole for $R < 10$ kpc in the DM distribution
[2]PE – pressure equilibrium cosmic rays, C – constant energy density in the halo 0.12 eV cm^{-3};
[3] local mid-plane density in cm^{-3}, best fit result from EGRET data
[4] local mid-plane density in cm^{-3}, best fit result from rotation curve

4. Conclusions

We have calculated the background gamma-emission from DM halo baryons in the form of dense compact MC. The best fit model is found to correspond to optically thick MC with only 10 % baryons contributing to the observed γ-ray background. This is a dynamically self-consistent model, in the sense that it is stable against large scale perturbations, and in addition explains the all-sky surveys in radio continuum, 21 cm, and X-ray emission as described by Kalberla & Kerp (1998). The parameters in this model are constrained thus in a narrow range.

We have studied also the gamma-emission from a DM halo with distribution in the form (3.4), and found that only models with a flattening $q \in (0.2, 04)$ can reasonably reproduce EGRET data, however in all cases the RMS deviation and the scatter is bigger than in our best fit model.

YS acknowledges partial support from German Science Foundation through Sonderforschungsbereich 191 and from the Organizing Committee.

REFERENCES

DE PAOLIS, F., INGROSSO, G., JETZER, PH. & RONCADELLI, M. 1995 *PR Lett.* **74**, 14–17
DE PAOLIS, F., INGROSSO, G., JETZER, PH. & RONCADELLI, M. 1999 *ApJ* **510** L103–L106
DIXON, D. D. ET AL. 1998 *New Astronomy* **3**, 539–561
DRAINE, B. T. 1998 *ApJ* **509**, L41–L44
FIEDLER, R. L., DENNISON, B., JOHNSON, K. J. & HEWISH, A. 1987 *Nature* **326**, 675–678
GERHARD, O. & SILK, J. 1996 *ApJ* **472**, 34–45
KALBERLA, P. M. W. & KERP, J. 1998 *A & A* **339**, 745–758
KALBERLA, P. M. W., SHCHEKINOV, YU. A. & DETTMAR, R.-J. 1999 *A & A* **350**, L9–L12
PFENNIGER, D. & COMBES, F. 1994 *A & A* **285**, 94–118
SCIAMA, D. W. 1999 *astro-ph/9906159*
WALKER M. & WARDLE, M. 1998 *ApJ* **498**, L125–L128

2. Formation - Destruction

Harvey Liszt and Peter Kalberla artistically posing for the photograph, with François Rostas reflecting about the last talk.

Experiments with Trapped Ions and Nanoparticles

By D. Gerlich, J. Illemann, AND S. Schlemmer

Faculty of Natural Science, Chemnitz University of Technology, 09107 Chemnitz, Germany

This contribution summarizes experimental work which has been performed predominantly in our laboratory using ion guides and specific traps for studying ions, molecules and dust particles under astrophysical conditions. After a short reminder of the basics of the technique and a brief discussion of our newest device, the nanoparticle trap, we shall review experimental results for low temperature gas phase collisions with H_2. In the last part we will summarize our present activities related to chemistry involving cold H atoms.

1. Introduction

Despite the fact that our knowledge on the role of hydrogen in space has significantly increased in recent years due to a combination of extensive new observations and astrophysical model calculations with fundamental theory and detailed innovative experiments, there are still many unsolved problems related to the interaction of H or H_2 with ions, radicals, surfaces and also photons. The most obvious example is the formation of H_2 itself; other examples include specific state-to-state cross sections, ortho-para transitions in H_2, H-D isotopic scrambling, formation and destruction of the H_3^+ molecule, or the role of hydrogen clusters and anions. In addition to gas phase reactions we will discuss in this paper our most ambitious goal, the detection of catalytic formation of H_2 molecules on an interstellar dust analogue localized in a cold trap.

2. Experimental: Ion guides and particle traps

2.1. *Inhomogeneous RF or AC fields*

From the point of view of experimental techniques, our research is predominantly based on the use of specific inhomogeneous, time-dependent, electrical fields, $\mathbf{E_0}(\mathbf{r},t) = \mathbf{E_0}(\mathbf{r}) \cdot \cos(\Omega t)$. The method to trap or guide ions and many applications in ion physics and ion chemistry have been documented thoroughly in Gerlich (1992). For practical applications, there are two very important facts to remember. (i) In good approximation the main influence of the oscillatory force can be described by the so called effective potential $V^* = q^2 \mathbf{E_0}(\mathbf{r})^2/4m\Omega^2$, where m and q is the particles mass and charge, respectively. (ii) In order to stay within the range of validity of this approximation, one has to operate under conditions such that $\eta < 0.3$, where η is the adiabaticity parameter defined by $\eta = 2q \mid \nabla \mathbf{E_0}(\mathbf{r}) \mid /m\Omega^2$. This basic principle is quite general and the effective potential can be used for confining electrons (Ω must be in the GHz range), ions (radio frequency range), clusters or charged microscopic particles (audio frequency, AC voltages). Typical applications include strong focusing lenses in accelerator physics, the well-known Paul trap invented in the 1950's or the Guided Ion Beam technique developed in the 1970s. For a detailed review of the historical development and of specific applications see Gerlich (1992). This and other confinement techniques in combination with laser cooling and detection schemes have been the foundation for a number of precision measurements (Schneider (1999)).

2.2. Temperature variable ion traps

Very important for studying gas phase reactions at temperatures and densities of interstellar interest are special traps with wide field free regions where the ions are cooled to the temperature of the cold electrodes or sourrounding walls. Examples include the temperature variable 22-pole trap and the ring electrode trap, which both have been described in detail in Gerlich (1994) and Gerlich (1995). These innovative devices have been utilized for experimental investigations of a variety of ion-molecule reactions. A proof of the unique sensitivity of these traps are the systematic studies performed on radiative association, a reaction mechanism of astrochemical importance (Gerlich & Horning (1992), Gerlich (1994)). In recent years, more versatile experimental setups, based on this technique, have been constructed by applying laser methods for ion preparation or detection (Schlemmer et al. (1999c), Schlemmer et al. (2000b)) or by integrating atomic beams Haufler et al. (2000). This latter apparatus is now being used for studying reactions between stored ions and a cold effusive beam of H atoms.

2.3. The NPMS technique

Based on the fact that the effective potential can be tailored for many purposes we have designed a special trap for observing and characterizing one single nanoparticles over long times. Whereas in the above mentioned case of a cold ion cloud a wide field-free region is required, here the precise localization of the object determines the field geometry. This is best achieved with a Paul trap, i.e., with a quadrupolar field the effective potential of which is harmonic. In practice we have used a rather complicated electrode arrangement in order to fulfill a variety of boundary conditions (Schlemmer et al. (1999a), Schlemmer et al. (2000a)). Briefly, the trap consists of two cones opposing each other at a distance of 6.6 mm. These two electrodes are surrounded by eight rods which approximate the usual ring electrode of a Paul trap. This open design has been chosen in order to obtain either a large solid angle for optical detection or to measure angular distributions of scattered light. Moreover the trap volume can be accessed through several ports with additional tools such as the particle source, the laser beam and an electron gun. The benefit of this new techniques lies in the combination of the following features: long time trapping of a single particle, isolation of the particle with respect to disturbing surfaces and gases , i.e., UHV- and low temperature conditions, very good localization of the particle (sub-μm). This makes it perfectly suited to sensitive optical detection (scattering, fluorescence, absorption, etc.).

One of the first obvious applications of this trapping technique is based on its mass resolving power in a mass range which is otherwise not accessible (NPMS: nanoparticle mass spectrometry). For this purpose, the oscillatory macro-motion of the particle is monitored using a collimated beam from a diode laser ($\sim 2.5\,\mathrm{mW/mm^2}$) which is directed near the center of the trap. The light scattered from the particle is collected by a lens, transferred outside the vacuum and transmitted to an avalanche photo diode. The eigenfrequencies are determined by Fourier transforming the detected signal. It is important to note that the particles trajectory has to remain close enough to the center of the quadrupolar field, where both the electrostatic and the effective potential are harmonic. In this case the derived eigen-frequencies, which are proportional to the particles q/m ratio, are independent on the amplitudes. Controlled charging/discharging of the particle in steps of single elementary charges leads to the absolute charge and mass. Thanks to the long term stability of our experimental setup the particles q/m-ratio can be followed over days.

First test measurements have been performed with 500 nm diameter SiO_2 spheres (nominal mass m $\approx 10^{-16}$ kg, $\approx 10^9$ SiO_2 molecules, $\approx 10^{11}$ u). For these particles a mass resolution $\Delta m/m \approx 10^{-4}$ has been achieved in a single ten second measurement. Recording the modulated signal over minutes improves the resolution to the ppm regime (for details see Schlemmer et al. (1999a)). Since the mass determination is based on a frequency measurement there are several possibilities to further improve the resolution significantly. Presently we extend the method to smaller particles by replacing the light scattering detection scheme with a LIF based method. For particles with a diameter of a few nm the secular frequencies are increasing from the few Hz- towards the kHz-regime reducing the time to determine q/m by FFT method. In addition low frequency noise which is hard to be rejected will no longer play a role. Also gravitational compensation is no longer necessary. In total we are confident that single molecule mass resolution will become feasible. Planned applications of this method in combination with hydrogen atoms and molecules will be discussed below.

Another result which is also important for understanding catalytic properties of interstellar dust particles, is the precise registration of the actual charge of the particle and its distribution on the surface. In general, photons, electrons, ions or suitable neutrals can be used to change the charge of a stored nanoparticle in a controlled way. In a recent experiment (Schlemmer et al. (1999a)) one single SiO_2 sphere (d = 500nm) has been bombarded with fast electrons. The individual events of charge jumps have been followed online step by step. From a series of such data quantitative values for the secondary electron emission yield have been derived.

Summarizing this experimental section we want to point out that confinement of charges in a fast oscillating field is a very versatile tool. Devices such as octopole beam guides, 22-pole ion traps, or nanoparticle traps open up a wide range of experiments which are important to understand the physical and chemical behavior of interstellar matter. In the following we summarize earlier applications of these methods to interactions of ions with the hydrogen molecule. In the last part we describe our present activities for determining sticking coefficients of H_2 and H on grain surfaces and for - finally - observing the formation of hydrogen molecules on dust grain equivalents.

3. H_3^+, $H^+ + H_2$ collisions, and deuterated variants

The H_3^+ molecular ion and the simple collision system $H^+ + H_2$ and isotopic variants play a specific role both in the chemistry of interstellar space and in our fundamental understanding of simple chemical reactions. Using the ion guiding and trapping technique, various aspects of this system have been studied over the last decades. Results include state-to-state differential cross sections (Gerlich (1977), Teloy (1978)), integral cross sections and thermal rate coefficients for isotopic scrambling (Gerlich (1982)) radiative and ternary association (Gerlich & Horning (1992)), rate coefficients for the growth of clusters (Paul et al. (1995)), and many specific results such as rate coefficients for ortho to para conversion (Gerlich (1990)). A variety of specific results from trapping experiments and from a merged beam arrangement can be found in Gerlich (1992).

Some of this experimental information already has been used in modeling certain aspects of the energy balance of the ISM; however it may be necessary that more detailed information has to be incorporated in modern models, for example, in order to understand the intensities of the H_2 ro-vibrational and pure rotational lines detected recently by ISO, or for predicting the correct branching ratios in reactions where D atoms compete with H atoms. Useful for such tasks are the numerical predictions which can be calculated from a dynamically biased statistical theory (Gerlich (1982)). This theory which has been developed some time ago (Gerlich (1977)), describes in great detail the above mentioned experimental results. A few details concerning the inclusion of dynamical constraints which have been derived from trajectory calculations, have been published in Gerlich et al. (1980). The statistical model also accounts in good approximation for the long range attraction and for restrictions imposed by the conservation of nuclear spin and parity. This is especially important for predicting correctly ortho to para transitions in the highly symmetric system $H^+ + H_2$ (Gerlich (1990)). One also should be aware that isotopic fractionation, e.g. in $D^+ + H_2$ collisions is strongly influenced by simple symmetry selection rules. From an experimental point of view, a renaissance of this simple statistical theory will become interesting for describing the results, we plan to measure for the $H_2^+ + H$ collision system.

The molecular H_3^+ ion plays an important role in the subsequent ion - molecule chemistry. Since formation via radiative association of $H^+ + H_2$ is negligible, the gas phase reaction of H_2^+ with H_2 leading to $H_3^+ + H$ is an important step in the H - H_2 balance. This reaction has been studied using the merged beams technique and also in a single beam arrangement using REMPI preparation of rotational states of H_2^+ (Glenewinkel-Meyer & Gerlich (1997)).

4. Collisions of H_2 with other important ions

There are a variety of other basic reactions involving hydrogen molecules which may or may not influence its thermal or chemical balance or which are important for forming molecules of special interest. Examples we briefly refer to include the hydrogen anion H^-, the first step in the formation sequence of hydrocarbons and the reactions leading to ammonia.

For the anionic system $H^- + D_2$ and $D^- + H_2$ the importance of tunneling has been shown experimentally by measuring in a guided ion beam experiment the absolute integral and differential cross sections for reactive scattering at energies below the barrier (Haufler et al. (1997)). Comparison of the integral cross sections for the two isotopic variants reveals a significant isotope effect. The measured barrier heights of 350±60meV for $H^- + D2$ and 330±60meV for $D^- + H_2$ indicate that, at low temperatures, scrambling via this reaction is as inefficient as via the neutral analogue, $H + D_2$.

The endothermic reaction $C^+ + H_2(j) \rightarrow CH^+ + H$ has been studied experimentally and theoretically in the threshold region and an interesting effect of reagent rotation has been resolved in the experiment which is consistent with the frozen nuclear spin (FNS) approximation, Gerlich et al. (1987). The large endothermicity of this reaction makes it extremely inefficient at low temperatures, as can be seen from the thermal rate coefficients published in this paper. Therefore, radiative association, i.e. direct formation of CH_2^+ via emission of a photon, becomes an important process in the formation of hydrocarbons (Gerlich & Horning (1992)). Precise values have been measured in the 22-pole trap at low temperatures for p-H_2, n-H_2 and n-D_2. The results can be found in Gerlich (1994).

The gas phase formation of ammonia starts by the reaction $N^+ + H_2 \rightarrow NH^+ + H$, followed by a chain of hydrogen abstraction reactions. This system is of special interest since it is almost thermoneutral and one does not yet know, whether an activation barrier hinders the NH^+ formation or whether it is actually endothermic by about 12 meV. The reverse reaction $NH^+ + H$ and the fine structure dependence will significantly increase our knowledge on this simple reaction which plays a key role in the interstellar formation of ammonia. Ion beam and ion trap results have been reported for $N^+ + D_2 \rightarrow ND^+ + D$ in Tosi et al. (1994).

5. Collisions involving H atoms

Despite the fact that H is the simplest atom, experiments with a well prepared ensemble of these radicals are still a challenge. In the following we report briefly on two related activities, going on in our laboratory. These activities are based on the development of an intense temperature variable H or D atom beam source. Combining this source with a special 22-pole trap device (Haufler et al. (2000)), allows us to study a variety of important and interesting gas phase processes under conditions of the ISM. The integration of the H atom beam into the nanoparticle trap machine will allow us to study surface mediated processes such as sticking and desorption or catalytic activities.

In the gas phase we will begin to study the interaction between H atoms and ions such as H_2^+, NH^+, CH^+, $C_2H_3^+$, $C_3H_2^+$, etc. For sufficiently large molecular ions it is expected that an H atom just "sticks", i.e., the collision complex gets stabilized via radiative emission very efficiently (Herbst & Le Page (1999)). A recent related issue is whether such charged or neutral molecules can be the missing link between atomic and molecular hydrogen. The catalytic cycle proposed is $XY^+ + H \rightarrow XYH^+$ followed by $XYH^+ + H \rightarrow XY^+ + H_2$.

A typical example which already may fulfill these assumption is the interaction of H atoms with $C_2H_3^+$ and $C_3H_2^+$. The reverse channel, i.e., collisions of $C_2H_2^+$ with H_2 have been extensively investigated, Gerlich (1993). Very recently the rotational dependence of $C_2H_3^+$ formation has been measured by infrared excitation of the acetylene ions (Schlemmer et al. (2000b)). Despite all these activities the details of the asymptotic potential surface are unknown and one does not yet know whether under interstellar conditions the chemical equilibrium $C_2H_2^+ + H_2 \leftrightarrow C_2H_3^+ + H$ is determined by the asymptotic energies or by some small barriers. The next more complicated system concerns the $C_3H_3^+$ complex. Low temperature studies of radiative association vs. hydrogen abstraction have revealed a very interesting temperature dependence in $C_3H^+ + H_2$ collision, Sorgenfrei & Gerlich (1994).

6. H atoms and NPMS

The overall importance of grain surfaces for processing interstellar matter is generally accepted. Nonetheless one has to admit that not much quantitative knowledge exists on the actual role, these nanometer or micrometer large particles play in the chemical balance, e.g. in the catalytic formation of H_2. For learning more about this fundamental process, interesting new experimental activities have been started in recent years some of which are discussed in this book. In the following we point out that, in comparison to these surface science activities, our NPMS technique follows a different approach which accounts more for the special properties of dust grains (Schlemmer et al. (1999b)). One important difference between macroscopic surfaces and nanoparticles is that quantum confinement affects the band structure, optical properties and also the chemical behavior. Other differences are caused by finite size effects and the large surface to volume ratio. Another important aspect is that in most cases interstellar dust particles are charged.

As reported above the mass resolving power we have achieved so far is already suited to study in average processes such as adsorption and desorption of H and H_2 on nanosurfaces. In this way, coefficients for sticking and desorption or parameters describing mobility, diffusion, reaction, and tunneling can be determined quantitatively. Based on our present precision, the adsorption of one single hydrogen atom on a 5 nm particle can be monitored simply as a 10^{-5} increase in mass. This corresponds to a jump in the secular frequency of the particles motion a factor of ten larger than the resolution. The ultimate goal of this NPMS technique is the direct observation of the catalytic activity of the grain by recording step by step the adsorption of individual H atoms and the desorption of hydrogen atoms or molecules.

Interesting questions are related to the finite size of the particle. For example the temperature of a structure consisting of 1000 atoms is raised by 15 K if the H-H recombination energy is distributed statistically among all degrees of freedom of the particle. If energy transfer is slow, local heating will lead to even higher temperatures followed by evaporation or may be also by chain reactions. Significant jumps in temperature can also be induced by absorbing a single photon or cosmic ray. Due to the small heat capacity it can be questioned in general, whether the nanoparticles "temperature" is in equilibrium with the "ambient temperature". There are various laboratory approaches to find realistic analogues to interstellar dust grains. In our trapping experiment the 500 nm SiO_2 particles will be replaced by smaller particles and other nanoparticles containing Si and C and other ingredients. Using molecular beam techniques the particles may also be coated with realistic icy mantles which have been identified on interstellar grains by μm observations.

In conclusion we hope that the combination of an H-atom beam with our new NPMS technique will provide quantitative data needed to explain the abundance of molecular hydrogen in the interstellar matter. However, we also will continue to search for efficient low temperature catalytic gas phase cycles, producing H_2 as well.

Acknowledgments: Most of this work has been funded by the Deutsche Forschungsgemeinschaft via the INK *Methods and material systems for the nanometer scale* and in the SPP *Star formation*.

REFERENCES

GERLICH, D. 1977 Ph.D. Thesis, University Freiburg, Germany.

GERLICH, D., NOWOTNY, U., SCHLIER, CH. & TELOY, E. 1980 *Chem. Phys.* **47**, 245.

GERLICH, D. 1982 in: *"Symposium on Atomic and Surface Physics"*, W. Lindinger, F. Howorka, T.D. Märk, and F. Egger (eds.), 304.

GERLICH, D., DISCH, R. & SCHERBARTH, S. 1987 *J. Chem. Phys.* **87**, 350.

GERLICH, D. 1989 *J. Chem. Phys.* **90**, 3574.

GERLICH, D. 1990 *J. Chem. Phys.* **92**, 2377.

GERLICH, D. 1992 *Adv. in Chem. Phys.* **LXXXII**, 1.

GERLICH, D. & HORNING, S. 1992 *Chem. Rev.* **92**, 1509.

GERLICH, D. 1993 *J. Chem. Soc., Faraday Trans.* **89**, 2199.

GERLICH, D. 1994 in: *"Molecules and Grains in Space"*, I. Nenner (ed.), AIP Press, New York, 489.

GERLICH, D. 1995 *Physica Scripta* **T59**, 256.

GERLICH, D. 1998 *Faraday Discuss.* **109**, 362.

GLENEWINKEL-MEYER, T. & GERLICH, D. 1997 *Israel J. of Chem.* **37**, 343.

HAUFLER, E., SCHLEMMER, S. & GERLICH, D. 1997 *J. Phys. Chem.* **101**, 6441.

HAUFLER, E., SCHLEMMER, S. & GERLICH, D. 2000 to be submitted to Rev. Sci. Instrum.

HERBST, E., LE PAGE, V., 1999 *Astron. Astrophys.* **344**, 310.

PAUL, W., LÜCKE, B., SCHLEMMER S., & GERLICH, D. 1995 *Int. J. Mass Spectrom. Ion Proc.* **150**, 373.

SCHLEMMER, S., ILLEMANN, J., WELLERT, S. & GERLICH, D. 1999a in: *"Trapped Charged Particles and Fundamental Physics"*, D. Schneider (ed.), AIP press, New York, 80.

SCHLEMMER, S., ILLEMANN, J., WELLERT, S. & GERLICH, D. 1999b in: "*The Physics and Chemistry of the Interstellar Medium*", V. Ossenkopf (ed.), CGA Publ.Co., Herdecke, 391.

SCHLEMMER S., KUHN T., LESCOP E., & GERLICH, D. 1999c *Int. J. Mass Spectrom. Ion Proc.* **185**, 589.

SCHLEMMER, S., ILLEMANN, J., WELLERT, S. & GERLICH, D. 2000a submitted to Science.

SCHLEMMER S., LESCOP E., V.RICHTHOFEN, J. & GERLICH, D. 2000b submitted to J. Chem. Phys.

SCHNEIDER, D. 1999 "*Trapped Charged Particles and Fundamental Physics*", AIP press, New York.

SORGENFREI, A. & GERLICH, D. 1994 in: "*Molecules and Grains in Space*", I. Nenner (ed.), AIP Press, New York, 505.

TELOY, E. 1978 in: "*Electronic and Atomic Collisions*", G. Watel (ed.), North-Holland, Amsterdam, 591.

TOSI, P., DMITRIJEV, O., BASSI, D., WICK, O. & GERLICH, D. 1994 *J. Chem. Phys.* **100**, 4300.

Dieter Gerlich and Danielle Dowek discussing the conference program.

Laboratory Studies of Molecular Hydrogen Formation on Surfaces of Astrophysical Interest

By Valerio PIRRONELLO[1], Ofer BIHAM[2], Giulio MANICÓ[1], Joe E. ROSER[3] AND Gianfranco VIDALI[3]

[1]Departimento di Metodologie Fisiche e Chimiche per l'Ingegneria, Universitá di Catania,
Viale Doria 6, 95125 Catania, Italy
e-mail: pirronello@dmfci.ing.unict.it

[2]Racah Institute of Physics, Hebrew University, Jerusalem, Israel

[3]Physics Department, Syracuse University, Syracuse, NY 13244, USA

We review laboratory studies of the formation of molecular hydrogen on surfaces and in conditions of astrophysical interest. Theoretical analysis and predictions are given on how experimental results can shed light on actual physico-chemical processes occuring in the interstellar medium. Preliminary measurements of H atom sticking are also shown.

1. Introduction

Molecular hydrogen is the most abundant molecule in the Universe. In space it plays two main roles that render it of incomparable importance:
- once formed it becomes a very efficient coolant that increases the rate of collapse of interstellar clouds contributing to shape galaxies and to regulate their dynamics;
- once ionized by UV photons or by cosmic rays H_2 enters and triggers all reactions schemes that form most of molecular species in the gas phase.

Molecular hydrogen has always been a source of puzzles for astrophysicists in spite of its simplicity. The mechanism of its formation in the extreme conditions encountered in the interstellar medium is certainly one of them. When a sufficient abundance of electrons and ions exists, ion-atom reactions may be quite effective in producing H_2 in the gas phase (Stancil & Dalgarno 1998; when, on the contrary, physical conditions allow almost only the presence of neutral hydrogen atoms, the radiative association of two of them is highly inefficient and cannot explain the observed abundances. The main reason is that the time-scale for the release of a consistent fraction of the formation energy (4.5 eV) through the emission of a photon via forbidden transitions is too long and the proto-molecule that is formed in a repulsive state almost invariably breaks up. To overcome this insurmountable difficulty it has been proposed, about forty years ago (Gould & Salpeter 1963), that interstellar dust grains play the role of catalysts.

There are two main mechanisms of heterogeneous catalysis on a solid: the Eley-Rideal and the Langmuir-Hinshelwood reactions (Zangwill 1984). The so-called Eley-Rideal mechanism ascribes the formation of a molecular species on the surface of a catalyst to the direct interaction of an adsorbed radical with an atom hitting it upon arrival from the gas phase; this mechanism has been positively identified experimentally only in a small number of systems, as, for instance, in the case of hydrogen and chlorine adsorbed on single crystal copper surfaces (Jackson et al. 1994, Rettner et al. 1995). In the so-called Langmuir-Hinshelwood mechanism, molecule formation occurs through a reaction between two atoms (or radicals) that are already adsorbed on the surface. Several processes have to take place for this mechanism to produce molecular hydrogen: (a) collision of an H atom with the solid; (b) accommodation and sticking of the H atom on the surface; (c) mobility of the adsorbed H atom; (d) encounter with a previously adsorbed H atom (via processes (a) and (b)); (e) recombination and (f) possible restoration of the product to the gas phase.

Because the efficiency of the Eley-Rideal mechanism increases linearly with the surface density of the adsorbed radical (i.e., with its coverage), it is the Langmuir-Hinshelwood mechanism that is considered to be the most relevant in interstellar conditions, where a relatively low coverage of H ad-atoms is expected.

The problem of the formation of H_2 is one of those in which there is an abundance of theoretical work (Gould & Salpeter 1963, Williams 1968, Hollenbach & Salpeter 1970, 1971, Hollenbach, Werner and Salpeter 1971, Smoluchowski 1981, 1983, Aronowitz & Chang 1985, Duley & Williams, 1986, Pirronello & Averna 1988, Duley 1996, just to quote a few of them) but a dearth of experimental ones. Due to intrinsic experimental difficulties, only few laboratory attempts have been performed to measure the efficiency of the catalytic process (King & Wise 1963, Schutte et al. 1976, Pirronello et al. 1997a, Pirronello et al. 1997b, Pirronello et al. 1999). Of course the best understanding of the problem necessarily needs the results of experimental investigations that simulate the occurrence of the process of H recombination in conditions close to those encountered in interstellar clouds and that provide the values of those parameters for more realistic theoretical descriptions.

In the next section we describe the experimental apparatus, the measurement procedures, a selection of recent results, and a discussion. In the third section a theoretical framework is given and a connection between our experimental results and previous theoretical work is made. In the last section we discuss the impact of these results to astrophysically relevant environments.

2. The experiments

From the experimental point of view, some pioneering work was performed in the early '60s (see e.g. King & Wise 1963) and '70s (Schutte et al. 1976). The very first measurements of the '60s suffered from experimental limitations in particular due to poor vacuum conditions. Later, Schutte et al. (1976), made measurements on the semiconductor surface of a bolometer using a flux of H atoms about two orders of magnitude higher than the one used today and with an estimated undissociated faction of molecular hydrogen coming from the source of $\sim 10\%$. They obtained that at a surface temperature of $\sim$ 3 K the hydrogen recombination efficiency on a surface with a H_2 layer is about 0.25 but it drops to 0.05 to 0.1 on a H_2-free surface. These measurements also suffer from the astrophysical point of view of limitations. Such limitation are mainly related with: the type of surface used (that of a semiconductor); the choice of the surface temperature ($\sim$ 3 K) that is "far" from the range between 10 K and 15 K (the lowest in which dust grains are found in clouds) especially in the sense that in this temperature range the effect of the finite residence time of H ad-atoms starts to become important; the use of high fluxes (not in comparison with interstellar fluxes that clearly cannot be attained in the laboratory), high in the sense that they do not allow to investigate easily H ad-atom coverages as low as a small fraction of a monolayer.

Only very recently the problem of the formation of molecular hydrogen has been investigated again (Pirronello et al. 1997a, Pirronello et al. 1997b, Pirronello et al. 1999) on realistic surfaces and in conditions close to those encountered in space.

In order to have a detailed description of all the important processes, one would be tempted to investigate the single processes that make the Langmuir-Hinshelwood mechanism work, such as atom sticking and mobility. This, however, would be a tremendously difficult task to be accomplished with amorphous and polycrystalline surfaces, which are the ones that better represent surfaces of solids in space. Measurements of the mentioned single processes have been successfully performed on well characterized crystalline surfaces, however, in conditions of little astrophysical interest (Zangwill 1984).

Chemically inert supersonic He beams at thermal energy are used to characterize the morphology, atomic structure and dynamics of the topmost layer of a solid surface or the structure and dynamics of atoms and molecules residing on top of a substrate (adsorbates) (Hulpke 1992, Vidali & Zeng 1996). The method is sensitive down to parts of 1% of one layer in coverage, chiefly because of the large He-ad-atom scattering cross-section (Poelsema & Comsa 1989). Unfortunately this method can be applied only when the sample surface is ordered enough on a $10 - 100 Å$ scale, so that there is a measurable amount of specular He beam reflection. Sticking of both molecular and atomic hydrogen on HOPG (Highly Oriented Pyrolytic Graphite) has been measured by Lin and Vidali (1996) by looking at the dimming of the specular reflection of a thermal helium beam due to the increase of diffuse scattering with increasing coverage of adsorbed species; for other cases see for instance Rettner et al. 1995 and Grovers et al. 1980. These are clear examples of a technique that has to be abandoned when studying polycrystalline or amorphous surfaces because the specular reflection of the control helium beam is far too small even in the absence of any adsorbed species. An even more difficult problem has to do with the measurement of diffusion of ad-atoms on surfaces; it can be directly probed only in the case of metals by means of Field Ionization Spectroscopy or Scanning Tunneling Microscopy (Zangwill 1984)

We decided to study the process of molecular hydrogen formation by means of the Langmuir-Hinshelwood mechanism considered as a whole, i.e., by depositing hydrogen atoms on a surface and by measuring the amount of recombined H_2.

Nonetheless, it is possible to obtain information on elementary atom-surface processes. For example, preliminary results on sticking (or more exactly on the residence time, assuming that the temperature dependence of sticking is due to the residence time) of H atoms on an amorphous carbon sample have been obtained (see Fig.1) by sending a beam of H atoms to the sample and measuring the change in the partial pressure from the case when the sample is warm (no sticking) to the case when is cold (substantive sticking). From the drop in pressure due to the adsorption of atoms by the surface, the sticking of atoms and molecules can be calculated. (Rettner et al., 1986)

2.1. *The experimental apparatus and methods*

A schematic top view of the experimental apparatus used by us at Syracuse University in the most recent series of measurements is shown in Fig. 2.

The apparatus consists of a Ultra-High Vacuum (UHV) chamber pumped by a cryopump and a turbo-molecular pump (operating pressure in the low 10^{-10} torr range). The sample is placed in the center of the UHV chamber and is mounted on a liquid helium continuous flow cryostat. By varying the flow of liquid helium and with the use of a heater located behind the sample, temperatures can be maintained in the range of 5-30 K. For cleaning purposes, the temperature of the sample can be raised to about 200 ° C (without liquid helium in the cryostat). The temperature is measured by an iron-gold/chrome thermocouple and a calibrated silicon diode placed in contact with the sample. Two triple differentially pumped atomic beam lines are aimed at the surface of the sample. Each has a radio-frequency cavity in which the molecular species is dissociated, cooled to $\sim$ 200 K by passing the atoms through a cooled aluminum channel, and then injected into the line. Dissociation rates are typically in the 75 to 90 % range, and are con-

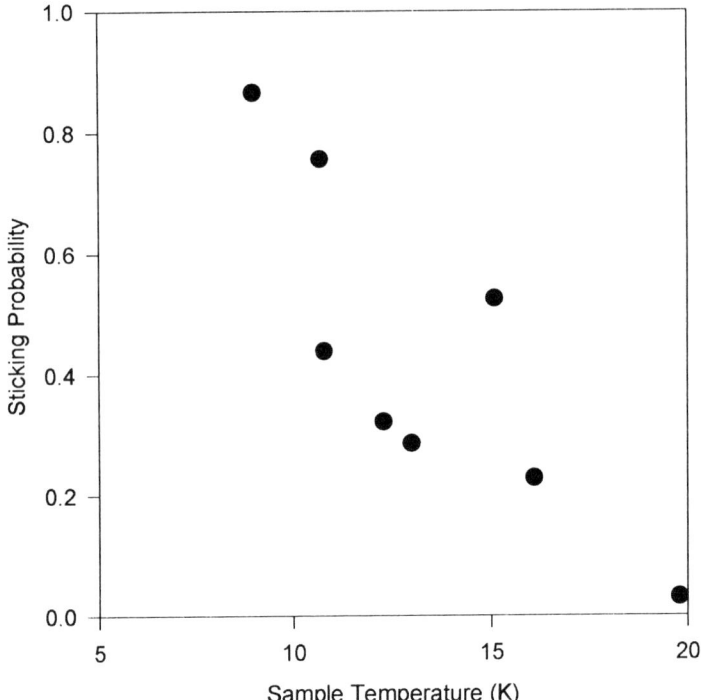

FIGURE 1. Probability of sticking of H atoms on amorphous carbon as a function of sample temperature

stant throughout a run. Estimated fluxes are as low as 10^{12} atoms $cm^{-2}s^{-1}$. A detailed description of the measurement methods employed is given in Vidali et al. 1998a.

The reason for using two beam lines and two isotopes (one line for H and the other for D) is dictated by the fact that, in preliminary runs using only one line, it became evident that the signal of H_2 (or D_2) formation was hidden in the background given by the undissociated fraction of molecules coming directly from the beam source. The possibility of using a second line is undoubtedly one of the most important features of this equipment.

By using H atoms in one line and D atoms in the other, we can look at the formation of HD on the surface, knowing that there are no other spurious sources of HD. The signal of HD is collected by a quadrupole mass spectrometer mounted on a rotatable flange. The experiment is done in two phases. First, H and D are sent onto the surface for a given period of time (from tens of seconds to tens of minutes). At this time any HD formed and released is detected. In the second phase, the sample temperature is quickly (~ 0.6 K/sec) ramped and the HD signal is measured. This latter experiment is called Temperature Programmed Desorption (TPD).

2.2. Experimental results

Two types of surfaces of astrophysical interest have been studied: a natural olivine mineral sample (a polycrystalline silicate mechanically polished until shiny (Pirronello et al. 1997a,b) and an amorphous carbon sample, prepared in an arc discharge (Pirronello et al. 1999). The production yield of HD is measured for different sample temperatures

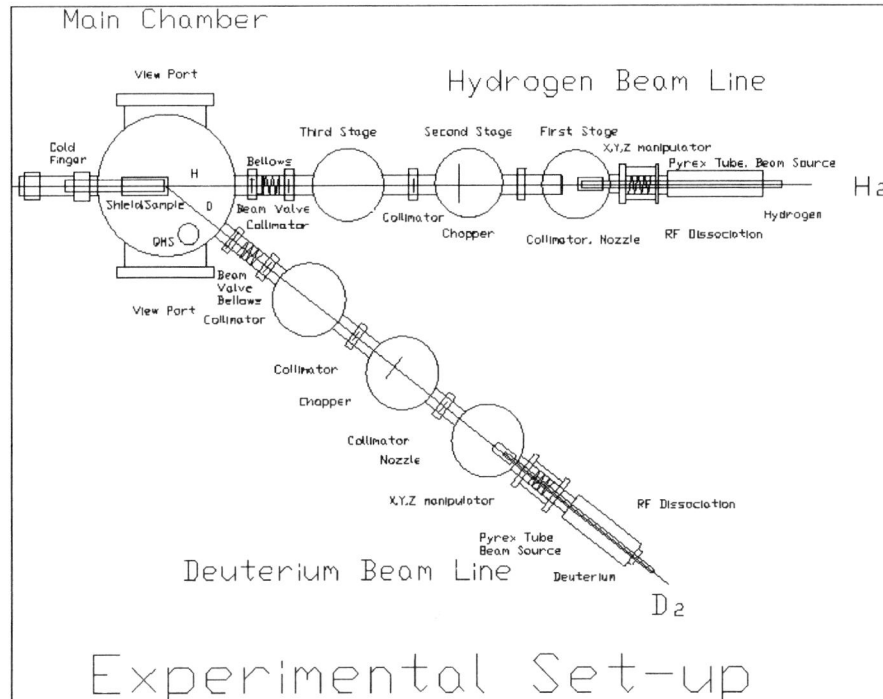

FIGURE 2. Schematic top view of the apparatus

both during and after the irradiation. In the latter case, a TPD experiment is done to obtain the yield of HD, as described below.

The so-called efficiency of the process, which is proportional to the ratio of the HD yield and the number of H and D atoms sent on the target, is computed by taking into account the fact that H and D yield also H_2 and D_2. The numerical values obtained are as high as 0.35 for olivine and 0.5 for amorphous carbon at the bottom irradiation temperatures of 5-7 K, but drop to significant lower values at temperatures of interstellar importance (i.e. above 10 K), see e.g. figure 2 in Pirronello et al. 1999. This efficiency depends, in general, also on the total fluency of H and D atoms.

As in the case for olivine, and at the lowest sample temperatures, most of HD detected is formed because of thermal activation during the heat pulse. Only a small fraction of HD is formed during the irradiation process, showing that, at least under our experimental conditions, prompt-reaction mechanisms (Duley 1996) or fast tunneling (Hollenbach & Salpeter 1971) are not that important.

The main result, in our opinion, is obtained by analyzing the kinetics of the desorption as a function of the coverage of H ad-atoms on the surface after irradiation performed at the lowest possible temperatures. The desorption spectra obtained from amorphous carbon surface are shown in Fig. 3.

The positions of the peak maxima that shift toward lower ramp temperatures (hence to shorter times from the beginning of the temperature ramp) clearly show that H and D atoms are still present on the surface when the warm-up begins. Hydrogen and deuterium ad-atoms are able to overcome the energetic barrier for diffusion and are set in motion at around ~ 10 K; shortly thereafter, they start encountering one another, recombining, leaving the surface as HD and producing the peak signal in the quadrupole. The kinetics of the process is said to be of the second order because the production and desorption

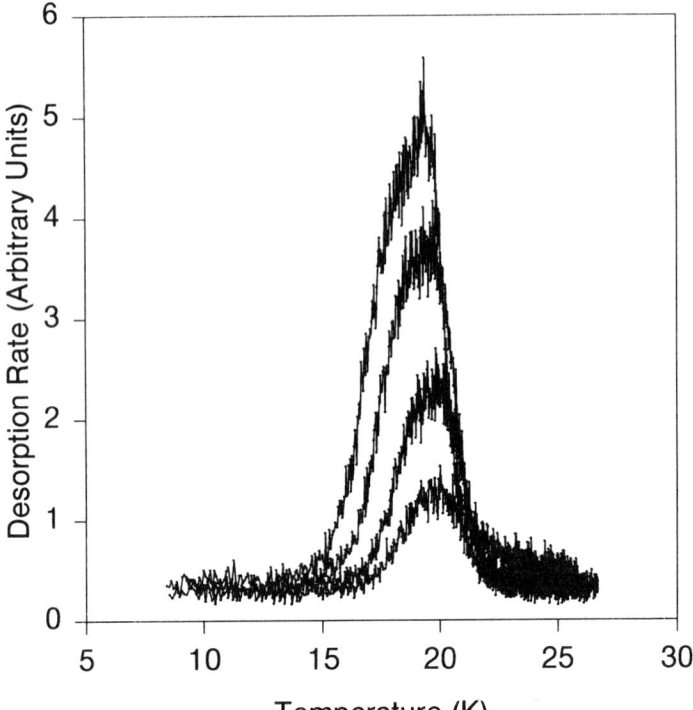

FIGURE 3. HD desorption rate vs. surface temperature of an amorphous carbon sample during TPD runs following adsorption of H and D on amorphous carbon at ~ 7 K for (bottom to top): 24, 48, 96, and 192 sec.

rates of the final products are proportional to the product of the surface concentrations of the reaction partners. The higher the coverage of H atoms, the shorter the average time before they encounter and form the HD molecule. This evidence is even clearer in the case of olivine (Pirronello et al. 1997b) where a lower coverage could be obtained. From this type of measurements we may safely conclude that tunneling does not assure high mobility (at least on polycrystalline and amorphous surfaces), as it has been assumed up to now following the influential papers of Hollenbach and Salpeter (1970, 1971), but that thermal activation is necessary.

This experimental evidence allowed us (Pirronello et al. 1997b) to propose a new expression for the formation rate of molecular hydrogen in interstellar clouds

$$R_{H_2} = \frac{1}{2}(n_H v_H A S t_H)^2 n_g \alpha \gamma\prime, \qquad (2.1)$$

where n_H and v_H are the number density and the speed of H atoms in the gas phase, respectively; A is the average cross-sectional area of a grain; t_H is the residence time of adsorbed H atoms on the surface, S is the sticking coefficient, n_g is the number density of dust grains, and α is the hopping rate of a single H ad-atom. $\gamma\prime$ takes into account the possibility that there is an activation energy for recombination. Notice that the quantity that is squared, i.e., $(n_H v_H A S t_H)$, is the average number N of H atoms adsorbed on the surface at a given time (the coverage).

It is worth noting that this expression is quadratic in the coverage of H ad-atoms and thus, it is qualitatively different from the one of Hollenbach et al. (1971), this latter being independent on the coverage of H ad-atoms, see below.

The physical rationale behind this expression is as follows. For an ad-atom that is localized on an adsorption site on the grain surface, α represents the number of hops it makes in the unit of time from one site to another and hence the number of sites it explores by a random walk in the unit of time; the probability for such an ad-atom to encounter another H ad-atom is, of course, proportional to the number of other H ad-atoms that are on the grain surface (i.e., to the total number N of H ad-atoms on the surface minus one, which is neglected in Eq.1 and is important only when very few H atoms are adsorbed). This probability is then $(N-1)\alpha$ for each single H ad-atom; the total rate of encounter between H ad-atoms on the surface is then $N(N-1)/2\alpha$ because there are N ad-atoms on the surface. The factor $1/2$ arises to avoid counting the number of encounters twice. The fraction that recombines is given by $\gamma\prime$. n_g gives the rate of H_2 recombination per unit volume in a cloud.

Theoretical papers on H_2 recombination have analyzed more sophisticated facets of the problem, taking into account both the possibility of a passive or an active surface where the catalytic process takes place. Incidentally we note that the work of Hollenbach and Salpeter (1970, 1971) still remains the reference point for any advancement in the field and any new expression for the formation rate of molecular hydrogen in clouds should be compared with the one they have obtained. They studied the formation of H_2 via the Langmuir-Hinshelwood mechanism as applied to interstellar conditions. They chose as a model surface that of a mono-crystalline ice layer (with some enhanced binding sites) and calculated sticking of H atoms from the gas phase in a semi-classical way. To obtain the mobility of adsorbed H atoms (H ad-atoms) they considered tunneling from one adsorption site to an adjacent one. They obtained a very high mobility of H ad-atoms consistently with the high diffusion of the wave packet describing quantum mechanically the H ad-atom in the periodic potential of the mono-crystalline ice surface. Hollenbach et al. (1971) obtained this simple expression for the steady state production rate of molecular hydrogen per unit volume in a cloud:

$$R_{H_2} = \frac{1}{2} n_H v_H A S \eta n_g, \quad (2.2)$$

where η is the probability that two H ad-atoms on the surface meet and recombine to form H_2. This equation substantially states that, for $\eta = 1$ (the value chosen by the authors to be consistent with the high mobility due tunneling), whenever two H atoms are adsorbed on a grain, a H_2 molecule is formed. This expression has been accepted and widely used to estimate the rate R_{H_2}. However, Smoluchowski (1981, 1983) obtained rather different results after a careful quantum mechanical treatment of the H ad-atom mobility on an amorphous ice surface. His results, orders of magnitude lower than those of Hollenbach and Salpeter, motivated the search for other possible mechanisms (Pirronello & Averna 1988, Averna & Pirronello 1991), that could be effective in regions where ice can accrete on grains.

3. The Theoretical Framework

Our expression and that of Hollenbach and Salpeter are different and may seem incompatible. However, it was shown by Biham et al. (1998) that Eqs. (2.2) and (2.1), are two limiting cases of the H recombination rate that are valid in two different regimes of mobility and/or coverage.

One possible approach to describe theoretically the processes of H_2 formation that take place on grains is to use rate equations. This formalism gives quantitatively, as a function of time, the population of H ad-atoms on grains by taking into account the contributions of sources and sinks (Biham et al. 1998). One should be aware of the difficulty that this approach runs into when very few adsorbate atoms on a very small grain, as discussed by Tielens & Hagen, 1982. For sake of simplicity, let's consider here what happens on a single grain; the final result can then be multiplied by the number density of grains to get the rate per unit time and unit volume.

The number of H ad-atoms on the surface N_H of a single grain as a function of time and the number of H_2 molecules produced in the unit of time (R_{H_2}) and immediately released in the gas phase are given by the rate equations

$$\dot{N}_H = FAS - pN_H - \alpha N_H^2 \tag{3.3}$$

$$R_{H_2} = \frac{1}{2} \alpha N_H^2 \tag{3.4}$$

where F is the flux of H atoms on the grain, i.e. $F = n_H v_H$; p is the desorption rate for H atoms, i.e. it is equal to the inverse of their residence time on the surface, i.e. $p = 1/t_H$. The first term in Eq. (3.3) represents the source of the H ad-atoms due to the incoming flux; the second term is the sink of H ad-atoms due to their evaporation (still as atoms) and the third term represents the rate at which H ad-atoms are lost due to the recombination into H_2. In Eq. (3.4) the factor $1/2$ reflects the fact that two H ad-atoms are needed to form one H_2 molecule.

In steady state conditions $\dot{N}_H = 0$ and the H population on the grain is given by $N_H = [-p + (p^2 + 4\alpha FAS)^{1/2}]/(2\alpha)$; the production rate of H_2 from a single grain then becomes

$$R_{H_2} = \frac{p^2 - p(p^2 + 4\alpha FAS)^{1/2} + 2\alpha FAS}{4\alpha}. \tag{3.5}$$

It is interesting to evaluate expression (3.5) in two limiting cases.

The first case is when the ad-atom evaporation rate (as atoms) is negligible compared to their recombination rate on the surface. This is the limit of $p^2 \ll 2\alpha FAS$. In this case it is possible to neglect the first two terms in the numerator on the right hand side of Eq. (3.5), finding

$$R_{H_2} = \frac{1}{2} FAS; \qquad p^2 \ll 2\alpha FAS, \tag{3.6}$$

namely, all H atoms that stick to the surface recombine and desorb as H_2 molecules; it is exactly the expression obtained by Hollenbach et al. (1971). In the other limit, $p^2 \gg \alpha FAS$, expanding the square root in Eq. (3.5) according to $(1+x)^{1/2} \simeq 1 + x/2 - x^2/8$ with $x = (4\alpha FAS)/p^2$, one gets:

$$R_{H_2} = \frac{1}{2} \frac{(FAS)^2}{p^2} \alpha = \frac{1}{2} (FASt_H)^2 \alpha \qquad p^2 \gg 2\alpha FAS. \tag{3.7}$$

that is exactly equal to the expression proposed by Pirronello et al. (1997b) based on the analysis of the experimental results summarized in the previous section.

Up to this point we have learned the following about the mechanism of formation of molecular hydrogen on realistic surfaces: mobility is not assured by tunneling but by thermal hopping; and the two expressions proposed to calculate the formation rate of molecular hydrogen per unit volume in a cloud at the steady state are both valid but in two different regimes of H ad-atom coverage.

What is really needed at this point is the numerical value of the parameters, such as alpha and t_H to be able to evaluate the actual rate of formation. In order to overcome this problem Katz et al. (1999) performed a simulation of the experiments using olivine

Material	E_{diff}	$E_{H_{des}}$	$E_{H_2 des}$	θ
Olivine	24.7	32.1	27.1	.33
Amorphous Carbon	44.0	56.7	46.7	.42

TABLE 1. Energy barriers in MeV for diffusion and desorption of atomic and molecular hydrogen

(Pirronello et al. 1997b) and amorphous carbon (Pirronello et al. 1999). After trying unsuccessfully to reproduce the desorption spectra of fig. 2 with the assumption that H_2 is immediately released in the gas phase upon formation, rate equations were changed to take into account the possibility that a fraction θ of the molecules formed remained on the solid. The parameters of the simulation are θ and the energy barriers for desorption and mobility of H and H_2. The results obtained by Katz et al. (1999) allow quantitative applications of the experimental measurements to astrophysically relevant situations are summarized in Table 1.

4. Astrophysical Implications

In Figure 4 we show the steady state formation rate of molecular hydrogen as a function of grain temperature obtained for a diffuse cloud (characterized by a density of 100 H atoms per unit volume and a gas kinetic temperature of 80 K) when using the values of the parameters deduced by Katz et al. (1999) for amorphous carbon grains. The additional hypothesis was made that at low temperature (when the mobility is very low and coverage becomes close to unity) the Eley-Rideal mechanism produces H_2 with an efficiency close to 100 %.

From Figure 4 it is clear that the rate drops very quickly with increasing grain temperature; this may pose a problem for all astrophysical environments in which grains are warm. An increase in the efficiency of recombination could come from the possible existence of a very rough or even porous structure of grains that would compensate the short lifetime of adsorbed species with an increased probability of trapping of H atoms.

This problem becomes relevant for small grains in diffuse clouds. In these clouds the equilibrium temperature of large grains is low enough to give rise to an efficient production of molecular hydrogen, but small grains (that have the highest exposed surface) suffer from relevant temperature fluctuations induced by impulsive energy releases following the absorption of UV photons.

Such spikes can increase the temperature of grains to several tens or hundreds K in times of the order of 10^{-10} seconds. The effect of such a sharp rise is to desorb H ad-atoms directly to the gas phase without allowing them to recombine. The spikes will reset the population of adsorbed hydrogen atoms on the surface of small grains to zero, leaving to the numerically smaller population of larger grains the duty of forming molecular hydrogen. The involvement of small grains in the production of H_2 is then dictated by the competition of two processes: a) collision and adsorption of H atoms; b) absorption of energetic UV photons. Only when the time-scale for adsorption

$$t_{ad} = (n_H v_H A S)^{-1} \tag{4.8}$$

is much shorter than the time for the absorption of a single UV photon (assuming that the grain behaves like a black body at UV wavelengths)

$$t_{UV} = (\sigma \Phi_{UV})^{-1} \tag{4.9}$$

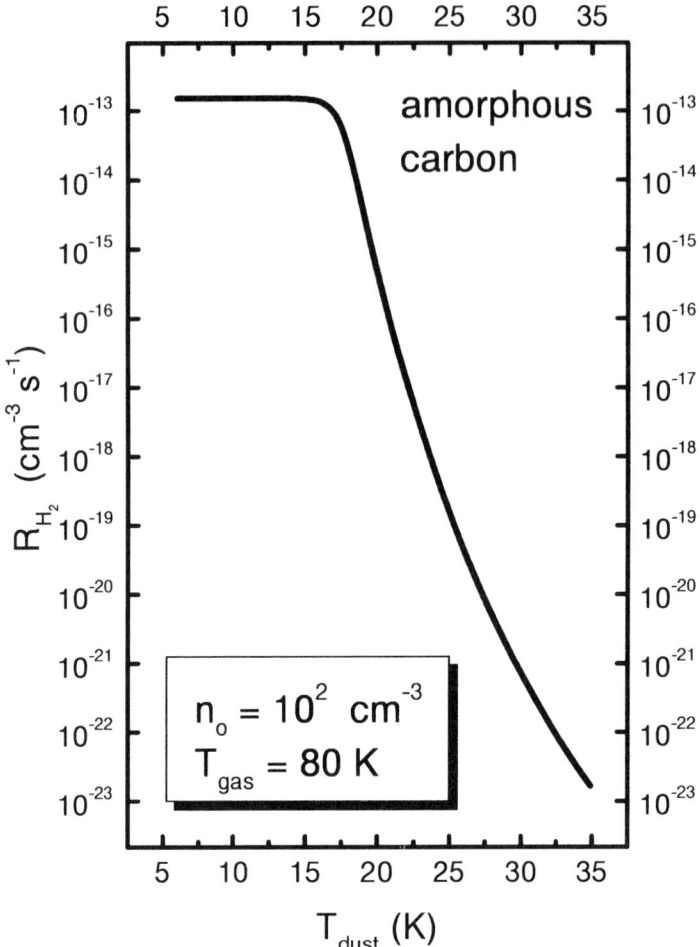

FIGURE 4. R_{H_2} versus T_d

where Φ_{UV} is the UV flux, then small grains will be able to participate to the production of molecular hydrogen.

In diffuse clouds which are permeated by almost unscreened UV radiation, formation of H_2 on small grains may be ruled out. In clouds of intermediate density, between diffuse and dense ones, there may be room for H_2 synthesis even on small grains.

Figure 5 shows the results of the first application of the reported experimental results and theoretical analysis to the chemical evolution inside a static diffuse cloud at the depth of $A_V = 1$ (density is equal to 100 H atoms per unit volume, gas kinetic temperature is 80 K) and grain temperature of 18 K. In this simple model (the chemical network includes only 130 species) it is calculated the temporal evolution of the relative concentration of molecular species starting with an initial composition that is totally atomic and allowing only molecular hydrogen to be formed by surface reactions on grains. All other species

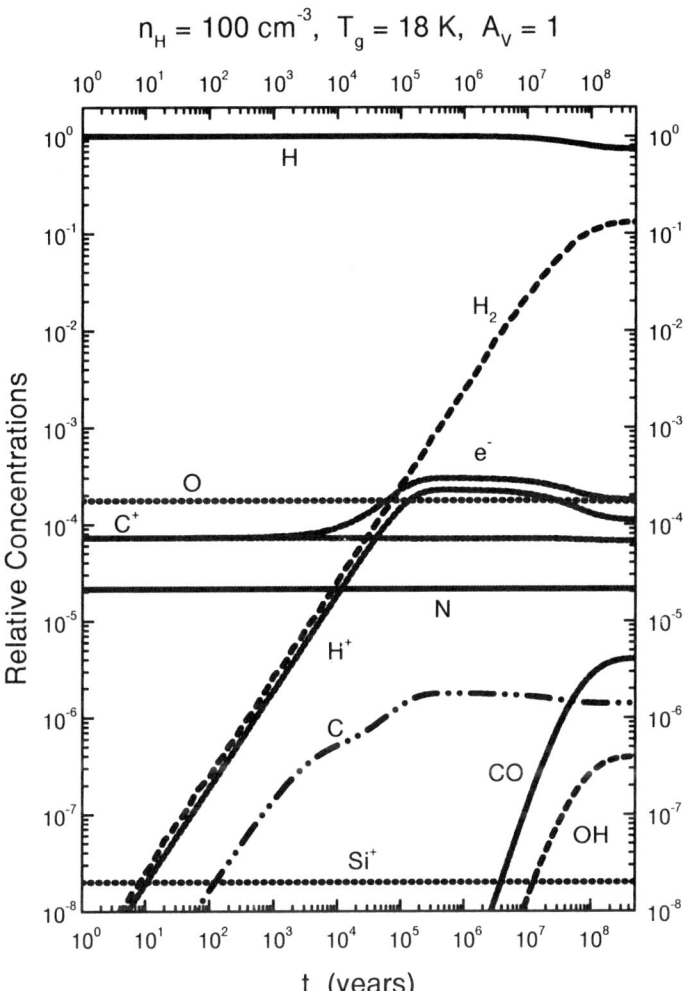

FIGURE 5. Chemical evolution in a diffuse cloud with dust temperature of $T_d = 18$ K

are obtained through gas phase reactions (using rate constants taken from the literature) that are triggered by molecular hydrogen involvement.

While more precise implications of chemical models have to wait for more sophisticated simulations involving at least other reactions on grains, from this simple model it is here worth to focus attention only on the parameters of the steady state. In the chosen conditions it is reached after about $2 \cdot 10^8$ years, already beyond the average lifetime of clouds. By generating simulations characterized only by different values of the grain temperature, it is possible to see (as one could already expect) that the entrance to the steady state (keeping fixed the values of the other parameters) is progressively delayed and the molecular hydrogen abundance is progressively decreased with increasing dust temperature.

At $T_d = 13$ K the model that uses our approach gives similar results of the model built using Hollenbach et al. (1971) expression to evaluate the H_2 production rate.

5. Conclusions

The combination of experimental and theoretical investigations has shed light on the fundamental problem of molecular hydrogen formation on realistic analogues of interstellar grains. We know now that mobility of ad-atoms on amorphous and polycrystalline surfaces is not assured by tunneling alone but chiefly by thermal hopping; depending on the coverage of H ad-atoms, recombination reactions occur at a rate that is either quadratic with coverage or independent of it. The exact solution to the steady state problem has been found and the combination of experimental results and theoretical analysis for some of the most important candidates of realistic analogues of interstellar dust give us confidence on the approach we have taken.

Much still remains to be done: measurements have to be performed on other important types of surfaces and the problem of the partition of H_2 formation energy between the grain and the molecule (kinetic and excitation) has to be addressed.

Financial support has been provided by the Italian Ministry for University and Scientific Research (MURST) under grant (to V. Pirronello) 21043088-98/225132010. Support for this work has been provided by NASA under grants (to G.Vidali) NAG5-4998 and NAG5-6822.

REFERENCES

ARONOWITZ S. & CHANG S. 1985 *Astrophys. J.* **293**, 243.

AVERNA, D. & PIRRONELLO, V. 1991 *Astron. Astrophys.* **245**, 239.

BIHAM, O., FURMAN, I., KATZ, N., PIRRONELLO, V. & VIDALI, G. 1998 *Mon. Not. R. Astr. Soc.* **296**, 869.

DULEY, W. W. 1996 *Mon. Not. R. Astr. Soc.* **279**, 591.

DULEY, W.W. & WILLIAMS, D.A. 1986 *Mon. Not. R. Astr. Soc.* **223**, 177.

GOULD, R. J. & SALPETER, E. E. 1963 *Astrophys. J.* **138**, 393.

GROVERS, T.R., MATTERA, L. & SCOLES, G. 1980 *J. Chem. Phys.* **72**, 5446.

HOLLENBACH, D. & SALPETER, E.E. 1970 *J. Chem. Phys.* **53**, 79.

HOLLENBACH, D. & SALPETER, E.E. 1971 *Astrophys. J.* **163**, 155.

HOLLENBACH, D., WERNER M.W. & SALPETER, E.E. 1971 *Astrophys. J.* **163**, 165.

HULPKE, E. 1992 Ed. *Helium Beam Scattering from Surfaces*, Springer Series in Surface Science, v.22 (Springer-Verlag).

JACKSON, B., PERSSON, M. & KAY, B. D. 1994 *J. Chem. Phys.* **91**, 5120.

KATZ,N., FURMAN,I., BIHAM,O., PIRRONELLO, V. & VIDALI, G. 1999 *Astrophys. J.* **522**, 305.

KING, A.B. & WISE, H. 1963 *J. Phys. Chem.* **67**, 1163.

LIN, J. & VIDALI, G. 1996 in *The Cosmic Dust Connection*, Ed. by M.Greenberg (Kluwer, 1996) p.323.

PIRRONELLO, V. & AVERNA D. 1988 *Astron. Astrophys.* **201**, 196.

PIRRONELLO, V., LIU, C., SHEN L. & VIDALI, G. 1997a *Astrophys. J.* **475**, L69.

PIRRONELLO, V., BIHAM, O., LIU, C., SHEN L. & VIDALI, G. 1997b *Astrophys. J.* **483**, L131.

PIRRONELLO, V., LIU, C., ROSER J.E & VIDALI, G. 1999 *Astron. Astrophys.* **344**, 681.

POELSEMA, B. & COMSA, G. 1989 *Scattering of thermal energy atoms from disordered surfaces* Springer Tracts in Modern Physics (Springer Verlag, New York).

RETTNER, C.T., DELOUISE, L.A. & AUERBACH, D.J. 1986 *J. Chem. Phys.* **85**, 1131.

RETTNER, C. T., MICHELSEN, H. A. & AUERBACH, D. J. 1995 *J. Chem. Phys.* **102**, 4625.

SCHUTTE, A., BASSI, D., TOMMASINI, F., TURELLI, F., SCOLES, G. & HERMAN, L. J. F. 1976 *J. Chem. Phys.* **64**, 4135.

SMOLUCHOWSKI, R. 1981 *Astrophys. Space Sci.* **75**, 353.
SMOLUCHOWSKI, R. 1983 *J. Phys. Chem.* **87**, 4229.
STANCIL, P.C. & DALGARNO, A. 1998 *Faraday Discussions* **109**, 61.
TIELENS, A.G.G.M. & HAGEN, W. 1982 *Astron. Astrophys.* **114**, 245.
VIDALI G., PIRRONELLO V., LIU, C. & SHEN L.Y. 1998a *Astrophys. Lett. Comm.* **35**, 423.
VIDALI., G., ROSER, J. LIU, C., PIRRONELLO, V. & BIHAM, O. 1998b *Proceedings of the NASA Laboratory Space Science Workshop*, (Cambridge, Mass. April 1-3 1998).
VIDALI, G. & ZENG, H. 1996 *Appl. Surf. Sci.* **92**, 11.
WILLIAMS, D.A. 1968 *Astrophys. J.* **159**, 935.
ZANGWILL, A. 1984 *Physics at Surfaces* Cambridge University Press.

Jacques Le Bourlot listening attentively to Eric Herbst, among a noisy crowd.

The Formation of H_2 and Other Simple Molecules On Interstellar Grains

By Eric Herbst[1]

[1]Department of Physics and Department of Astronomy, Ohio State University, Columbus, OH 43210, USA

Molecular hydrogen is formed on interstellar grains by two main processes. In the first, or Langmuir-Hinshelwood, mechanism, hydrogen atoms land on a grain and diffuse over the surface by either tunneling or hopping until they find each other. In the second, or Eley-Rideal, mechanism, hydrogen atoms landing on grains are fixed in position. Reaction occurs only when a gaseous hydrogen atom lands atop an adsorbed one. Based on new experimental results concerning the rate of diffusion of H atoms on interstellar-like surfaces, it is clear that the rate is significantly slower than estimated in the past. The range of temperatures over which diffusive formation of H_2 occurs is correspondingly reduced although sites of strong binding can raise the upper temperature limit. The surface formation of molecules heavier than hydrogen is still not well understood.

1. Introduction

It is almost certain that H_2 and a variety of other molecules are formed on the surfaces of low-temperature interstellar dust particles. On these surfaces, binding sites for adsorbates exist interspersed among regions of higher potential. On a grain of typical radius 0.1μ there are roughly 10^6 such binding sites, onto which neutral gas-phase molecules stick with high efficiency. The binding energy, or energy required for desorption (E_D), depends on the surface and on the adsorbate. For example, the binding energy of H atoms on olivine (a silicate-type material) has just been measured to be 372 K by Katz et al. (1999), who also measured the binding energy of H on amorphous carbon to be 658 K. The binding energies for H_2 on these surfaces were measured by Katz et al. (1999) to be 314 K and 542 K, respectively. Heavier and larger species tend to possess larger values of E_D (Tielens & Allamandola 1987; Hasegawa et al. 1992). At a temperature of 10 K, evaporation can occur only for the lightest species, and heavier molecules can desorb only via uncertain non-thermal means. At higher temperatures, evaporation becomes possible for a wider variety of species.

On a homogeneous surface, there are two possible mechanisms for surface chemistry. The better known one, due to Langmuir and Hinshelwood, depends on diffusion of weakly bound adsorbates. The species can either tunnel under barriers between sites (if they are light) or thermally hop over these barriers. The rate of diffusion is governed by the barrier between binding sites (E_b) and the distance between sites. The barrier against diffusion is typically less than E_D. Lighter species (especially H atoms) tend to diffuse more readily, and dominate the chemistry if abundant. Once in the same binding site, two species, unless prevented from doing so by a large activation energy barrier, can either associate into a larger molecule or react to form more than one product. Association occurs via the transfer of excess energy into the grain. Some of the excess energy can be used to eject the newly-formed molecule from the grain. The diffusion mechanism can occur over a variety of temperatures, the upper limit of which is determined by a competition between diffusion and evaporation. Depending on the barrier against diffusion of the adsorbate and whether or not tunneling occurs, there will be a minimum temperature below which diffusion cannot readily occur. Under this temperature, surface chemistry is still possible but will have to occur via the so-called Eley-Rideal mechanism. Here, surface adsorbates wait patiently until gas-phase species fall over them to initiate chemical reaction.

2. Additional Complications

In addition to uncertainties as to which mechanism dominates under interstellar conditions, dust chemistry is plagued by a variety of complications. The first has to do with the size and topology of the grain particles. Arguments have been made that a large fraction of dust particles is much smaller than the standard size (Weingartner & Draine 1999). The chemistry on these small particles, which can also be regarded as large molecules, is likely to be different from that occurring on larger grains. Energetic events, such as cosmic ray, X-ray, and UV photon bombardment, as well as exothermic chemical reactions, are likely to heat up small grains sufficiently to remove weakly bound surface species, so that primarily strongly bound species remain. This analysis would argue for an Eley-Rideal mechanism only, but the chemistry occurring on small particles is likely to be highly dependent on the shape and chemical nature of the particles, as is the case for gas-phase reactions (Herbst & Le Page 1999). If larger dust particles are highly irregular, in the sense of being fluffy, porous, or chemically diverse, then the chemistry is likely to be very complex and the existence of yet another mechanism emerges. In this mechanism, adsorbates can diffuse over large areas of the surface but are eventually trapped by unusual sites of high binding. The trapped adsorbates cannot evaporate easily, and await reaction via a diffusive partner or a partner from the gas landing nearby. This hybrid mechanism can raise the upper temperature limit of the simple diffusive mechanism.

A second complication is caused by the time-dependent chemical nature of the grain surface. In the initial stages of a cloud, the surface is either composed of some silicate-type material, some carbonaceous material, or a mixture of these. It is on these species that molecular hydrogen is produced. But, at least in denser clouds, water ice is also thought to be formed reasonably quickly and, unlike H_2, does not evaporate easily from the surface, but builds up monolayers. The rate of diffusion for adsorbates then changes because different surface-adsorbate combinations lead to different barriers against diffusion (E_b). For example, Hasegawa et al. (1992) used $E_b = 100$ K for H atoms on dirty ice whereas Katz et al. (1999) measured this energy to be 290 K on olivine and 511 K on amorphous carbon.

A third complication is caused by our relative lack of knowledge of branching fractions between associative and normal product channels for surface reactions. For example, the reactants H and H_2CO can produce two sets of products:

$$H + H_2CO \longrightarrow H_3CO, H_2 + HCO \tag{2.1}$$

and the relative amounts of each product set as a function of surface are not known.

Finally, there are mathematical problems associated with modeling the diffusive surface chemistry of species on small particles with only a few reactive species at any given time. As first noted by Tielens (1995, unpublished), the use of differential ("rate") equations to determine average surface concentrations for species, the technique of most modelers, can result in error. On the other hand, the Monte Carlo technique advocated by Tielens & Hagen (1982) and Charnley (1998) has heretofore proved to be difficult to use in a complete gas-grain chemical model of interstellar clouds. Caselli et al. (1998) and Shalabiea et al. (1998) have advocated some modifications of the rate equations to mimic most of the results obtained by Monte Carlo results for simple systems.

3. Monte Carlo Calculations of Surface H_2 Formation

The standard diffusive model for molecular hydrogen formation on grain surfaces requires that two H atoms find one another before evaporation occurs. We have used a Monte Carlo calculation (Caselli et al. 1998) to follow H_2 formation for a cloud in which the gas is at a density of 10^4 cm^{-3} and is overwhelmingly atomic H in nature. The Monte Carlo technique follows the evolution of a single grain probabilistically, with large fluctuations at short times. One million accretion events were allowed, corresponding to an elapsed time of less than a year. A desorption energy of 350 K was assumed for the H atoms, while the barrier against diffusion was varied from a low of 100 K to a high of 300 K. Tunneling from site to site was excluded, following Katz et al. (1999). At the highest barrier, which is close to the measured value for olivine, the production of molecular hydrogen via diffusion was found to occur with at least a nominal efficiency over the small temperature range 9 - 13 K. Lowering the diffusion barrier does not affect the upper limit, but lowers the lower limit. At still lower temperatures, the Eley-Rideal mechanism can be operative. The observation of molecular hydrogen in regions where the grain temperature is expected to be significantly higher than 13 K suggests that a significant amount of molecular hydrogen is formed on grain surfaces with a larger E_D, e.g., amorphous carbon, that a hybrid mechanism with impurity sites of high binding is operative, or that H_2 production is confined to high density clumps. More calculations are needed to decide among these possibilities.

4. Comparison of Two Modeling Techniques

The unmodified rate equation method works worst at late stages of a cloud, when much of the atomic hydrogen has been converted into molecular hydrogen. Thus, the need to modify the rate equations is most urgent not for the study of molecular hydrogen formation, but for the study of the formation of heavier species at times after most H_2 is produced. Of the two major published modifications to the rate method (Caselli et al. 1998), the first concerns the production of molecular hydrogen only while the second concerns the synthesis of all hydrides. The more important correction is the latter, in which the diffusion rate of atomic hydrogen on a grain is reduced so that the rate of reaction per hydrogen atom does not exceed the rate of accretion of H atoms onto a grain. Care must be taken that the reduced rate of diffusion is not so slow that evaporation occurs instead of reaction. The slower the actual diffusion rate of H, the less the need to artificially slow it down.

In addition to these modifications, it is necessary to modify the rate of a reaction involving H with a small activation energy if the reaction is still sufficiently rapid that it exceeds the accretion rate of H. This possibility can occur if a reaction partner is very abundant on a grain surface - the most obvious example is surface CO. Its reaction with H has a small activation energy ($\approx$ 1000-3000 K) but the high concentration of surface CO still renders the reaction reasonably efficient. Once the rate of reaction per H atom exceeds the accretion rate of H atoms, it must be reset. Such considerations are important in the sequential conversion of CO into methanol (CH_3OH) via H atom addition on grain surfaces (Charnley et al. 1997):

$$CO \to HCO \to H_2CO \to H_3CO \to CH_3OH. \qquad (4.2)$$

In addition to the association of H and CO, the reaction between H and H_2CO also possesses a small amount of activation energy.

Using all of the above modifications, Caselli & Herbst (in preparation) have compared the modified rate method with the Monte Carlo approach for an interstellar cloud at temperatures ranging from 10-20 K with a gas phase, constant in composition, consisting of H, O, and CO in varying proportions. We allow these species to adsorb onto (and possibly evaporate from) dust particles for 1000 yr and determine the amounts of H_2, O_2, CO_2, H_2O, H_2CO, and CH_3OH produced during this period. A variety of barriers against diffusion have been used for all species. In general, the two approaches are in excellent agreement with one another. Exceptions occur for the largest barriers against diffusion, in which case the diffusive mechanism breaks down and the two approaches are both incorrect since they do not contain the Eley-Rideal mechanism. The results are very sensitive to temperature. The formation of methanol is a case in point - at lower temperatures, more methanol is produced than at higher temperatures despite the fact that activation energy barriers exist to the hydrogenation of CO into methanol. The reason is that H atoms tend to evaporate at higher temperatures, so that hydrogenation becomes much less efficient.

5. Conclusions

The formation of molecular hydrogen on grain surfaces is still poorly understood because the detailed characteristics of the surfaces are poorly constrained, time-dependent, and quite variable from one likely material to another. Depending on these characteristics and the physical conditions, H_2 formation can occur via the diffusive (Langmuir-Hinshelwood) mechanism, a hybrid diffusive mechanism, in which one of the reactants is trapped in a site of especially strong binding, and the Eley-Rideal mechanism, in which all surface species are stationary. The relative roles of the latter two mechanisms should be studied more closely.

The formation of heavier molecules on grains is highly uncertain. In addition to the uncertainties plaguing the analysis of H_2 formation, the additional complication exists that diffusive chemistry when low H abundances are present in the gas cannot be treated accurately with the standard rate equations. Needed modifications to these equations are most severe when diffusion of H is at its most rapid.

We would like to acknowledge the support of the National Science Foundation (US) for our program in astrochemistry.

REFERENCES

CASELLI, P., HASEGAWA, T. I. & HERBST, E. 1998 *Astrophys. J.* **495**, 309–316.

CHARNLEY, S. B. 1998 *Astrophys. J.* **509**. L121–L124.

CHARNLEY, S. B., TIELENS, A. G. G. M. & RODGERS, S. D. 1997 *Astrophys. J.* **482**, L203–L206.

HASEGAWA, T. I., HERBST, E. & LEUNG, C. M. 1992 *Astrophys. J. Suppl. Ser.* **82**, 167–195.

HERBST, E. & LE PAGE, V. 1999 *Astron. & Astrophys.* **344**, 310–316.

KATZ, N., FURMAN, I., BIHAM, O., PIRRONELLO, V. & VIDALI, G. 1999 *Astrophys. J.* **522**, 305–312.

SHALABIEA, O. M., CASELLI, P. & HERBST, E. 1998 *Astrophys. J.* **502**, 652–660.

TIELENS, A. G. G. M. & ALLAMANDOLA, L. J. 1987 in *Interstellar Processes* (ed. D. J. Hollenbach & H. A. Thronson, Jr.), pp. 397–470. Reidel.

TIELENS, A. G. G. M. & HAGEN, W. 1982 *Astron. & Astrophys.* **114**, 245–260.

WEINGARTNER, J. C. & DRAINE, B. T. 1999 *Astrophys. J.* **517**, 292–298.

The interaction of H atoms with interstellar dust particles: Models

By Victor SIDIS
L. JELOAICA†, A. G. BORISOV AND S. A. DEUTSCHER

Laboratoire des Collisions Atomiques et Moléculaires
(Unité Mixte de Recherche: CNRS – Université Paris Sud, No. 8625)
Bâtiment 351, Université Paris Sud, 91405 Orsay Cedex, France

Two topics of relevance for H_2 formation in the interstellar medium are considered: (i) the interaction of H and H-H with a model-graphite surface (Coronene: $C_{24}H_{12}$), and (ii) H^- formation by charge transfer in the interaction of H with a model-silicate surface ($MgO\{100\}$ representing forsterite: $Mg_2SiO_4\{100\}$). The first topic is related to the frequently invoked Langmuir-Hinshelwood and Eley-Rideal mechanisms for H_2 formation near carbonaceous zones of interstellar dust grains. *Ab initio* calculations based on Density Functional Theory are used. The second topic proposes a new scenario in which the efficient production of H^- ions would subsequently enable the formation of H_2 via associative detachment. It stems from recent work of the authors on charge transfer between neutral atoms and ionic insulators.

1. Introduction

The mechanism of H_2 formation in the interstellar medium (ISM) is still an open problem. Owing to the temperature and density conditions existing in the ISM, 3-body recombination and radiative association processes in the gas phase are unable to account for actual H_2 abundances. The existence of dust particles in the ISM has attracted the attention of astrophysicists as plausible catalysts or mediators of H-H recombination in space (Hollenbach and Salpeter (1970), Hollenbach and Salpeter (1971)). Current knowledge of interstellar dust particles (IDPs) indicates that they have both a carbonaceous and a silicate composition. This has in particular stimulated the investigation of the role graphitic bonds may have on H_2 formation as a result of elementary interactions between H atoms and graphite-like surfaces or platelets. The interaction of H atoms with silicates is a much younger field of investigation.

The present contribution to the field of "H_2 in space" concerns the interaction of H-atoms with two types of surfaces: a graphite surface and a silicate surface. These interactions are studied theoretically from different points of views.

2. Characterization of the adsorption of H and H-H on graphite

The interaction of H-atoms with graphite is involved in the reaction

$$H + H - graphite \longrightarrow H_2 + graphite$$

through two mechanisms:
• the Langmuir-Hinshelwod mechanism whereby two adsorbed and mobile H atoms interact to recombine and desorb as H_2 molecules, and
• the Eley-Rideal (ER) abstraction mechanism whereby an H atom from the gas phase impinges on a chemisorbed H atoms to finally yield desorbed H_2.

† also: Natnl. Inst. for R & D of Material Physics, Bucuresti-Magurele, P.O.B Mg–7, Romania

2.1. Other work

Most of the previous theoretical work in the field is devoted to the characterization of the H–graphite interaction, especially: chemisorption.† Until the work of Jeloaica and Sidis (1999a) most of the available knowledge on this interaction came from *semi-empirical* calculations based on *approximate Hartree-Fock*-type schemes (*e.g.* CNDO, MNDO, INDO, MINDO/3)† and employing *minimal basis set* expansions. The work of Fromhertz et al. (1993) seems to be the most accomplished in this category. Extended-Hückel-type calculations as well as perturbative calculations have also been reported.† Given the approximations made in all these calculations, they can only be considered as qualitative. Moreover, they do not mutually agree in general.

Considering the H-H–graphite interaction, much less theoretical information is available. In order to perform molecular dynamics calculations on the above ER mechanism, Parneix and Bréchignac (1998) have tentatively built an H-H–graphite potential energy surface based on the H–graphite results of Fromhertz et al. (1993). In the meantime, and while the H-H–graphite calculations presented below where underway, calculations based on density functional theory were announced by Farebrother and Clary (1998); results of these calculations are presented in another contribution of these proceedings (Williams et al. (2000)).

2.2. Present work

Our aim was to go beyond previous approaches by the use of a non-empirical (*ab initio*), non-perturbative, quantum mechanical treatment that incorporates electron correlation (as opposed to Hartree-Fock), spin polarization and extended expansion basis sets and that also considers substrate relaxation to some extent. This ambitious task can be envisaged today within the framework of density functional theory (DFT).

Our calculations are based on a cluster model of the graphite (0001) basal surface. Large polycyclic aromatic hydrocarbon (PAH) molecules like coronene ($C_{24}H_{12}$) which we consider here have often been advocated for this purpose. The H atoms serve only to passivate the pending bonds of the C_{24} cluster.

In practice, the calculations are carried out using the ADF‡ computer code. The exchange-correlation density functional used in this work includes the local functional of Vosko et al. (1980) and the PW91 (Wang and Perdew (1991)) generalized gradient correction. The latter has been chosen because it is (a) non-empirical, and (b) satisfies best the known formal properties of the exact exchange-correlation functional (Burke et al. (1995)). The Kohn-Sham (KS, Kohn and Sham (1965)) molecular orbitals are expanded as linear combinations of atom centered Slater type orbitals (STO); double-ζ+polarisation (DZP) (Jeloaica and Sidis (1999a)) and triple-ζ+polarization (TZP) bases have been used.‡ In these calculations spin polarization of the electrons is allowed for by letting the ↑ and ↓ spin densities be different (unrestricted KS). This is an important feature to be taken into account since, in general for H–graphite, and especially in the dissociation region of H···H–graphite, the electron spins are not paired.

The calculations consider perpendicular approaches at three sites of the central ring of the C_{24} cluster: site $\mathcal{A}$, on top of a C atom; site $\mathcal{B}$, above the middle of a C-C bond; and site $\mathcal{C}$, above the center of the hexagon.

---

† Cf. Fromhertz et al. (1993) and Jeloaica and Sidis (1999a) for an extensive list of references.
‡ ADF: Amsterdam Density Functional (Scientific Computing and Modelling, Vrije Universiteit, Amsterdam, The Netherlands) - ADF 2.0.1 and ADF 2.3 Versions (Baerends et al. (1973), TeVelde and Baerends (1992)).

2.2.1. The H–graphite interaction

Our first results (Jeloaica and Sidis (1999a)) showed quite different adsorption characteristics from those reported before. Two zones of adsorbate-substrate distances z were found: an inner zone ($z^{in} \leq 1.8\,\text{Å}$) and an outer zone ($z^{out} > 1.8\,\text{Å}$). The inner zone, where substantial mixing and electron transfer occur, may be considered as a *chemisorption* zone. On the other hand, the outer zone, where charge transfer is negligible, may be considered as a *physisorption* zone.

For a planar substrate, the H atom can hardly chemisorb: at the $\mathcal{C}$-site, the potential energy as a function of z is repulsive; at the $\mathcal{B}$-site, it has a shoulder on a repulsive wall; and, at the $\mathcal{A}$-site it exhibits a sort of "metastable chemisorption position" with a well, located around $z^{in}(\mathcal{A}) = 1.25\,\text{Å}$, hanging at $0.3\,\text{eV}$ on a repulsive wall (see figure 3 of Jeloaica and Sidis (1999a)).

The calculations of Jeloaica and Sidis (1999a) also showed the existence of a potential well (having nearly the same characteristics at the $\mathcal{A}$-, $\mathcal{B}$-, and $\mathcal{C}$-sites) in the outer zone. This well ($\approx 0.07\,\text{eV}$) was tentatively related to physisorption. Although it is not commonly believed that Van der Waals minima can be correctly described by generalized gradient functionals, arguments were developed which supported the latter identification. For comparison, the physisorption well deduced from experimental H–graphite scattering data (Ghio *et al.* (1980)) is $\approx 0.04\,\text{eV}$.

Allowance for substrate relaxation¶ at the $\mathcal{B}$- and $\mathcal{C}$-sites does not change the results in a significant way. On the other other hand, dramatic changes appear in the inner zone upon relaxation at the $\mathcal{A}$-site (figures 1 and 2): a true chemisorption well builds up and the minimum energy path exhibits a protruding barrier (of height $0.144\,\text{eV}$) between the inner and the outer zones (at $z(\mathcal{A}) = 1.9\,\text{Å}$). As also observed by Fromhertz *et al.* (1993), the chemisorption well at the $\mathcal{A}$-site builds with a substantial pull out (tetrahedrization) of the nearest C neighbour (see figure 5 of Jeloaica and Sidis (1999a)). The results also indicate that the chemisorbed H atoms would have no tendency to migrate parallel to the surface; in particular the path along C-C bonds, suggested earlier for H migration would lead to desorption of the H-atoms.

New calculations involving a TZP-STO basis have recently been undertaken (Jeloaica and Sidis (1999b)), especially to measure the sensitivity of the results against basis set size, and to check and correct for eventual effects of basis set superposition errors (BSSE). The results remain qualitatively the same as the DZP ones (Jeloaica and Sidis (1999a)), but are modified in two ways: (i) the outer well is dramatically reduced but still exists ($0.008\,\text{eV}$), and it is displaced outward; (ii) both chemisorption well and barrier are moved up energetically (figure 1). The evolution of the H–graphite interaction potential upon displacement of the nearest C atom from the surface (figure 2) aids to understand figure 1.

On the basis of the above results, we are inclined to say that, in the outer zone, the characterization of the H-*physisorption* to graphite with the employed computation method is debatable. As concerns the characteristics of H-*chemisorption*: the height and thickness of the barrier along the minimum energy path of figure 1 as well as the the strong reconstruction of the graphite surface required to stick the H-atom in the inner zone are such that *rather high temperatures are required to chemisorb H on graphite*.

¶ At the $\mathcal{A}$-site, the C atom is allowed to move perpendicular to the basal plane, at the $\mathcal{B}$-site the two nearest C atoms are allowed to move simultaneously perpendicular to the basal plane, and, at the $\mathcal{C}$-site the six C atoms of the central hexagon are allowed to move simultaneously perpendicular to the basal plane.

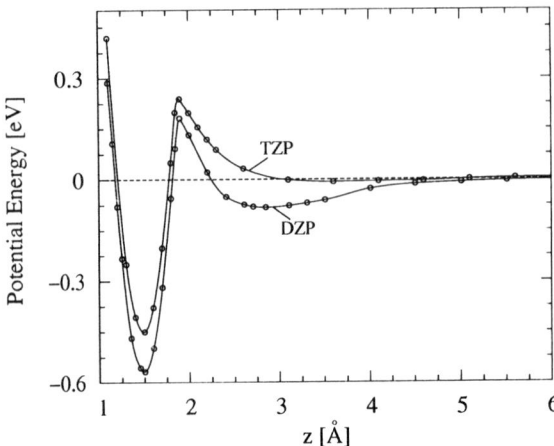

FIGURE 1. Minimum potential energy of the H–graphite system at the $\mathcal{A}$-site when the nearest C atom is allowed to relax. The results of two different basis sets are compared.

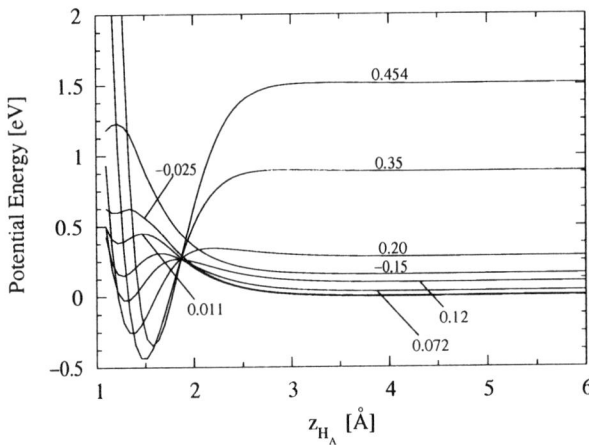

FIGURE 2. Potential energy curves of the H-H–graphite system on top of the $\mathcal{A}$-site for several Z_C values (as indicated) at a large distance (5Å) between the H atoms.

2.2.2. The H-H–graphite interaction

The TZP calculations have been extended to the case when two H atoms approach along an axis perpendicular to the model graphite surface.

At large distances (in the outer zone defined in § 2.2.1) we have carried out minimum energy path calculations, at the $\mathcal{A}$-, $\mathcal{B}$- and $\mathcal{C}$-sites. The altitudes of the H-atoms are varied and the nearest C neighbours are allowed to relax.‡ A sort of physisorption well (with a depth ≈ 0.025 eV) is found at a distance of the closest H atom to the surface (called H_A), $z_{H_A} \approx 3$ Å. These characteristics are little dependent on the site; the substrate has no tendency to relax and the H_2 molecule stays at its equilibrium distance ($r_e = 0.75$ Å). The computed well depth is nearly twice smaller than the one determined by Mattera et al. (1980) from experimental molecule-surface scattering data.

In the inner zone, the $\mathcal{A}$-site was found to be the only one able to chemisorb an H-

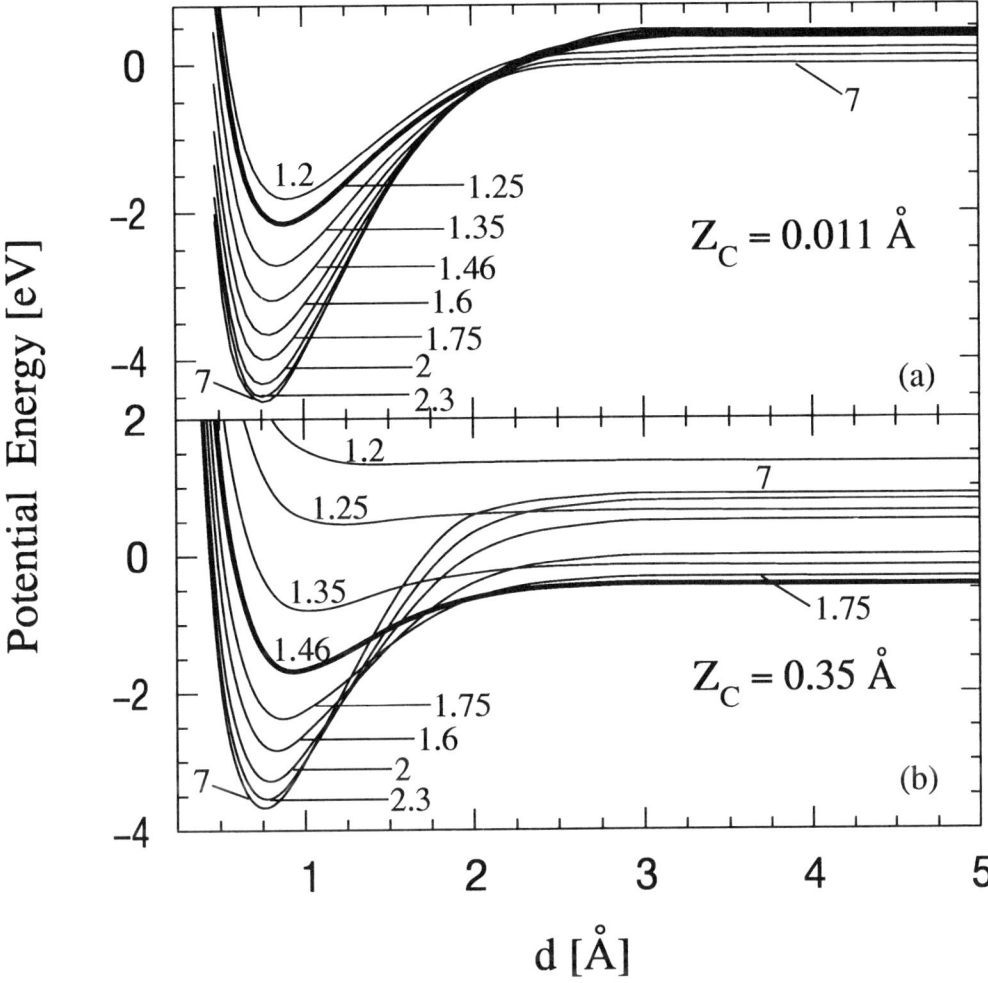

FIGURE 3. Cuts in the H–H–graphite potential energy surfaces for particular values of Z_C, on top of the $\mathcal{A}$-site, as a function of the distance d between the H atoms. The parameter of the cuts is z_{H_A}. The bold line in each pannel indicates the path that starts from the H chemisorption position when the other H atom is far away (cf. figure 2).

atom (§ 2.2.1, figures 1 and 2). Thus, we have restricted our study to this site. For a given distance of approach z_{H_A} of the H_A atom to the basal plane and a given distance Z_C by which the nearest C atom is displaced from the surface we have computed the potential energy as a function of the H_A-H_B distance. Figure 3 shows samples of the results for $Z_C = 0.011$ Å and $Z_C = 0.35$ Å respectively; the former corresponds to nearly no relaxation and the latter corresponds to the bottom of the chemisorption well along the H–graphite minimum energy path of figure 1 (see also figure 2).

Figure 3 shows that *the H-H recombination in an on-top encounter at the $\mathcal{A}$-site is an easy reaction*. This is seen in these figures by following (from right to left) the minimum energy path for the approach of the initially remote H_B atom to the H_A atom stuck to the surface at the minimum of the chemisorption well. (As already discussed in § 2.2.1 and shown in figure 2 the latter well may be considered as a "metastable chemisorption position in a hanging well" for the smallest Z_C values [figure 3a]). This path shows no barrier and descends steeply towards the formation of the H_B-H_A molecule away from the surface. The way in which the H_2 well is distorted when the molecule is close the surface lets one anticipate, on qualitative grounds, that the desorption should be accompanied by vibrational excitation of the product.

3. On H$^-$ formation in the interaction of H atoms with silicates

Our interest in this topic is motivated by the search for new ways to form H_2 in space.‖
Why H$^-$? The associative detachment reaction

$$H^- + H \longrightarrow H_2 + e^-$$

is a very efficient way of making H_2 in the gas phase. It has a large cross section (few Å^2) at eV energies (Phaneuf (1992)) and is likely to have a Langevin-type rate constant at low temperatures (see also Launay et al. (1991)). Thus, it is of interest to explore various ways through which H$^-$ can be produced efficiently in media of astrophysical relevance. The dust particles found in such media are now recognized as objects of great interest, yet their role in the production of H$^-$ has not been investigated so far.

3.1. From dust via silicates to MgO

Why silicates? Apart from carbonaceous material and water ice, interstellar dust particles contain silicates, typically olivines and pyroxenes. An end member of the olivine family, forsterite (Mg_2SiO_4), is thought to be a principal component of interstellar dust.

A computer simulation by Watson et al. (1997) of the structure and stability of forsterite surfaces indicates that the more stable ones are arranged such that silicon is as far from the surface as possible. The most stable surface is the {100} one; it has alternating layers of MgO and SiO_2 *terminating with an MgO layer*. The study also concludes that structure and energy of this surface are very similar to that of the {100} surface of MgO. Moreover, the energy band gaps of the two materials differ by less than 1 eV (Nitsan and Shankland (1976)).

As a first benchmark, we have, hence, calculated the H$^-$ production from neutral hydrogen atoms on an MgO{100} surface.

3.2. The model

Solid MgO is an ionic insulator, built of ions of alternating charge $\pm Q$ in a cubic crystal lattice; $Q \approx 2$ (Illas et al. (1993), Sousa et al. (1993)) and the lattice constant is 4.21 Å. It has recently been established that neutral atoms (H, O, F) are very efficiently converted into negative ions during grazing scattering from the surfaces of ionic insulators such as alkali halides (Auth et al. (1995), Auth et al. (1998)) and metal oxides, in particular MgO (Ustaze et al. (1997), Esaulov et al. (1997), Maazouz et al. (1997), Ustaze et al. (1998), Esaulov et al. (1999)). The negative ion yield in these interactions can even be larger than that obtained with metals (e.g. Al, Au). This phenomenon appeared at first sight surprising in view of the large energy difference between the electron affinities of the considered neutrals (0.75 eV, 1.47 eV, and 3.4 eV, respectively) and the binding energies of electrons in the valence band of ionic insulators (figure 4a). It has been explained theoretically (Borisov et al. (1996), Borisov and Sidis (1997), Auth et al. (1998)) by a *diabatic energy level confluence* as outlined below.

The valence band electrons of ionic insulators are localized at the anionic sites of

‖ This topic has been suggested to one of us (V.S.) by D. Field (Univ. Aarhus, Denmark) at the 6th ECAMP conference (Siena, Italy, July 1998); cf. his contribution at the H_2 in Space International Conference (Paris, France October 1999).

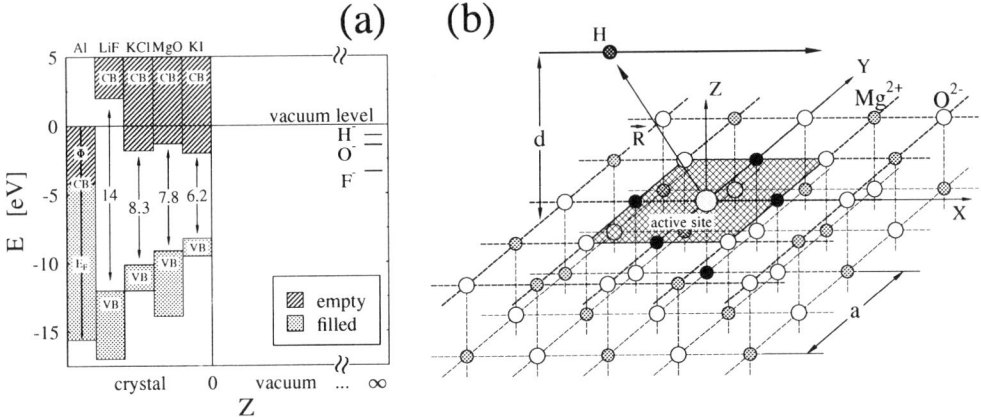

FIGURE 4. (a) Energies of valence (VB) and conduction (CB) bands of several ionic crystals, and affinity levels of the free projectiles. Fermi energy E_F and work function Φ of Al are shown for comparison. (b) Collision system: Cluster around active O^{2-} (centre) embedded in a grid of ± 2 point charges. The projectile moves in the X–Y plane in X direction at altitude $Z = d$.

the crystal. After an electron transfer has taken place, a +1 charge (electron hole) is left on the surface at an anionic site. The attractive Coulomb interaction between this positive charge and the outgoing H^- leads to a dramatic reduction of the energy difference between reactants and products, making a negative ion formation possible. The described mechanism is reminiscent of ionic-covalent interactions in gas phase atom-atom and atom-molecule chemi-ionization processes and the related harpoon mechanism (Levine and Bernstein (1987)). The negative ion formation process can, thus, be viewed as *a succession of binary interactions* (Auth et al. (1995), Borisov et al. (1996), Borisov and Sidis (1997)) of the form

$$H + O^{2-} \longrightarrow H^- + O^-$$

between the projectile and a given anionic site ("active site").

The other Mg^{2+} and O^{2-} ions are considered as spectator point charges (figure 4b). We fixed the point charges of the ions on the MgO lattice sites to ± 2 which is known to be a good approximation (Mejías et al. (1995), Zuo et al. (1997)).

This model has been the basis of quantitative work on F^- and H^- formation in small angle scattering from an MgO surface (Deutscher et al. (1999a), Deutscher et al. (1999b)) In practice, we consider the binary collision of the projectile with a $(Mg_5O)^{8+}$ cluster embedded in a lattice of 778 point charges which guarantees a proper account of the Madelung potential. The cluster consists of 10 electrons and six positive ion cores: one central O^{8+} ion surrounded by five ℓ-dependent Mg^{2+} pseudopotential cores. While freedom is given in these calculations to the electrons to spread around the Mg^{2+} pseudopotential cores, they actually remain localized around the O center, albeit in a rather diffuse way.

3.3. Results

Results have first been obtained (Deutscher et al. (1999b)) at high energies ($\geq 100\,eV$) and a rather small incidence angle to the surface plane (3.5°) for comparison with experimental data of Ustaze et al. (1997). As already found for other cases (Borisov and Sidis (1997), Deutscher et al. (1999a)), the computed $H \to H^-$ capture rate is quite high and reaches 100% at keV energies. Yet the comparison with experiment reveals the existence of an important loss channel mainly due to electron detachment towards the vacuum or the conduction band of the surface. The related probability of electron loss per site was estimated to be as large as 98%, and after inclusion of this loss rate on a phenomenological level the experimental data were well reproduced (Deutscher et al. (1999b)).

These calculations were extended to lower energies. Our procedure is well suited for small incidence angle scattering conditions and, therefore, exploratory calculations were performed for incidence angles in the range of $3.5°$ to $20°$. Considering only the *capture* process, the H→H$^-$ conversion is sizable (about a few permil at 10 eV collision energy), and it grows considerably with increasing collision energy ($\sim 10\%$ at 100 eV). Discussing these results, care should be taken with respect to the loss channel which should decrease the actual negative ion fraction compared with the calculated ones. It can be expected that the H$^-$ level, although at large separations in resonance with the MgO conduction band, will be shifted into the band bap by the image potential in front of the surface, thus reducing electron loss. For small energy, large angle collisions the electron loss should, hence, be less efficient than what was found for fast, grazing ones where dynamic electron stripping from the H$^-$ ion can occur during short ranged interaction with anion sites.

It is noteworthy that each time an electron capture of this sort occurs the resulting H$^-$ ion is slower than the initial H projectile by ≈ 8 eV. A succession of capture-then-loss processes in multiple interactions of the H atom with the surface can (a) significantly decelerate the outgoing H or H$^-$ species and (b) enrich the surrounding with electrons.

4. Conclusion

New information on the interaction of H atoms with objects modelling interstellar dust particles have been reported. The calculations reported in section 2 open a possibility of investigating theoretically the dynamics of H–H recombination near graphitic zones of IDPs. At the present stage of the calculations it is the initial sticking stage of H to graphite which seems the most problematic owing to its energetic requirements. The calculations reported in section 3 open new perspectives related to basic processes near silicate zones of IDPs.

We acknowledge financial support from the "Programme National de Physique et Chimie du Milieu Interstellaire". Thanks are due to P. Parneix and P. Bréchignac for having drawn our interest to graphite, and to D. Field for discussions on silicates. S.A.D. acknowledges support from the European Union under contract ERBFMBICT971983. L.J. acknowledges support from the French Ministry of Education, Research, and Technology for a grant within the French-Romanian Network for Training and Research.

REFERENCES

Auth, C. Borisov, A. G. & Winter, H. 1995 *Phys. Rev. Lett.* **75**, 2292.
Auth, C. Mertens, A. Winter, H. Borisov, A. G. & Sidis, V. 1998 *Phys. Rev. A* **57**, 351.
Baerends, E. J. Ellis, E. D. & Ros, P. 1973 *Chem. Phys.* **2**, 41.
Borisov, A. G. Sidis, V. & Winter, H. 1996 *Phys. Rev. Lett.* **77**, 1893.
Borisov, A. G. & Sidis, V. 1997 *Phys. Rev. B* **56**, 10628.
Burke, K. Perdew, J. P. & Levy, M. 1995 in *Modern Density Functional Theory: A Tool for Chemistry* (eds. J M. Seminario & P. Politzer) vol. 2, Elsevier.
Deutscher, S. A. Borisov, A. G. & Sidis, V. 1999a *Phys. Rev. A* **59**, 4446.
Deutscher, S. A. Borisov, A. G. & Sidis, V. 1999b *NIM B* **157**, 61.

Esaulov, V. A. Gnizzi, O. Guillemot, L. Maazouz M. Ustaze, S. & Verruchi, R. 1997 *Surf. Sci.* **380** L521.

Esaulov, V. A., Guillemot, L. Lacombe, S. 1999 *Comment At. Mol. Phys.* **34**, 273.

Farebrother, & Clary, D. C. in *MOLEC XII, European Conf. on the Dynamics of Molecular Collisions* (Abstracts) 1998, p.73.

Fromhertz, T. Mendoza, C. & Ruette, F. 1980 *MNRAS* **263**, 851.

Ghio, E. Mattera, L. Salvo, C. Tommasini, F. & Valbusa, U. 1980 *J. Chem. Phys.* **73**, 5561.

Hollenbach, D. H. & Salpeter, E. E. 1970 *J. Chem. Phys.* **53**, 79.

Hollenbach, D. H. & Salpeter, E. E. 1971 *Ap. J.* **163**, 155.

Ilas, F. Lorda, A. Rubio, J. Torrance, J. B. Bagus, P. S. 1993 *J. Chem. Phys.* **99**, 389.

Jeloaica, L. & Sidis, V. 1999a *Chem. Phys. Lett.* **300**, 157.

Jeloaica, L. & Sidis, V. 1999b in *DFT99 8th International Conference on the Applications of Density Functional Theory to Chemistry and Physics* (Abstracts), also to be published.

Kohn, W. & Sham L. 1965 *Phys. Rev. A* **140**, 1133.

Launay, J.M. Ledourneuf, M. & Zeippen, C. J. 1991 *A&A* **252**, 842.

Levine, R. D. & Bernstein, R. 1987 *Molecular Reaction Dynamics and Chemical Reactivity* Oxford.

Maazouz, M. Guillemot, L. Lacombe, S. & Esaulov, V. A. 1997 *Phys. Rev. Lett.* **77**, 4265.

Mattera, L. Rosatelli, F. Salvo, C. Tommasini, F. Valbusa, U. & Vidali, G. 1980 *Surface Sci.* **93**, 515.

Mejías, J. A. Márquez, A. M. Sanz, J. F. Fernańdéz-García, M. Ricart, J. M. Sousa, C. & and Illa, F. 1995 *Surf. Sci.* **327**, 59.

Nitsan, U. Shankland, TJ (1976) *Geophys. JR Astr. Soc.* **45**, 59; *Eos* **57**:3, 160.

Parneix, P. & Bréchignac, Ph. 1998 *A&A* **334**, 363.

Phaneuf, R.A. 1992 *Nuclear Fusion* (Supplement) **2**, 75.

Sousa,C. Ilas, F. Bo, C. & Poblet, J. M. 1993 *Chem. Phys. Lett.* **215**, 97.

TeVelde, G. & Baerends, E. J. 1992 *J. Comp. Phys.* **99**, 84.

Ustaze, S. Verruchi, R. Lacombe, S. Guillemot, L. & Esaulov, V. A. 1997 *Phys. Rev. Lett.* **79**, 3526.

Ustaze, S. Guillemot, L. Verruchi, R. Lacombe, S. & Esaulov, V. A. 1998 *NIM B* **135**, 319.

Vosko, S. H. Wilks, L. H. & Nussair, M. 1980 *Can. J. Phys.* **58**, 1200.

Wang, Y. & Perdew, J.P. 1991 *Phys. Rev. B* **43**, 8911.

Watson, G. W. Oliver, P. M. Parker, S. C. 1997 *Phys. Chem. Miner.* **25** 70.

Williams, D. A. *et al.* 2000, in this issue.

Zuo, J. M. O'Keeffe, M. Rez, P. & Spence, J. C. H. 1997 *Phys. Rev. Lett.* **78**, 4777.

Pierre Cox listening to Guillaume Pineau des Forêts, with Jacques Le Bourlot onlooking.

The energetics and efficiency of H_2 formation on the surface of simulated interstellar grains

By
David A. Williams[1], David E. Williams[2], David Clary[2],
Adam Farebrother[2], Andrew Fisher[1], Jon Gingell[2],
Richard Jackman[3], Nigel Mason[1], Anthony Meijer[2],
James Perry[2], Steven Price[2], AND Jonathan Rawlings[1].

[1]Department of Physics and Astronomy, University College London, Gower Street, London WC1E 6BT

[2]Department of Chemistry, University College London, Gower Street, London WC1E 6BT

[3]Department of Electronic and Electrical Engineering, University College London, Gower Street, London WC1E 6BT

This paper reports the theoretical and experimental work on H_2 formation on interstellar dust mimics. These studies are being carried out under the auspices of the UCL Centre for Cosmic Chemistry and Physics.

1. Introduction

The purpose of this article is to report on the current state of work at the UCL Centre for Cosmic Chemistry and Physics, a consortium of scientists at University College London addressing problems of chemistry arising in astronomy. All the work currently in progress in this consortium is concerned with H_2 formation on surfaces, and it consists of both theoretical and experimental programmes.

The Centre was formed a few years ago when it was realised that advances in both experimental and theoretical techniques now make it possible to address in a realistic manner some problems of longstanding and fundamental interest in astronomy. The expertise at UCL, both in theory and experiment, is very strong on surface reactions; the current motivation from astronomy also emphasises the gas/dust interaction (Williams 1998). It was decided, therefore, to undertake a long-term and coordinated programme on surface processes of relevance to astronomy. Of course, the most fundamental interaction is that leading to H_2 formation on dust. There is currently some important experimental and theoretical work being carried out in this particular area, and much of this work has been reported at this meeting. Nevertheless, it was felt that the UCL consortium could make a useful contribution without simply replicating the experiments and calculations of others.

2. Motivation

The astrophysical requirement for H_2 production was set a quarter century ago by Jura (Jura 1975). He used Copernicus satellite data along lines of sight to bright stars that included both optically thin and thick lines of the H_2 B-X band. His interpretation of these data was that in diffuse clouds nearly every H atom arriving at a grain must leave as part of an H_2 molecule. Since the 1970s, our understanding of the interstellar medium has changed enormously. While the diffuse cloud requirement for H_2 production has remained essentially unchanged, we need to know much more about the process of

H_2 formation if we are to extend the ideas into a range of situations in interstellar and circumstellar space. These regions offer a much wider spread of physical conditions than those represented by diffuse clouds. To what extent can the canonical diffuse cloud rate be applied in these other situations?

There are a number of factors affecting H_2 formation about which our ideas have changed considerably since the 1970s. For example, it is now less clear what is the amount of grain surface area per unit volume. The interpretation made by Draine (1984) of infrared emission was of a grain size distribution extending to much smaller sizes than had been assumed hitherto. Also, it became recognised that interstellar particles, like some collected interplanetary particles, (e.g. Sandford 1989), might have a very open, porous structure. Such particles exhibit much more surface area than the canonical interstellar particles, often assumed to be spherical or spheroidal. However, it remains unclear to what extent very small particles - or small thermally independent units in a larger composite particle - can contribute to surface chemistry, for they may undergo rather frequent temperature excursions following the absorption of single UV photons (Sellgren 1984). Such excursions may lead to desorption before reaction between adsorbed species can occur. The chemical nature of the dust has also become less clear since the 1970s; the additional information from ISO and other facilities has shown a much richer chemical variety than was previously assumed for dust. Also, the range of densities and temperatures in which interstellar and circumstellar molecules are found has extended enormously from the canonical diffuse cloud values of 100 H cm^{-3} and 100 K. Now, molecules in general and H_2 in particular may be found in conditions ranging from $\gtrsim 1$ H nucleus cm^{-3} and temperatures $\lesssim 4000$ K. Regions with these parameters contain most of the non-stellar baryonic matter in the Universe.

It is also evident that we need to understand much more fully than we currently do the nature of the H_2 formation process, not only to explain the abundance of H_2 in diffuse clouds, but in other situations, too. The list of what we really need to know - from the astrophysical modellers point of view - is extensive.

The fundamental concern is the nature of the interaction between an H-atom and a dust grain. Is this simply physisorption, or does chemisorption occur? To answer this question, the physical and chemical nature of the surface must be specified. If the nature of the interaction is specified then the next quantity of interest is the sticking probability, for which the nature of the surface is very important. After sticking, are adsorbed H-atoms mobile on the surface, and - if so - what is their mobility? Does the interaction between two H-atoms on interstellar dust grains occur most readily in the Eley-Rideal (prompt) or the Langmuir-Hinshelwood (diffusion) modes? Such mechanisms will have rather different dependences on the nature of surface and on gas and grain temperatures (Duley 1996). When either reaction occurs, what is the energy budget of the reaction? Is the nascent molecule ro-vibrationally excited, and does it contribute to infrared emission detected in regions of high excitation? How much kinetic energy is carried by the newly formed H_2? Is this a significant contribution to interstellar heating? If some energy from the reaction forming H_2 is deposited in the surface, does this affect the local temperature and drive desorption of nearby species? Only when answers to all of the above questions are available will it be possible to construct an overall H_2 formation rate for any given physical situation and set of grain types, and to describe in detail the role of H_2 in the Universe.

The programme we have started at UCL is intended to throw some light on all of these questions, particularly the energetics of the reaction

$$H + \text{H-grain} \rightarrow H_2(v'', J'') + \text{K.E.} + \text{grain} \qquad (2.1)$$

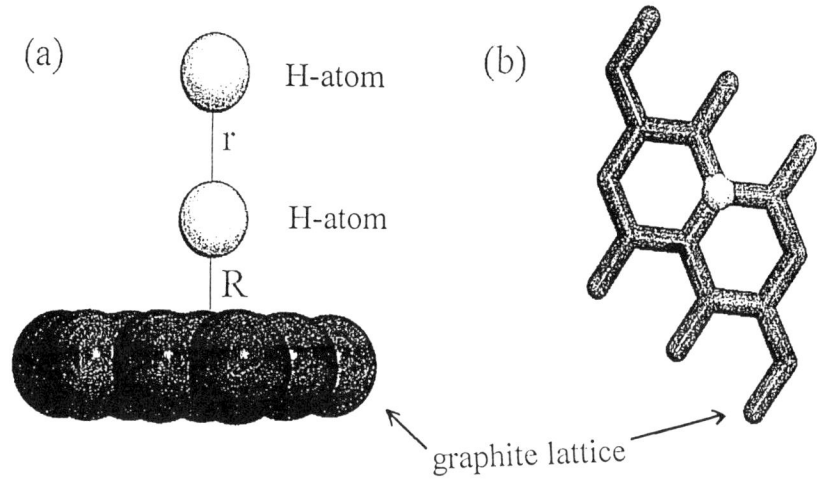

FIGURE 1. The model system. (a) Side view illustrating the coordinate system used in the calculations of Farebrother et al., and (b) shows the location of binding site.

3. Theory

A very detailed modelling of the H_2 formation process on graphite has been carried out by Farebrother et al. (1999). The Eley-Rideal mechanism is explored, in a collinear process H-H-C, where the C atom is in the graphite surface. The Eley-Rideal mechanism implies that surface motions are of less importance than in the Langmuir-Hinshelwood case, and therefore the surface is held rigid in this calculation. The problem as treated by Farebrother et al. is therefore reduced to a two-dimensional problem (see Figure 1). A portion of the graphite surface is included in the calculation, with periodic extension in all directions. The surface is assumed to be flat and rigid. Fully quantum mechanical calculations are made to determine the potential energy surface of the H-H-graphite system, using the time-independent Schrödinger equation solved in both the Local Density Approximation (LDA) and in Generalized Gradient Approximation (GGA) of Density Functional Theory. The super cell contains 18 C atoms plus two H atoms, and the spacing between layers in the vertical dimension is set at $40a_o$, to reduce interactions with image particles. The potential energy surface, fitted to 250 calculated points, shows no barrier to Eley-Rideal reaction (see Figure 2).

Test calculations of possible binding sites for H atoms on graphite show that of three possible sites (above a lattice C-atom, above the midpoint between neighbouring C-atoms, or above the centre of a hexagon) the first of these sites is the most stable (in agreement with Jeloaica and Sidis 1999), - see Figure 1. The binding energy of H to the surface is found to be 1.89 eV, implying that the Eley-Rideal reaction is exothermic by 2.9 eV.

Farebrother et al. carried out quantum scattering calculations of the H + H-graphite

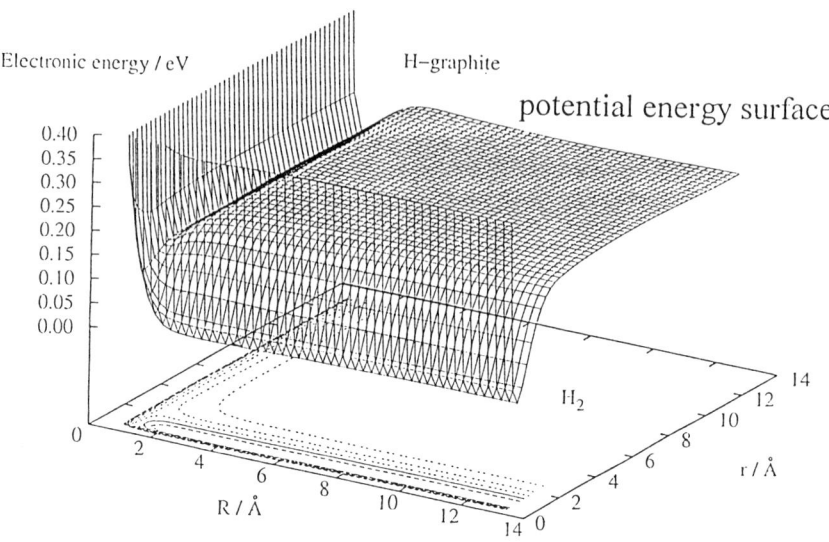

FIGURE 2. The potential energy surface for the H_2-graphite system.

system, using the R-matrix propagator method, solving the close coupling equations for 39 channels for a given energy. The scattering matrix then gives information about scattering into the v'' states of the product H_2, and the overall probability of reaction.

The total reaction probability is high, close to unity for energies $\gtrsim 0.01$ eV, and > 0.1 for energy $\lesssim 0.001$ eV (see Figure 3). The vibrational distribution favours high v'' states (see Figure 4, which shows that the level $v'' = 2$ is favoured, and that levels 3, 1, 4, 5 are also significantly populated, as well as the ground level $v'' = 0$).

The implications of these results for astrophysics are significant. It appears that the formation of H_2 on graphite surfaces by the Eley-Rideal process would be efficient over a wide range of interstellar conditions. This is rather different from the traditional view (Hollenbach, Werner and Salpeter 1971). The overall rate, of course, will depend on the fraction of grains that have graphitic surfaces, and there may also be a contribution from the Langmuir-Hinshelwood mechanism.

The most striking aspect of these results is the high vibrational excitation to be expected in the product molecule formed in the Eley-Rideal mechanism on graphite. Such excitation was predicted on the basis of very simple considerations (Duley & Williams 1986), and this conclusion now seems to be strengthened. This excitation may have several consequences in the interstellar medium. Firstly, relaxation from these high levels should contribute to the infrared emission, and may be detectable. At present, the highest vibrational level detected is $v'' = 3$ (Federman et al. 1995). Secondly, the internal energy may overcome barriers and endothermicities in the reaction of H_2 with atoms and ions, opening up new channels in interstellar chemistry, as was pointed out by Stecher & Williams (1972). Thirdly, the ionization of H_2 by photons available in the neutral interstellar medium ($h\nu \leq 13.6$ eV) becomes possible for sufficiently high vibra-

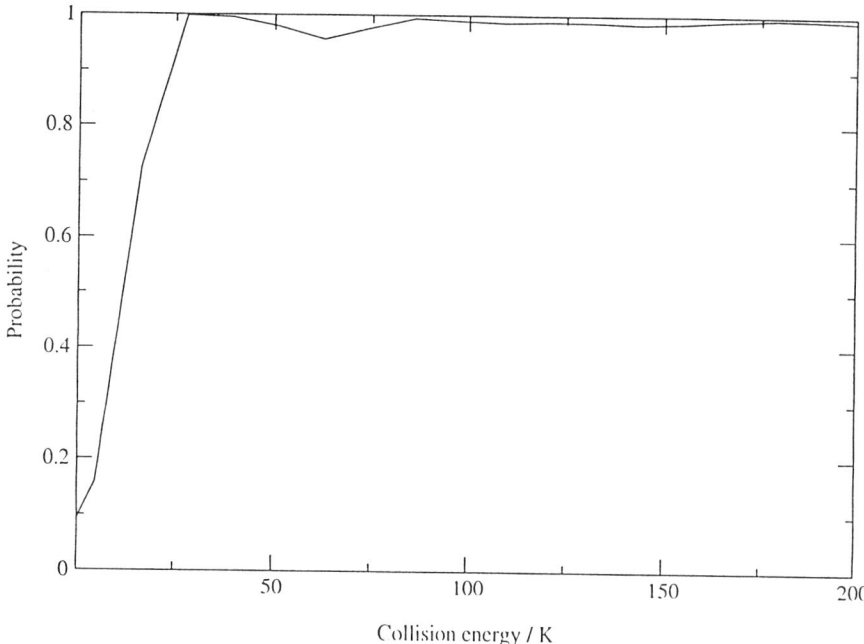

FIGURE 3. The total reaction probability for H + H- graphite → H_2 + graphite.

tional excitation ($v'' \geq 4$), so the ionization balance in localized regions may be affected (Stecher and Williams 1975). Fourthly, excited H_2 may contribute to the heating of the interstellar medium through collisional relaxation (Stecher & Williams 1973).

The calculations of Farebrother et al. are for the collinear case. A generalization to the non-linear approach of the H atom is being performed by Clary and Meijer (work in progress). This fully quantum mechanical calculation is being made using a time-dependent method and a semi-empirical (LEPS) description of the surface, which remains rigid.

The assumed rigidity of the surface in these calculations precludes a calculation of energy adsorbed by the lattice. It has been considered that this energy may be able to cause non-thermal effects on the grain surface, such as localized desorption of weakly bound species (Duley and Williams 1993). It is worth reporting here that a detailed calculation of the H_2 formation process on amorphous ice has been made by Takahashi et al. (1999a,b), using a classical description in a molecular dynamics formulation. The predicted vibrational distribution in the product H_2 was also found to be high in this calculation. The energy deposited in the amorphous ice was, however, rather small, at about 5% of the H_2 bond energy. The desorption of adsorbed CO by the resulting non-thermal pulse has been found to be negligible (Takahashi and Williams 2000).

4. Experiment

The aims of the UCL experiment are (i) to measure the reaction efficiency of H_2 formation from H atoms on a substrate, (ii) to determine the rovibrational energy distribution in the newly formed molecules, and (iii) to measure the kinetic energy of the desorbed

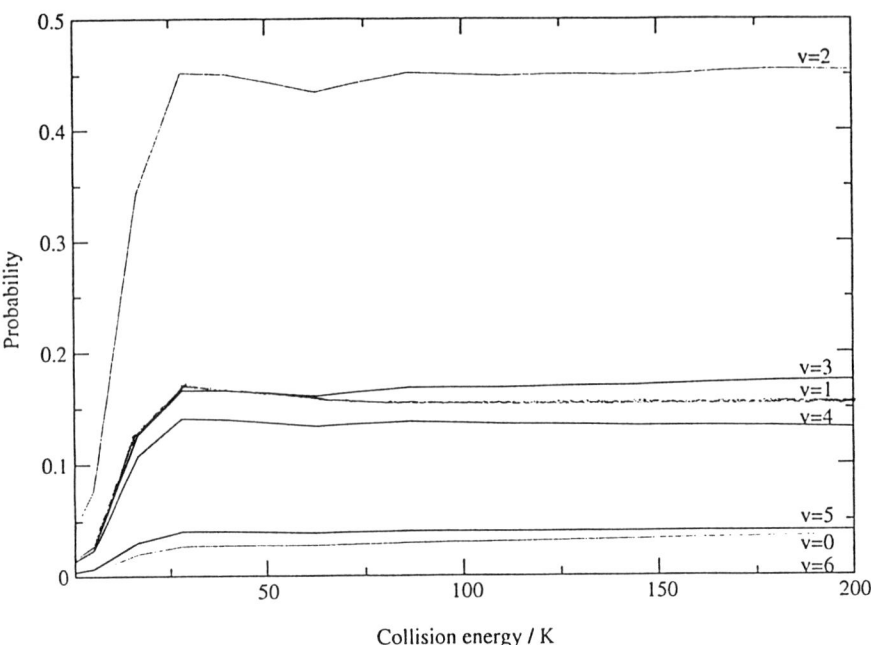

FIGURE 4. The vibrational distribution in the product H_2 molecule.

H_2. The experiments will be carried out for a range of physical conditions and for a variety of substrates. The work at UCL will therefore supplement and extend the excellent work being carried out in the Syracuse laboratory (Pirronello et al. 1997a; Pirronello et al. 1997b; Biham et al. 1998; Pirronello et al. 1999; Katz et al. 1999).

Our experiments are performed with equipment that is illustrated schematically in Figure 5. Detailed descriptions of each part of the equipment can be found at the laboratory web site http://www.ucl.ac.uk/~uccacdl.

In brief, the system operates at ultrahigh vacuum, about 10^{-10} torr. This is still a high pressure compared to most of the interstellar gas, but is the best that can practically be achieved. A beam of H-atoms is produced via a microwave discharge and cooled to about 100 K by passage through a cooling block whose temperature is controlled by a cryogenic cold-head. The cooled beam is then pulsed by a tuning fork chopper that provides an aperture of 5 mm that is open for 5 ms every 200 ms. The beam then enters the reaction chamber and impinges on the simulated dust grain target.

The targets are prepared in the Department of Electronic and Electrical Engineering at UCL and work is concentrated initially on carbonaceous materials. Carbon films will be deposited from a plasma of H_2 and CH_4, and the sp^2:sp^3 ratio of the resulting material will be controlled, and materials of differing ratio will be employed. The first targets used are polycrystalline diamond and HOPG.

The targets are mounted on a cryogenic cold-head which will maintain the grain simulated material at 10 K. The usable surface areas are 1.5 cm^2 and the films are 100 micrometres thick. The targets are flashed to 1300 K for a few seconds to drive off any contaminants, and this process is carried out *in situ*. The target and heater are mounted on a ruby heat shield and attached to the cold-head. The ruby conducts heat well at low

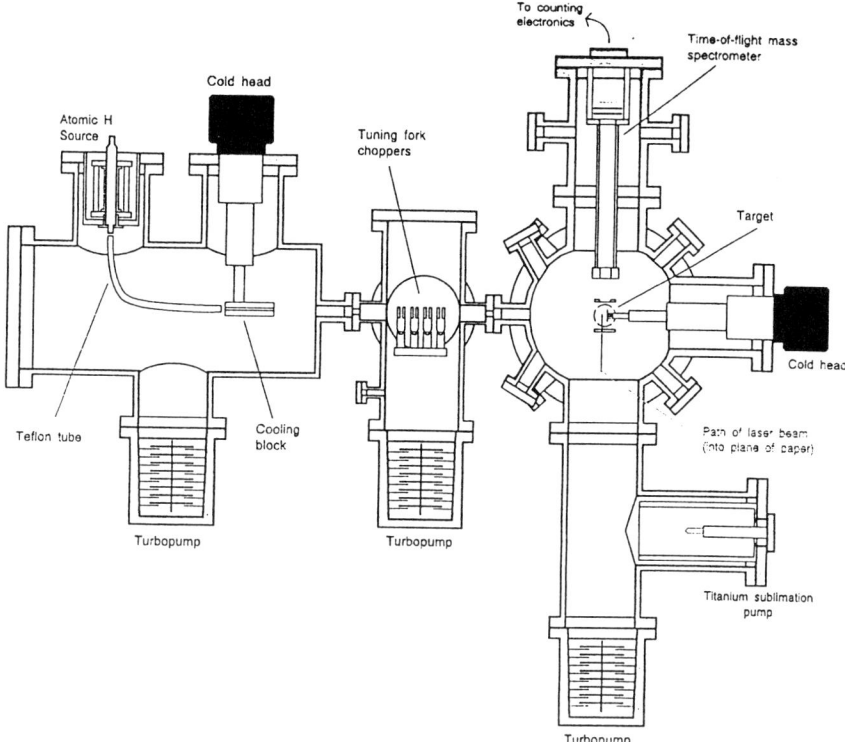

FIGURE 5. Schematic diagram of the experimental set-up.

temperatures, but insulates at high temperatures. Therefore, the flash heating does not penetrate the cold head.

The product H_2 is then ionized by a pulsed laser, in (2 + 1) Resonance Enhanced Multi-photon Ionization. The laser system consists of a tuneable dye laser pumped by a Nd:YAG laser, and radiation in the red part of the spectrum is tripled in frequency to output in the UV. The system is tuned so that excitation by frequency-tripled radiation from individual v'', J'' levels of the ground electronic state, X, of the product H_2 occurs to $v' = 0$, $J' = 0$ of the E, F $^1\Sigma_g^+$ state which is 12.4 eV above ground. Then a further photon ionizes the excited H_2, to give H_2^+. Thus, ionization occurs only from selected internal energy states of the product H_2. The resulting H_2^+ is then detected in a time-of-flight mass spectrometer, tuned to monitor H_2^+ ions exclusively.

The experiment, therefore, measures directly the formation efficiency into individual v'', J'' levels of the ground electronic state. Integration of these state-selected results will give the overall efficiency. Time-of-flight measurements of the laser-ionized H_2 will give a direct determination of the kinetic energy of the desorbed H_2. Consideration of the total energy budget will then give the amount of energy transferred to the substrate surface.

At the time of writing, all parts of the system are tested, and operating as planned. First results from the experiment are expected soon.

5. Conclusion

The theory programme has produced results for the H_2 formation efficiency and vibrational distribution, for the case of a collinear approach to a graphite surface. This calculation will be generalized to a 3D interaction, so that rotational excitation, in addition to vibrational excitation, can also be considered. These theoretical results provide the ability to interpret the experimental results that will be obtained. Experiments will be performed using a range of substrates.

REFERENCES

Biham, O., Furman, I., Katz, N., Pirronello, V. & Vidali, G. 1998 *MNRAS* **296** 869

Draine, B. T. & Anderson, N. 1984 *ApJ* **292** 494

Duley, W. W. 1996 *MNRAS* **279** 591

Duley, W. W. & Williams, D. A. 1993 *MNRAS*, **260** 37

Duley, W. W. & Williams, D. A. 1986 *MNRAS*, **219** 177

Farebrother, A. J., Meijer, A. J. H. M., Clary, D. C. & Fisher, A. J. 2000 *Chem. Phys. Lett.*, in press

Federman, S. R., Cardelli, J. A., van Dishoeck, E. F., Lambert, D. L., & Black, J. H. 1995 *ApJ.* **445**, 325

Hollenbach, D. J., Werner, M. W. & Salpeter, E. E. 1971 *ApJ.* **163** 165

Jeloaica, L. & Sidis, V. 1999 *Chem. Phys. Lett.* **300** 157

Katz, N., Furman, I., Biham, O., Pirronello, V. & Vidali, G. 1999, *ApJ.* **522** 305

Pirronello, V., Liu, C., Shen, L. & Vidali, G. 1997a *ApJ.* **475** L69

Pirronello, V., Biham, O., Liu, C., Shen, L. & Vidali, G. 1997b *ApJ.* **483** L131

Pirronello, V., Liu, C., Roser, J. E. & Vidali, G. 1999 *A&A* **344** 681

Sandford, S. A. 1989 in "*Interstellar Dust*", eds. L. J. Allamandola, A. G. G. M. Tielens, Kluwer, Dordrecht, pp. 403–413.

Sellgren, K. 1984 *ApJ* **277** 623

Stecher, T. P. & Williams, D. A. 1972 *ApJ* **177** L145

Stecher, T. P. & Williams, D. A. 1973 *MNRAS* **161** 305

Stecher, T. P. & Williams, D. A. 1975 *Astrophys. Sp. Sci.* **32** 211

Takahashi, J., Masuda, K. & Nagaoka, M. 1999a *ApJ* **520** 742

Takahashi, J., Masuda, K. & Nagaoka, M. 1999b *MNRAS* **306** 22

Takahashi, J. & Williams D.A. 2000 *MNRAS* in press

Williams, D. A. 1998 *Faraday Disc.* **109** 1

Probing the connection between PAHs and hydrogen (H, H$_2$) in the laboratory and in the interstellar medium

By C. JOBLIN[1], J.P. MAILLARD[2], I. VAUGLIN[3], C. PECH[1] AND P. BOISSEL [4]

[1] CESR-CNRS, 9 Av. du Colonel Roche, 31028 Toulouse, France

[2] IAP-CNRS, 98 bis Bd. Arago, 75014 Paris, France

[3] Observatoire de Lyon, 69561 Saint Genis Laval Cdx, France

[4] LPCR-CNRS, Bât 350, Université Paris Sud, 91405 Orsay Cdx, France

Polycyclic aromatic hydrocarbons (PAHs) could play an important role in interstellar chemistry. In particular, it is important to evaluate their possible contribution to the formation of H$_2$. To address this question, recent laboratory results and new observations are presented. Although still preliminary, these first results are very encouraging. First, the photodissociation of PAHs isolated in ion traps and exposed to UV light involves the loss of pairs of hydrogen atoms which are likely to form H$_2$ molecules. Second, the PAH and H$_2$ emission observed in the photodissociation region associated with the young stellar object S106-IR was found to coincide at some positions. This suggests a coupling between the interstellar PAH and H$_2$ populations. More results are expected in the near future.

1. Introduction

The Infrared Space Observatory (ISO) has recently showed the ubiquity of the emission features at 3.3, 6.2, 7.7, 8.6, 11.3 and 12.7 μm in the interstellar medium (First ISO Results 1996, Boulanger 1999). Amongst the carriers for these features, polycyclic aromatic hydrocarbons (PAHs) are the best candidates as far as an excitation mechanism and a reasonable spectral agreement are concerned. A lot of infrared spectroscopy has been performed in the laboratory since the initial proposal by Léger & Puget (1984) and Allamandola et al. (1985) to find the laboratory species whose spectrum match the interstellar spectrum (Szczepanski & Vala 1993, Joblin et al. 1995, Cook et al. 1998, Allamandola & Hudgins 1999, Hudgins & Allamandola 1999, etc...).

The role of PAHs in interstellar chemistry may be fundamental. These species may contain up to 20% of the cosmic carbon (Boulanger 1999) and therefore could largely contribute to surface reactions including the formation of H$_2$. Another possibility would be to form H$_2$ by the photodissociation of PAHs induced by UV light. Léger et al. (1989) have shown that the absorption of a UV photon can sufficiently heat the molecule to provide the ejection of a fragment by photo-thermo-dissociation. Jochims et al. (1994, 1999) have studied the photodissociaion of PAHs submitted to (hard) UV photons. However, the experiment was performed at a rather high dissociation rate ($\sim 10^4$ s^{-1}), whereas in the interstellar medium dissociation is expected to occur at a much smaller rate, in direct competition with the IR radiative cooling (typically 10 s^{-1} according to Allamandola et al. 1989).

Slow dissociation processes can be studied in the laboratory by trapping PAHs in ion cells under the conjugated action of a magnetic field and an electrical field (see for instance Boissel et al. 1994). The results are presented in section 2 and a new experimental set-up called PIRENEA and specifically devoted to these studies is described.

In section 3, we address the question of the coupling of PAHs with hydrogen in the interstellar medium using observations of the S106 star formation region and in particular of the photodissociation region (PDR) between the young massive object S106-IR (S106 IRS4 in Gehrz et al. 1982) and its parent cloud. Thanks to the IR capabilities of the Canada-France-Hawaii telescope, we have observed the emission band of interstellar PAHs at 11.3 μm and the 1–0 S(1) rotation-vibrational fluorescent line of H_2. The spatial distribution of PAHs and H_2 can be generally understood if both populations are excited by UV photons from the central star. Evidence for secondary effects such as a formation of H_2 driven by PAHs is also discussed.

2. Photodissociation of PAHs as a source of H_2 in the laboratory

The photodissociation of PAHs has been studied using Fourier transform mass spectrometry (FTMS) in ion cyclotron resonance (ICR) cells. These studies concern PAH cations but the derived molecular constants are expected to be also valid for neutral species (Leach 1987). Boissel et al. (1994) have shown that the fragmentation can be probed very close to its threshold when the trapped ions are exposed to the radiation of a Xe arc lamp. This threshold is reached by building up the internal energy through the sequential absorption of several visible photons.

In these conditions, the pyrene cation ($C_{16}H_{10}^+$) undergoes dissociation through the loss of two hydrogen atoms. Several other small PAHs were also found to lose an even number of hydrogen atoms (Ekern et al. 1998). In the case of the coronene cation ($C_{24}H_{12}^+$) the whole series of -2H,-4H, -6H, -8H, -10H and -12H fragments leading to the completely dehydrogenated species C_{24}^+ was observed at high irradiation of the Xe lamp. The fragmentation of larger PAHs (up to 200 carbon atoms) was reported by Joblin et al. (1997). These species were produced by laser ablation of a sample of pyrolysed coronene ($C_{24}H_{12}$). At low laser irradiance, the coronene dimer appears at mass 596 a.m.u. (M=596), as expected, the additional peaks being due to ^{13}C isotopes (Fig. 1). For the trimer, the spectrum is slightly more complicated due to the presence of a condensed form at M=888 and a more linear one at M=892. Increasing the laser irradiance leads to a complex pattern of photodissociation products. However, it has been noticed that all the mass spectra can be synthetized using fragments derived from the parent species by the loss of even numbers of hydrogen atoms (lower part of Fig. 1).

All these studies strongly suggest that the photodissociation of compact and large PAHs preferentially involves the loss of pairs of hydrogen atoms. From an energetic point of view, it is tempting to conclude that these pairs recombine during the photodissociation process to form H_2. However, more laboratory studies have to be done in particular for large PAHs. This is one of the prime goal of the PIRENEA experiment.

The PIRENEA experiment is a FTMS set-up equipped with an ICR cell in a 5 Tesla superconducting magnet. This experiment has been specifically developed for interstellar chemistry studies in particular whith large molecules or small aggregates. To approach the physical conditions of the interstellar medium, the trapped ions are isolated under ultra-high vacuum and in a cold environment (~ 10 K). The experiment is now in the testing phase and we expect the first scientific results in the coming months. In particular, the photodissociation of large PAHs will be studied with a better control of the internal energy compared to previous studies (Joblin et al. 1997). An interesting point will be to compare the stability of the fragments against photodissociation. The most stable ones, those which need a higher energy to be dissociated, are likely to be real analogues of interstellar PAHs. The direct detection of H_2 produced in situ by the photodissociation

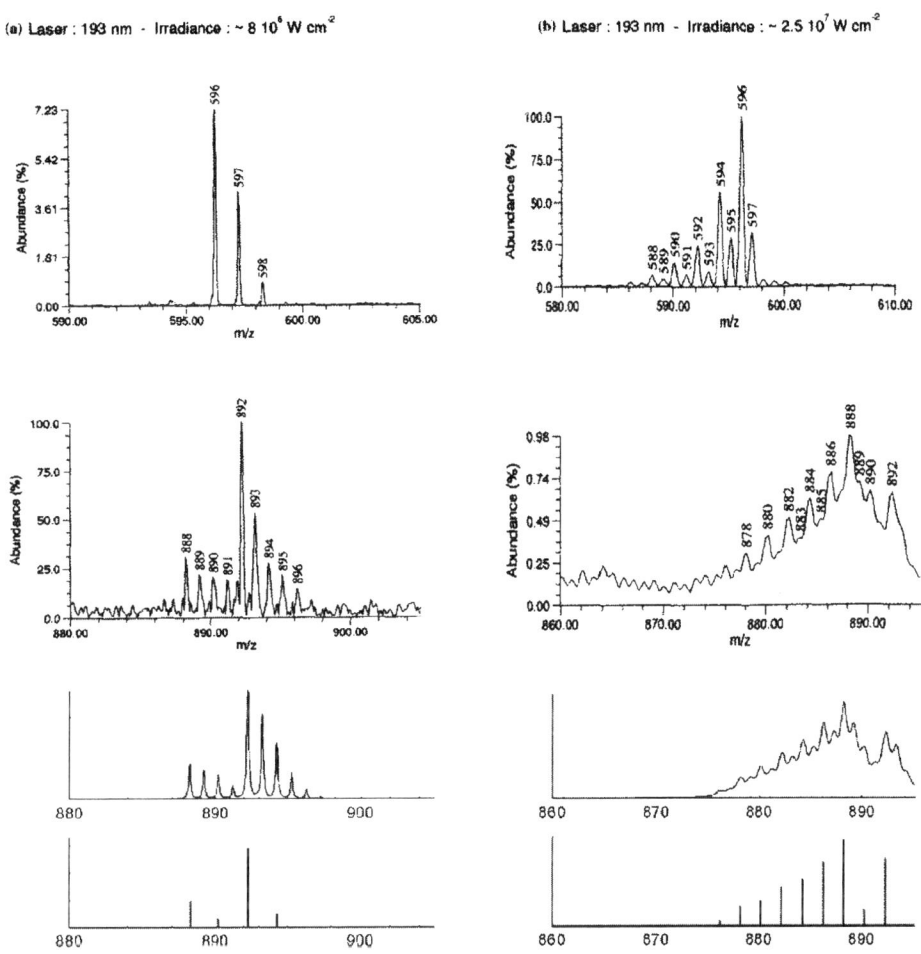

FIGURE 1. Mass peaks of the coronene dimer ($C_{48}H_{20}$; M=596) and trimer ($C_{72}H_{28}$; M=892) produced with 193 nm laser ablation (upper part). For the trimer, the laboratory spectra are compared with synthetized spectra (third row). The latter are obtained by taking into account the ^{13}C isotopes and by including several fragment species whose relative abundances are given in the bottom charts. At low laser irradiance (left column (a)), the satellite peaks mainly reveal the presence of a condensed form of the trimer ($C_{72}H_{24}$; M=888). Increasing the irradiance (right column (b)) clearly induces photodissociation by loss of pairs of hydrogen atoms (H_2?) as only species with even masses are present.

of PAHs is expected to be difficult in PIRENEA. This question will certainly require a theoretical approach. Another important aspect is that of the reactivity of hydrogen (H and H_2) with more or less dehydrogenated PAHs. Indeed, if PAHs contribute to the formation of H_2 when exposed to UV photons, they should be continuously rehydrogenated for this process to be efficient. Snow et al. (1998) and LePage et al. (1999) have shown that small PAH cations react rapidly with hydrogen to form protonated species. Finally, as discussed by F. Pauzat in her contribution to this conference, theoretical studies have suggested the formation of H_2 by reaction of atomic hydrogen with PAH cations (Cassam-Chenaï et al. 1994) or with protonated PAH cations (Bauschlicher 1998). This could be tested in our experiment as soon as it will be equipped with an atomic hydrogen source.

3. Emission of PAHs and H_2 in the S106 star formation region

S106 is a star formation region which contains about 160 stars (Hodapp & Rayner 1991) but the dynamics of the region is essentially dominated by the very massive young stellar object S106-IR (Eiroa et al. 1979). Our prime interest is to study the correlation between the emission of PAHs and of H_2 in the photodissociation region formed by the interaction of this star with its parent molecular cloud. One advantage of this object is to be extended and rather close which allows detailed spatial studies. Its distance is usually taken as 600 pc according to the measurements by Staude et al. (1982) although Rayner (1994) has determined a distance of 1200 pc.

The observations were performed using spectro-imagery (0.3"/pixel) at the IR focus of the Canada-France-Hawaii (CFH) telescope in July 1998. The 11.3 μm PAH band was observed using the C10μ camera (Merlin et al. 1996) at three positions of the CVF: 11.25, 10.75 and 11.75 μm to properly separate the emission in the band from the continuum emission. The H_2 1–0 S(1) emission line was imaged with the BEAR Fourier transform spectrometer (Maillard 1995) using a narrow-band filter centered at 2.12 μm. The BEAR instrument offers a very high spectral resolution providing essential information on the gas dynamics. In these observations, the spectral resolution was 14 km s^{-1}. The measured gas line was fitted with a gaussian profile in order to precisely derive its position, width and integrated intensity.

3.1. Results

The comparison of the H_2 and PAH emission maps leads to several distinct regions (Fig. 2).

(a) A large condensation at the western side where the PAH and H_2 emission fronts are shifted relative to each other

(b) A bright arc at the north in the H_2 emission where the PAH emission is weak

(c) An extended H_2 emission at the close south of S106-IR which is not observed in the PAH emission except for the bright clump right at the west of the star

(d) A long clumpy arc at the south where the PAH and H_2 emissions correlate quite well.

The dispersion of the velocity field is not important with a maximum difference of 10 km s^{-1} between the south arc (d) and the rest of the nebula (Fig. 3). The width of the H_2 line is given by the instrumental resolution (14 km s^{-1}) in most of the field. It reaches however a value of 20 km s^{-1} at the close vicinity of the S106-IR source and at the borders of the large hole present at the south-west. Observations in the Brγ line, not reported in this paper, clearly showed that this hole was created by a young companion star whose winds have disrupted the surrounding molecular cloud. Turbulence might be quite important in this area and account for the increased H_2 line width.

3.2. Discussion

The large condensation at the west (position a) corresponds to the typical case of an edge-on photodissociation region such as the well-known Orion Bar (Tielens & Hollenbach 1985). There are clearly separated layers of ionized gas (H^+), atomic gas where the PAH emission occurs and finally molecular gas. In the atomic region, H_2 is dissociated

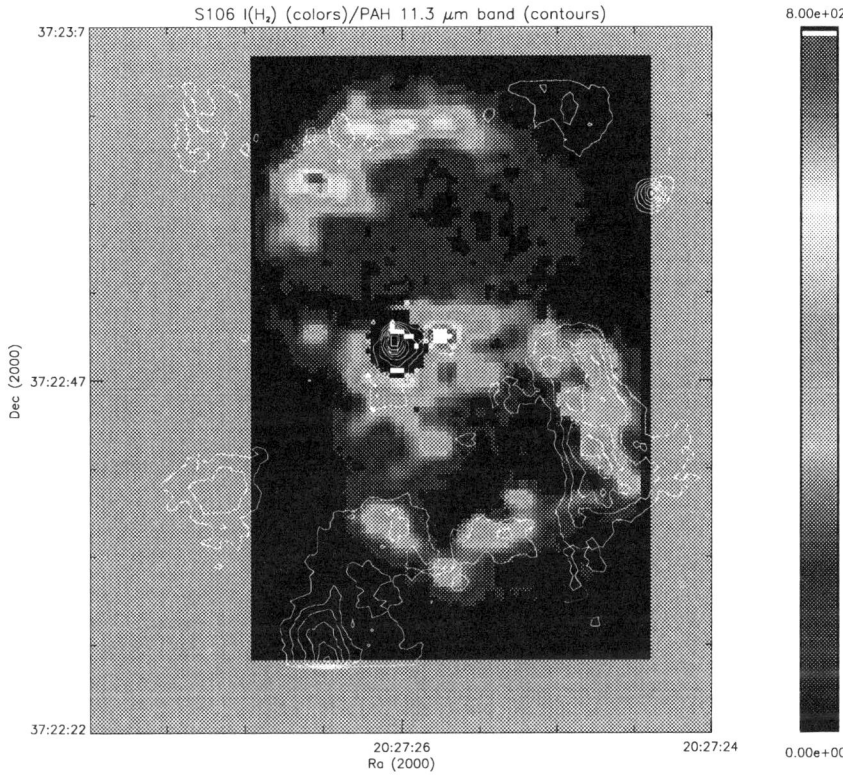

FIGURE 2. Comparison of the emission in the H_2 1–0 S(1) line at 2.12 μm (colors) with the emission in the 11.3 μm PAH band (contours). The gas line measured with the BEAR instrument at a spectral resolution of 14 km s^{-1} was fitted with a gaussian profile whose integrated intensity is shown here (arbitrary units). The 11.3 μm PAH band measured with the C10μ camera in CVF mode was derived by subtracting the emission in the continuum.

by the FUV photons. Only at an optical depth of 1-2 magnitudes, there is equilibrium between H_2 formation and dissociation. The emission in the near-IR vibrational bands of H_2 traces this region. At position (a), the H_2 emission seems therefore to be mostly radiative fluorescent emission.

The emission at position (b) is more difficult to interpret. It is particularly strong compared to the emission in the south lobe especially considering that we have not tried to correct for differential extinction: A_V=12 and 8 for the northern and southern lobes respectively (Felli et al. 1984). The fact that the PAH emission is weak in the northern region suggests either that the UV field is weak or that PAHs are not abundant. A plausible explanation could be that denser regions are present at the north in which gas-phase PAHs are depleted due to mutual sticking or coagulation on large grains (Boulanger et al. 1994). This is consistent with the results of theoretical models which show that H_2 emission can increase when the UV field decreases if the density is higher (Sternberg & Dalgarno 1989, Black & van Dishoeck 1989, Burton et al. 1990, Draine & Bertoldi 1996). In this case, collisions can play an important role. In particu-

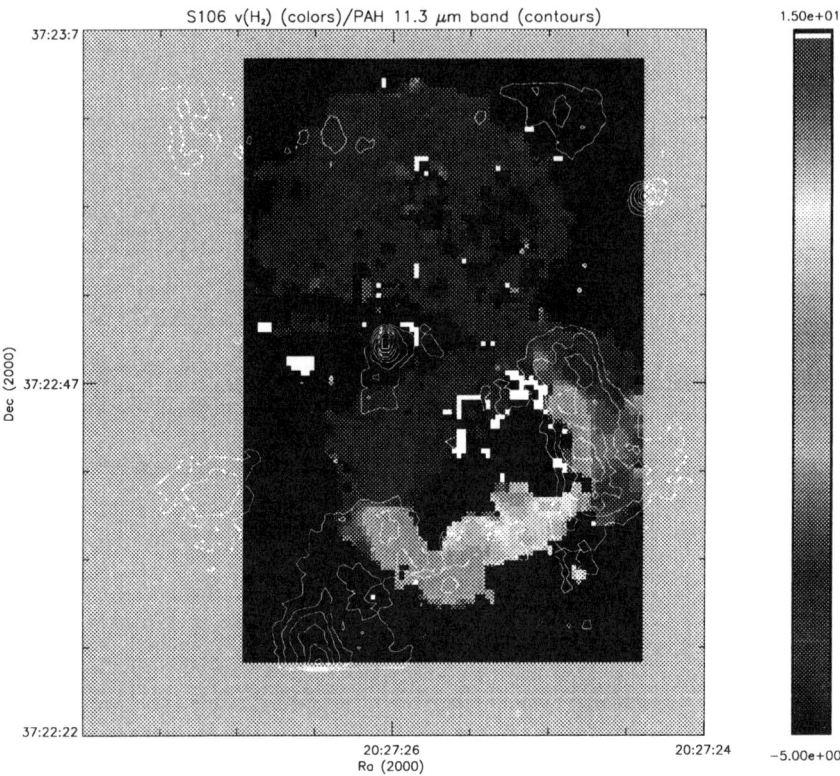

FIGURE 3. Comparison of the gas velocity field (km s^{-1}) derived from the H$_2$ 1–0 S(1) line (colors) with the emission in the 11.3 μm PAH band (contours). The velocity of the H$_2$ line was determined by a fit with a gaussian profile. It was corrected from the proper motion of the source derived from the observation of the Brγ line on the central star.

lar, the 2–1 S(1)/1–0 S(1) H$_2$ line ratio is expected to be weaker compared to the value of 0.55 calculated in the case of a pure fluorescent cascade by Sternberg et al. (1989). Longmore et al. (1986) measured the spectrum at 2 μm for a position just north of S106-IR. The authors showed that the line ratio was lower than 0.55. They suggested a possible contribution of shocks to the excitation of H$_2$ which would be more pronounced for geometrical positions which are directly exposed to the wind. This could happen for positions (b) and (c) although it is difficult to distinguish between a high-density radiative excitation and a thermal excitation in shocks. However, the narrow width measured for the H$_2$ line as well as the small dispersion in the velocity field argue against an excitation by high-velocity shocks.

Finally, position (d) is the one which gives the best spatial coincidence between the PAH and the H$_2$ emission. This coincidence might just be a geometrical effect. It could also be a place where the role of PAHs in H$_2$ formation is enhanced. We notice that the gas velocity in this region is higher (∼8 km s^{-1}) compared to the rest of the nebula (Fig. 3). This is likely due to a driving motion generated by the collimated redshifted jet from the young companion star as revealed by the Brγ line observations (cf. & 3.1).

4. Conclusions

In this paper, we have evaluated the possibility for PAHs to contribute to the formation of H_2 using both laboratory experiments and observations. This work is clearly in the beginning stage although the first results are quite encouraging.

Preliminary results obtained with ion trap experiments strongly suggest that the photodissociation of large and compact PAHs exposed to UV photons leads to the formation of H_2. These studies are one of the prime goal of the new experimental set-up called PIRENEA.

We have also searched for the coupling of PAHs with H_2 directly in the interstellar medium, using the BEAR FTS imager and the C10μ camera at the CFH telescope. The emission of PAHs and of H_2 was found to coincide at some positions within the PDR associated with the S106-IR young massive stellar object. We have shown that such results are difficult to interpret as several factors including the UV field intensity, the density and the presence of shocks can interfer. More observations are clearly required before providing any observational evidence for the contribution of PAHs to the formation of H_2. For instance, we have recently observed with BEAR the 2–1 S(1) H_2 line to better constrain the excitation mechanism(s) of the H_2 emission in the S106 region. Other observations are also planned, in particular to study reflection nebulae which are simpler PDRs where shocks are absent.

We would like to especially acknowledge G. F. Mitchell and P. Cox for their contribution to the BEAR observations and P. de Pesloüan for her help in the data reduction. We also thank P. Merlin and J.-P. Dubois for assisting in the C10μ camera operation.

REFERENCES

FIRST ISO RESULTS 1996 *Astron. Astrophys.* **315** (2).

ALLAMANDOLA, L., TIELENS, A. G. G. M. & BARKER, J. R. 1985 *Astrophys. J.* **290**, L25–L28.

ALLAMANDOLA, L., TIELENS, A. G. G. M. & BARKER, J. R. 1989 *Astrophys. J. Series* **71**, 733–775.

ALLAMANDOLA, L. J. & HUDGINS, D. M. 1999 *Astrophys. J.* **511**, L115–L119.

BAUSCHLICHER, C. W. 1998 *Astrophys. J.* **509**, L125–L127.

BLACK, J. H. & VAN DISHOECK, E. F. 1989 *Astrophys. J.* **322**, 412–449.

BOISSEL, P., LEFÈVRE, G. & THIÉBOT, PH. 1994 in *Molecules and Grains in Space* (ed. I. Nenner). Conf. Proceedings **312**, pp. 667–674. AIP, New York.

BOULANGER, F., PRÉVOT, M.L. & GRY, C. 1994 *Astron. Astrophys.* **284**, 956–970.

BOULANGER, F. 1999 in *Solid Interstellar Matter: The ISO Revolution* (eds. L. d'Hendecourt, C. Joblin & A. Jones). Les Houches Series **11**, pp. 20–31. EDP Sciences, Les Ulis.

BURTON, M. G., HOLLENBACH, D. J. & TIELENS, A. G. G. M. 1990 *Astrophys. J.* **365**, 620–639.

CASSAM-CHENAÏ, P., PAUZAT, F. & ELLINGER, Y. 1994 in *Molecules and Grains in Space* (ed. I. Nenner). Conf. Proceedings **312**, pp. 543–547. AIP, New York.

COOK, D.J., SCHLEMMER, S., BALUCANI, N., WAGNER, D. R., HARRISON, J. A., STEINER, B. & SAYKALLY, R. J. 1998 *J. Phys. Chem. A* **102** 1465–1481.

DRAINE, B. T. & BERTOLDI, F. 1996 *Astrophys. J.* **468**, 269–289.

EIROA, C., ELSÄSSER, H. & LAHULLA, J. F. 1979 *Astron. Astrophys.* **74**, 89–92.

EKERN, S. P., MARSHALL, A. G., SZCZEPANSKI, J. & VALA, M. 1998 *J. Phys. Chem. A* **102**, 3498–3504.

FELLI, M., STAUDE, H. J., REDDMANN, T., MASSI, M., EIROA, C., HEFELE, H., NECKEL, T. & PANAGIA, N. 1984 *Astron. Astrophys.* **135**, 261–280.

GEHRZ, R. D., GRASDALEN, G. L., CASTELAZ, M., GULLIXSON, C., MOZURKEWICH, D. & HACKWELL, J. A. 1982 *Astrophys. J.* **254**, 550–561.

HODAPP, K.-W. & RAYNER, J. 1991 *Astron. J.* **102**, 1108–1117.

HUDGINS, D. M. & ALLAMANDOLA, L. J. 1999 *Astrophys. J.* **513**, L69–L73.

JOBLIN, C., BOISSEL, P., LÉGER, A., D'HENDECOURT, L. & DÉFOURNEAU, D. 1995 *Astron. Astrophys.* **299**, 835–846.

JOBLIN, C., MASSELON, C., BOISSEL, P., DE PARSEVAL, P., MARTINOVIC, S. & MULLER, J. F. 1997 *Rapid Comm. in Mass Spectrom.* **11**, 1619–1623.

JOCHIMS, H. W., RÜHL E., BAUMGÄRTEL, H., TOBITA, S. & LEACH, S. 1994 *Astrophys. J.* **420**, 307–317.

JOCHIMS, H. W., BAUMGÄRTEL, H. & LEACH, S. 1999 *Astrophys. J.* **512**, 500–512.

LEACH, S. 1987 in *Polycyclic Aromatic Hydrocarbons and Astrophysics* (ed. A. Léger, L. d'Hendecourt & N. Boccara), pp. 99–127. Dordrecht. Reidel.

LÉGER, A. & PUGET, J. L. 1984 *Astron. Astrophys.* **137**, L5–L8.

LÉGER, A., BOISSEL, P., DÉSERT, F.X. & D'HENDECOURT, L. 1989 *Astron. Astrophys.* **213**, 351–359.

LE PAGE, V., KEHEYAN, Y., SNOW, T. P. & BIERBAUM, V. M. 1999 *Int. J. Mass Spectrom.* **185/186/187**, 949–959.

LONGMORE, A. J., ROBSON, E. I. & JAMESON, R. F. 1986 *Mon. Not. R. Astr. Soc.* **221**, 589–598.

MAILLARD, J.P. 1995 in *Tridimensional Optical Spectroscopic Methods in Astrophysics* (ed. G. Comte & M. Marcelin). IAU Coll. 149, **71**, pp. 316–327. ASP Conf. Series.

MERLIN, P., SIBILLE, F. & VAUGLIN, I. 1996 *User's Manual for the 10 micron camera C10μ*. available at http://www-obs.univ-lyon1.fr/ir.

RAYNER, J. 1994 in *Infrared Astronomy with Arrays: The next generation* (ed. I.S. McLean), pp. 185. Dordrecht. Kluwer.

SNOW, T. P., LE PAGE, V., KEHEYAN, Y. & BIERBAUM, V. M. 1998 *Nature* **391**, 259–260.

STAUDE, H. J., LENZEN, R., DYCK, H. M. & SCHMIDT, G. D. 1982 *Astrophys. J.* **255**, 95–102.

STERNBERG, A. & DALGARNO, A. 1989 *Astrophys. J.* **338**, 197–233.

SZCZEPANSKI, J. & VALA, M. 1993 *Astrophys. J.* **414**, 646–655.

TIELENS, A. G. G. M. & HOLLENBACH, D. 1985 *Astrophys. J.* **291**, 722–746.

3. Observations and Models

Dahbia Talbi in the foreground and David Wilgenbus in the background.

Non stationary C-shocks: H_2 emission in molecular outflows

By Guillaume PINEAU DES FORÊTS[1] AND David FLOWER[2]

[1] Observatoire de Paris, DAEC, UMR 8631 du CNRS, F-92195 Meudon, France

[2] Physics Department, The University, Durham DH1 3LE, UK

Shock waves in outflows are generated by the impact of jets, associated with low-mass star formation, on the surrounding molecular gas. These shocks give rise to a strong H_2 rovibrational emission spectrum which has been observed by the ISO satellite in several star formation regions. The dynamical time scales associated with these outflows are estimated to be a few thousand years and can be, in some regions, as short as a few hundred years. On the other hand, the time required to reach steady state for a C-shock is about 10^4 years. Under such circumstances, the shocks are unlikely to have attained a state of equilibrium, and a time dependent approach has to be considered. Non stationary C-shocks are found to exhibit both C- and J-type characteristics. The H_2 rotational excitation diagram can provide a measure of the age of the shock; in the case of the outflow observed in Cepheus A West by the ISO satellite, the shock age is estimated to be approximately 1.5×10^3 yr.

1. Time scales

1.1. Steady state shocks

Shocks propagating in the interstellar medium are expected to modify profoundly the local physical and chemical conditions. Even in the simplest case of planar shocks, the structure of the shock can take a number of different forms, from 'jump' or J-type structure, in which changes in density, velocity and temperature occur quasi-discontinuously, to 'continuous' or C-type, where the variations take place smoothly over a much larger distance scale. This structure, which depends on the physical conditions in the pre-shock gas, has been studied by a number of authors (see Draine & McKee, 1993, for a review). In the absence of a magnetic field, or in the presence of a weak magnetic field, the ionized and neutral fluids remain fully coupled throughout the flow, and a J type shock occurs. On the other hand, if the transverse magnetic field strength is sufficiently large, and hence the Alfvén speed in the ionized fluid, the ions are heated and compressed a long way upstream of the neutrals, and the steady state solution is a C-type shock (Mullan 1971, Draine 1980). These two types of shocks differ, not only in their distance scales, but also in their time scales. If we define the flow time as the time required by a neutral fluid particle to flow through the shock, the flow time is then:

$$t_{flow} = \int \frac{dz}{v_n}, \qquad (1.1)$$

where v_n is the velocity of the neutral fluid.

In Fig. 1, we show the comparison of the steady state temperature and density profiles as a function of the flow time for the two types of shocks. Both models have the same initial energy input, namely the same shock velocity ($v_s = 20$ km s^{-1}) and pre-shock density ($n_H = n(H) + 2n(H_2) = 10^4$ cm^{-3}). The only difference is in the strength of the pre-shock transverse magnetic field: $B = 10$ μG leading to a J-shock, and $B = 100$ μG to a C-shock structure.

Both C- and J-type shocks involve an initial rise in the temperature, followed by

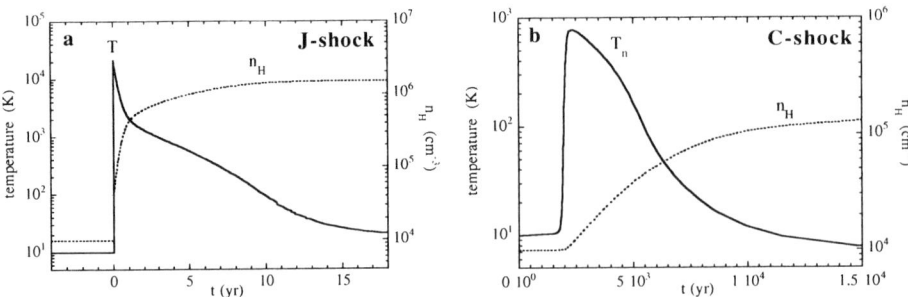

FIGURE 1. Temperature and density profiles as a function of the flow time for a J-shock (a) and a C-shock (b) of velocity $v_s = 20$ km s^{-1} propagating into a medium in which $n_H = 10^4$ cm^{-3} with a transverse magnetic field $B_0 = 10\mu$G (a) and $B_0 = 100\mu$G (b). 'Pre-shock' is to the left and 'posts-hock' to the right in this Figure.

radiative cooling and compression of the gas towards its posts-hock state. However, the quasi-discontinuous temperature rise associated with a J-type shock ensures that the maximum temperature is attained adiabatically, and it is consequently much higher than in a C-type shock of the same speed. Thus, whilst the C-type shock gives rise to a maximum temperature $T_n^{\max} \approx 800$ K of the neutral gas, $T_n^{\max} \approx 2 \times 10^4$ K in the corresponding J-type shock

The high maximum temperatures in J-type shocks give rise to much more rapid radiative cooling, by rovibrational transitions of molecular hydrogen, than in the corresponding C-type shocks. Consequently, the time of flow through the J-type shock is only 15 yr, as compared with 8000 yr for the C-type shock, this means that the time t_{equ} required, for such a C-shock, to reach a steady state is $\approx$ 8000 yr. The corresponding widths are 10^{13} cm and 10^{17} cm, respectively. Evidently, H$_2$ emission lines of high excitation energy are much more intense, relative to those of low excitation energy, under J-type shock conditions.

More generally, t_{equ} scales as (see Wilgenbus et al. in this volume):

$$t_{equ} \approx 8 \times 10^4 \text{ yr } [10^4 \text{ cm}^{-3}/n_H] [10 \text{ km s}^{-1}/v_s], \tag{1.2}$$

All steady state models explicitly assume that the action of the perturbation has to be maintained for at least this length of time in the region being modelled.

1.2. Outflows

Molecular outflows in regions of low-mass star formation have been the objects of increasing observational and theoretical study in recent years (see the Proceedings of the IAU Symposium 182: Reipurth & Bertout 1997). The bow shocks which are observed in these regions probably have C-shock characteristics over at least part of their extent. These outflows are also characterized by broad emission lines of SiO (see, for example, Gueth, Guilloteau & Bachiller 1998) and a rich spectrum of H$_2$ rovibrational transitions (Davis & Eislöffel 1995; Gredel 1994, 1996; Wright et al. 1996; Neufeld, Melnick & Harwit 1998). Schilke et al. (1997) interpreted the SiO emission in molecular outflows in terms of the sputtering of silicate grains in C-shocks and subsequent chemical processing in the gas phase. The model predicts that sputtering occurs in C-shocks when the ion-neutral drift speed is sufficiently large.

More difficult is the interpretation of the H$_2$ emission lines. In previous work, a combination of different shock types with different densities and velocities has often been invoked to match the observed flux of the H_2 rovibrational lines (see, for example, Wright et al. 1996 and Hartigan et al. 1996 for Cepheus A). Furthermore, all previous work has been based on the assumption of steady state. However, the dynamical ages of the outflows being modelled are estimated as being of the order of 1000 yr, which is insufficient for the C-shocks to attain steady state. In view of these short timescales, a time dependent approach has to be considered.

2. The temporal evolution of C-shocks

The present work builds on the study of Schilke et al. (1997). An important extension is to allow, in the current model, for the temporal evolution of the shock. The dynamical ages of the structures are insufficient for the shocks to attain steady state. Chièze, Pineau des Forêts & Flower (1998) have shown that, under these circumstances, the shock retains both C- and J-characteristics. Although the use of the exact time-dependent model is precluded, in the present case, by the computational demands of the additional physical processes which are treated, a quasi-time-dependent approach is followed which has been validated by comparison with the results of the exact calculations (Flower & Pineau des Forêts, 1999).

A total of 120 chemical species comprising H, He, C, N, O, Si, S and Fe are connected by a set of 860 reactions. Thus, 180 simultaneous, first order, highly non-linear differential equations (comprising 11 equations for the dynamical variables, 120 chemical rate equations, and 49 equations for the H_2 rovibrational levels) had to be solved. This task was performed by the GEAR differential equation solver, in the implementation of Hindmarsh (1974).

Here, we shall be concerned principally with the emission by H_2. The rate coefficients for the rotational excitation of H_2 have been updated and derive from the work of Flower and his collaborators (see the review by Flower in this volume). The treatment of the level populations of molecular hydrogen differs from previous work (e.g. Draine, Roberge & Dalgarno 1983) not only by the use of more recent collisional data but also by abandoning the statistical equilibrium assumption. In shocks in which discontinuities occur, the assumption that the level populations adjust instantaneously to the local temperature and density, implicit in the statistical equilibrium approximation, is not valid.

The dynamical importance of the chemistry should be emphasized: the chemical rate equations should also be solved in parallel with the dynamics, particularly the ion-molecule chemistry. The much lower degree of ionization, when chemical reactions are included, leads to weaker ion-neutral coupling, thereby enhancing the width of the shock wave and the time required to re-establish equilibrium, in the posts-hock gas. In particular, the maximum value attained by the kinetic temperature of the neutral fluid is lower when chemical reactions are included (Pineau des Forêts & Flower 1997).

To illustrate the temporal evolution of a C-shock, we adopted a value for the density of the pre-shock gas, $n_H = 10^4$ cm^{-3}, which is within the range of the estimates of densities in molecular outflows. The initial value of the magnetic induction, $B = 100$ μG, was obtained from the frequently-used scaling relation $B(\mu G) = [n_H$ (cm$^{-3})]^{0.5}$. A shock speed of 25 km s^{-1} was adopted; this value is within the range of velocities derived from radio emission line profiles, particularly of SiO, and is just sufficient to cause sputtering of refractory grain materials, such as silicates (Field et al. 1997). Other information on initial conditions is to be found in Schilke et al. (1997).

The time sequence of the temperature profiles (charged and neutral fluids) are presented in Fig. 2. Note that, contrary to Fig. 1, the pre-shock gas is to the right. The relative importance of the J-component increases towards earlier times and the maximum

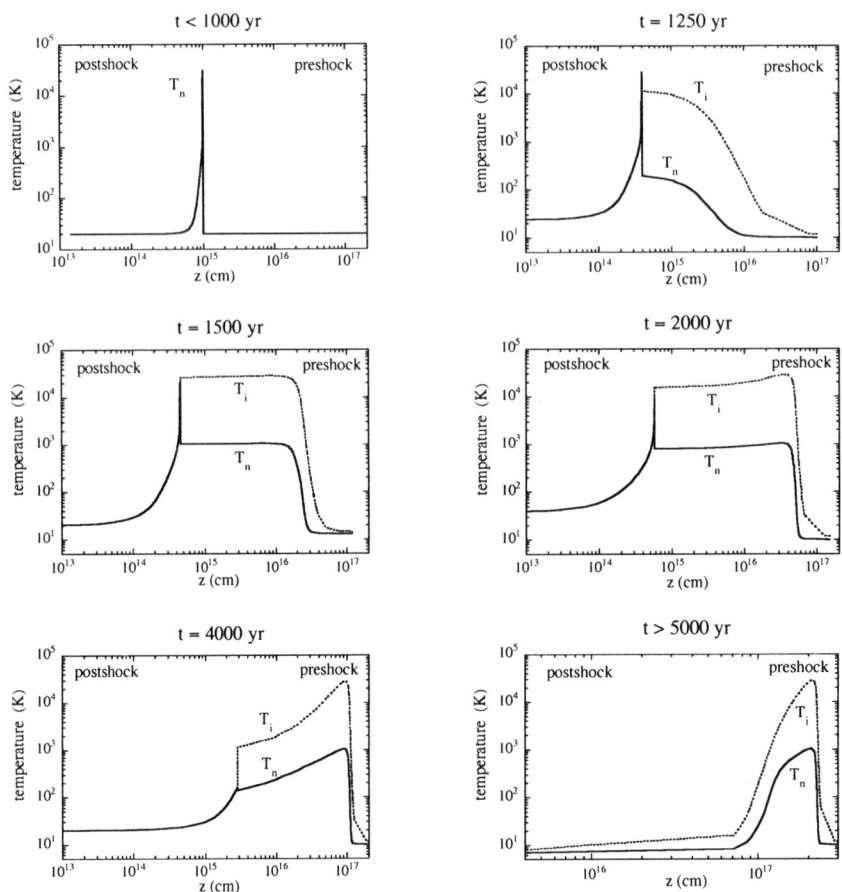

FIGURE 2. Time sequence of the temperature profiles of the charged (broken curves) and neutral (full curves) fluids as functions of position for a range of values of the age of the shock. The shock of velocity $v_s = 25$ km s^{-1} propagates into a medium in which $n_H = 10^4$ cm^{-3} and $B_0 = 100\mu$G. 'Pre-shock' is to the right and 'posts-hock' to the left in this Figure.

temperature T_n^{max} attained by the neutral fluid decreases from $\approx 2 \times 10^4$ K to ≈ 1000 K as the age increases; this has consequences for the relative intensities of the H$_2$ emission lines. Indeed, the prospect opens to estimating the age of the structure by fitting to the observed H$_2$ line intensities. Note also the rapid decrease of the shock width as a function of the age of the shock: from 10^{15} cm (≈ 100 AU) at $t_{flow} = 1000$ yr, to 10^{17} cm (≈ 0.03 pc) for $t_{flow} > 4000$ yr.

Wright et al. (1996) have observed the intensities in Cepheus A West of transitions within the ground vibrational state of H$_2$ extending to the rotational level $J = 11$, together with the intensities of a number of rovibrational lines, also in emission. In Fig. 3, the results of Wright et al. (1996) are plotted in the form of an excitation diagram. Also plotted are the results of models at three values of the flow time in the range $10^3 \leq t_{flow} \leq 4.0 \times 10^3$ yr. It may be seen that the shapes of the theoretical curves change markedly over this range of evolutionary times. At the earliest time, the contribution of the J-component to the transitions of lower excitation energy is seen to be too weak, whilst at the latest time the contribution of the C-component to the transitions of higher excitation energy is too weak. Only at an intermediate time $t_{flow} = 1.5 \times 10^3$

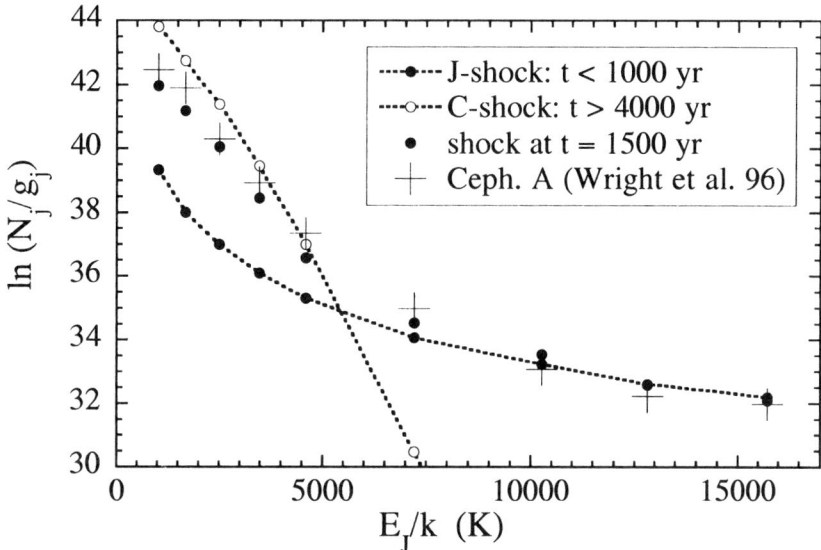

FIGURE 3. The column densities N_J divided by the statistical weights $g_J = (2J+1)(2I+1)$ of the emitting rovibrational states J of H_2 plotted against the excitation energy E_J of the emitting level. The values observed in Ceph. A West are plotted as crosses. The model results are given for three evolutionary times, t_{flow}: 10^3 yr (full circles joined by a dotted line); 1.5×10^3 yr (full circles) and 4.0×10^3 yr (open circles joined by a dotted line).

yr is the agreement with the observations satisfactory. Thus, the possibility exists of estimating the age of the shock wave by fitting to the observed intensities of H_2 emission lines of a large range of excitation energies.

3. Concluding remarks

The interpretation of ISO observations of Herbig-Haro complexes is likely to be influenced by allowing for the time-dependence. We note that, for Cepheus A West, Wright et al. (1996) derive two different temperatures from their analysis of observations of H_2 lines in GGD37: 700 K and 11 000 K. Their interpretation is in terms of two steady state C-type shocks with differing parameters. However, the present work suggests that, under the conditions which are likely to prevail within outflows in Herbig-Haro complexes, shocks are unlikely to have attained steady state: these outflows are believed to be sufficiently young for the shocks to be still evolving towards steady state.

We have presented a model computation of the time dependent structure of a shock wave and shown that, at early times, shocks exhibit both C-type and J-type characteristics. The existence of a hot and of a warm component is then explicable in terms of a single structure. The J-component of the shock wave, which appears naturally at early evolution times, is essential to account for the intensities of the higher excitation lines, particularly those arising from vibrationally excited levels.

We have compared the computed intensities of the H_2 infrared rotational and rovibrational lines with those observed in Cepheus A West by ISO. We find that an excitation diagram for the rovibrational transitions provides an effective discriminant of the evolutionary age of the shock wave. Considering all the rovibrational transitions that have been observed in Cepheus A West, and bearing in mind the very wide range of excitation energies involved, we believe the level of agreement with the model for $t = 1.5 \times 10^3$ yr to be encouraging.

However, significant improvements remain to be made in the models of such regions. In particular, whilst we have investigated plane parallel shock waves, observations of outflows tend to show the presence of bow shocks, which are believed to have J-shock characteristics at the apex and C-shock characteristics in the wake (Smith & Brand, 1990, Tedds et al. in this volume). The geometry of the shock wave could play a crucial role, notably with regard to the line profiles of other tracers of shocks, such as NH_3, SiO and CH_3OH (Flower & Pineau des Forêts, 1999).

A further constraint on the models may be provided by the observed ortho- to para-H_2 ratio, which is another indicator of the shock structure and perhaps of the shock age (Wilgenbus et al., 2000; see also the review by Wilgenbus et al. in this volume).

We are grateful to the Royal Society and the CNRS for financial support under the European Science Exchange Programme.

REFERENCES

Chièze J.-P., Pineau des Forêts G., Flower D. R., 1998, MNRAS, **295**, 672

Davis C. J., Eislöffel J., 1995, A&A, **300**, 851

Draine B. T., 1980, ApJ, **241**, 1021

Draine B. T., Roberge W. G., Dalgarno A., 1983, ApJ, **264**, 485

Draine B. T., McKee C. F., 1993, ARA&A, **31**, 373

Field D., May P. W., Pineau des Forêts G., Flower D. R., 1997, MNRAS, **285**, 839

Flower D. R., Pineau des Forêts G., 1999, MNRAS, **308**, 271

Gredel R., 1994, A&A, **292**, 580

Gredel R., 1996, A&A, **305**, 582

Gueth F., Guilloteau S., Bachiller R., 1998, A&A, **333**, 207

Hartigan P., Carpenter J.M., Dougados C., Skrutskie M., 1996, AJ, **111(3)**, 1278

Hindmarsh A. C., 1974, Lawrence Livermore Lab. Report UCID-30001

Mullan D. J., 1971, MNRAS, **153**, 145

Neufeld D. A., Melnick G. J., Harwit M., 1998, ApJ, **506**, L75

Pineau des Forêts G., Flower D. R., 1997, in 'Herbig-Haro Flows and the Birth of Low-Mass Stars', IAU Symposium 182, B. Reipurth & C. Bertout ed., Kluwer (Dordrecht)

Reipurth B., Bertout C., 1997, IAU Symposium 182, 'Herbig-Haro Flows and the Birth of Low-Mass Stars', Kluwer (Dordrecht)

Schilke P., Walmsley C. M., Pineau des Forêts G., Flower D. R., 1997, A&A, **321**, 293

Smith M.D., Brand P.W.J.L., 1990, MNRAS, **245**, 108

Wilgenbus D., Cabrit S., Pineau des Forêts G., Flower D. R., A&A, 2000, submitted

Wright C.M., Drapatz S., Timmermann R., van der Werf P.P., Katterloher R., de Graauw Th., 1996, A&A, **315**, L301

The ortho:para-H_2 ratio in C- and J-type shocks

By David WILGENBUS[1], Sylvie CABRIT[1], Guillaume PINEAU DES FORÊTS[2] AND David FLOWER[3]

[1] Observatoire de Paris, DEMIRM, UMR 8540 du CNRS, 61 Avenue de l'Observatoire, F-75014 Paris, France

[2] Observatoire de Paris, DAEC, UMR 8631 du CNRS, F-92195 Meudon Principal Cedex

[3] Physics Department, The University, Durham DH1 3LE, UK

We investigate the ortho:para-H_2 ratio (OPR) in both C- and J-type planar shocks, paying attention to H_2 excitation and dissociation processes. We relate the predictions of the models to observational determinations of the OPR. As an illustration, we consider ISO and ground-based observations of HH 54. Neither planar C-type nor planar J-type stationary shocks appear to be able to account fully for these observations.

1. Introduction

The H_2 molecule exists in two forms, according to the value of its total nuclear spin (0 or 1) and the corresponding rotational quantum numbers, even (para-H_2) or odd (ortho-H_2), respectively. In local thermodynamic equilibrium, the ortho:para-H_2 ratio is a monotonic increasing function of the temperature: very low at low temperature (< 0.01 at 25 K) and reaching asymptotically 3 (i.e. the ratio of nuclear spin degeneracies) at high temperature (≈ 2.9 at 200 K). As we shall see below, the ortho:para-H_2 ratio is an important parameter in studies of molecular clouds, as it depends strongly on the past history of the gas.

In cold molecular gas, the dominant mechanism for para-ortho conversion is spin exchange with protons or ions like H_3^+ or H_3O^+. For these reactions, we adopted the values of the rate coefficient k_{H^+,H_2} tabulated by Gerlich (1990). The abundance of these ions in well-shielded cloud interiors is $n(H^+)+n(H_3^+)+n(H_3O^+) \approx 10^{-6}\left[n_H(\text{cm}^{-3})\right]^{0.5}$ cm^{-3}, where $n_H = n(H) + 2n(H_2)$. The conversion timescale is then

$$\tau_{conv}(\text{cold}) \approx 10^6 \text{ yr } \left[10^4 \text{ cm}^{-3}/n_H\right]^{0.5}, \quad (1.1)$$

which is of the same order as the cloud contraction timescale ($10^5 - 10^7$ yr). Thus, it is difficult to establish whether the value of the ortho:para-H_2 ratio in a cold cloud has reached its equilibrium value or remains close to its initial value, which is unknown.

In warm gas, reactive collisions between H_2 and H (which have an activation energy) can occur. The corresponding reaction rate coefficient is $k_{H,H_2} = 8 \times 10^{-11} e^{-3900/T}$ cm^{-3} s^{-1} (Schofield 1967). The two important parameters are then the abundance of atomic hydrogen and the kinetic temperature. Gas may be heated by the passage of shock waves. In molecular clouds, where the degree of ionization is very low, shocks can be of two different types, depending on the strength of the magnetic field. When the field is weak, neutrals and charged particles are well coupled and the shock is of the J-type (Jump-type): the temperature and density undergo discontinuous changes. When the magnetic field is high enough to partially decouple the charged and neutral fluids, the charged particles are accelerated, compressed and heated upstream of the neutrals. The neutral fluid is also accelerated, as it is collisionally coupled to the ions, and there is no discontinuity in the temperature nor the density: the shock is of the C-type (Continuous-type). See Mullan (1971) and Draine (1980) for more details.

If we assume that H is produced in molecular shocks mainly by the chemical reactions O(H_2,H)OH and OH(H_2,H)H_2O (which is the case for C-type shocks: see section 2.1), and we neglect for simplicity the initial H present in pre-shock gas, then $n(H) \approx 2n(O) \approx 7 \times 10^{-4} n_H$ and the timescale for para-ortho conversion is

$$\tau_{conv}(600 \text{ K}) \approx 3.8 \times 10^4 \text{ yr } [10^4 \text{ cm}^{-3}/n_H], \quad (1.2a)$$

$$\tau_{conv}(1400 \text{ K}) \approx 920 \text{ yr } [10^4 \text{ cm}^{-3}/n_H], \quad (1.2b)$$

$$\tau_{conv}(T \geq 12000 \text{ K}) \approx 80 \text{ yr } [10^4 \text{ cm}^{-3}/n_H]. \quad (1.2c)$$

Conversion will only be complete if τ_{conv} is shorter than the shock duration, t_{shock}, which is of the order of (see section 2.2):

$$t_{shock} \approx 5000 \text{ yr } [10^4 \text{ cm}^{-3}/n_H] \quad \text{for C-type shocks,} \quad (1.3a)$$

$$t_{shock} \approx 20 \text{ yr } [10^4 \text{ cm}^{-3}/n_H] \quad \text{for J-type shocks.} \quad (1.3b)$$

Therefore, in C-shocks, we expect substantial conversion once the temperature reached in the shock wave is of order 1000 K. In J-shocks, however, an extra source of atomic H (i.e. collisional dissociation) will be needed to have $\tau_{conv} < t_{shock}$. We discuss this point in more detail in section 2.2.

The ortho:para-H_2 ratio has been measured by means of both absorption and emission lines. Lacy et al. (1994) found a ratio $N(J=1)/N(J=0) < 0.8$ toward NGC 2024 IRS2, using the v=0-1 S(0) and S(1) lines in absorption. This ratio of column densities is representative of the OPR at low temperatures, at which only the $v = 0, J = 0, 1$ levels are significantly populated. The observed ratio is consistent with OPR=0.2, the equilibrium value at the kinetic temperature deduced from CO lines, $T_{CO} = 45$ K.

Prior to the launch of ISO, only rovibrational emission lines were observable from warm (shock-heated) gas in bipolar outflows. Gredel (1994) observed several HH objects (including HH 54) in the v=1-0 $2\mu m$ lines. He inferred excitation temperatures in the range 2000–2700 K in these objects and OPR ≥ 2.5, consistent with the high-temperature value of 3.0. Since then, pure rotational lines have become accessible, thanks to ISO, and non-equilibrium values of the ortho:para-H_2 ratio have been measured in a few cases, for example, in HH 54 by Neufeld et al. (1998) and Cabrit et al. (1999) (using ISOSWS and ISOCAM, respectively). The excitation temperatures were found to lie between 500 and 700 K, for levels with $E/k_B \leq 7000$ K, and the observed OPR reached values as low as 1.0. These observations give rise to the questions:

- Is it possible to infer the initial OPR in dark clouds from observations of non-equilibrium values in shocks?
- Why is the OPR observed in rovibrational lines greater than in pure rotational lines ?

2. The Ortho:Para-H_2 Ratio in interstellar shocks

2.1. Shock model

We have modelled both C- and J- type planar shocks propagating in molecular gas. The model integrates the equations for the populations of the 49 rovibrational levels of H_2 up to $E/k_B = 20000$ K in parallel with the hydrodynamical equations and makes use of the most recent data for collisional excitation of H_2 by H, He, and H_2 (Le Bourlot et al., 1999; Flower, these Proceedings). The cooling due to species other than H_2, namely C, O, C$^+$, H_2O, CO, OH and NH_3, is also taken into account; see Flower & Pineau des Forêts (1999) for a more complete description. The rate of cooling

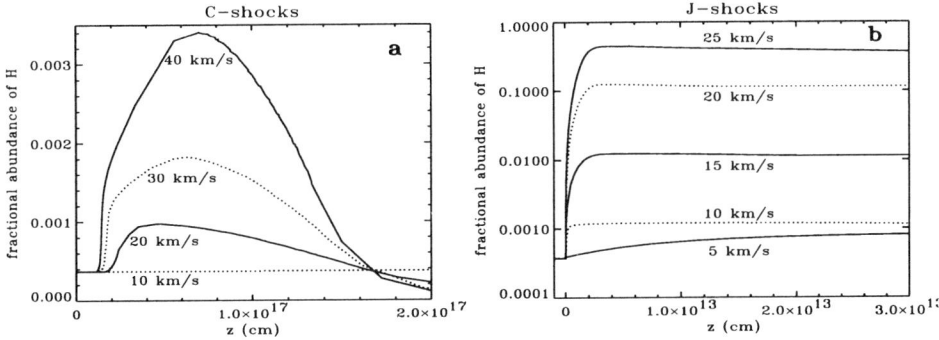

FIGURE 1. Fractional abundance of H in (a) C-type and (b) J-type shocks with an initial density $n_H = 10^4$ cm^{-3} and an initial magnetic induction $B_0 = 100$ μG (a) and $B_0 = 10$ μG (b).

and, for C-type shocks, the degree of ion-neutral coupling determine the shock profile and are also crucial to the rate of para-ortho conversion in the shock wave. In the latter context, another important parameter is the abundance of atomic hydrogen, determined by the chemistry and by collisional dissociation of H_2. The model includes 120 chemical species and a network of 864 reactions, for which the rate equations are solved in parallel with the hydrodynamical equations. We adopted a rate coefficient for dissociation of H_2 in collisions with drifting ions which is roughly 5 times lower than the upper limit of Draine et al. (1983); see Wilgenbus et al. (2000).

Figure 1 shows the fractional abundance of H, $n(H)/n_H$, in C- and J- type shocks for an initial (pre-shock) density $n_H = 10^4$ cm^{-3}. In C-type shocks with $v_s > 10$ km s^{-1}, atomic H is mainly produced by the chemical reactions O(H_2,H)OH and OH(H_2,H)H_2O, providing two H atoms per oxygen atom in the pre-shock gas. At shock speeds $v_s \geq 30$ km s^{-1}, collisional dissociation of H_2 by drifting ions further increases the abundance of atomic hydrogen, but by a factor of 3 at most. In J-type shocks with $v_s \geq 15$ km s^{-1}, the temperature is high enough (above 10^4 K) that collisional dissociation of H_2 by H and H_2 can considerably enhance the production of atomic H.

A comparison with previous models, particularly those of Timmermann (1998), is made by Wilgenbus et al. (2000). They conclude that Timmermann's C-shock models overestimate the fractional abundance of atomic hydrogen and hence the maximum ortho:para-H_2 ratio reached in post-shock gas.

2.2. Results

We have made a systematic study of para-ortho conversion in C- and J-type shocks, for ranges of the pre-shock density and of the shock speed. The initial ortho:para-H_2 ratio is unknown and was treated as a free parameter. A grid of C-type shock models was computed for
- initial densities $n_H = n(H) + 2n(H_2) = 10^3, 10^4, 10^5, 10^6$ cm^{-3},
- initial ortho:para ratios (o/p)$_{init}$ = 0.01, 1.0, 2.0, 3.0,
- shock speeds v_s = 10, 20, 30, 40 km s^{-1} (except for $n_H = 10^6$ cm^{-3}, when v_s = 10, 15, 20, 25, 30 km s^{-1}),
- initial magnetic induction $B_0(\mu G) = \sqrt{n_H (\text{cm}^{-3})}$.

A grid of J-type shock models was computed for
- $n_H = 10^3, 10^4, 10^5$ cm^{-3},
- (o/p)$_{init}$ = 0.01, 1.0, 2.0, 3.0,

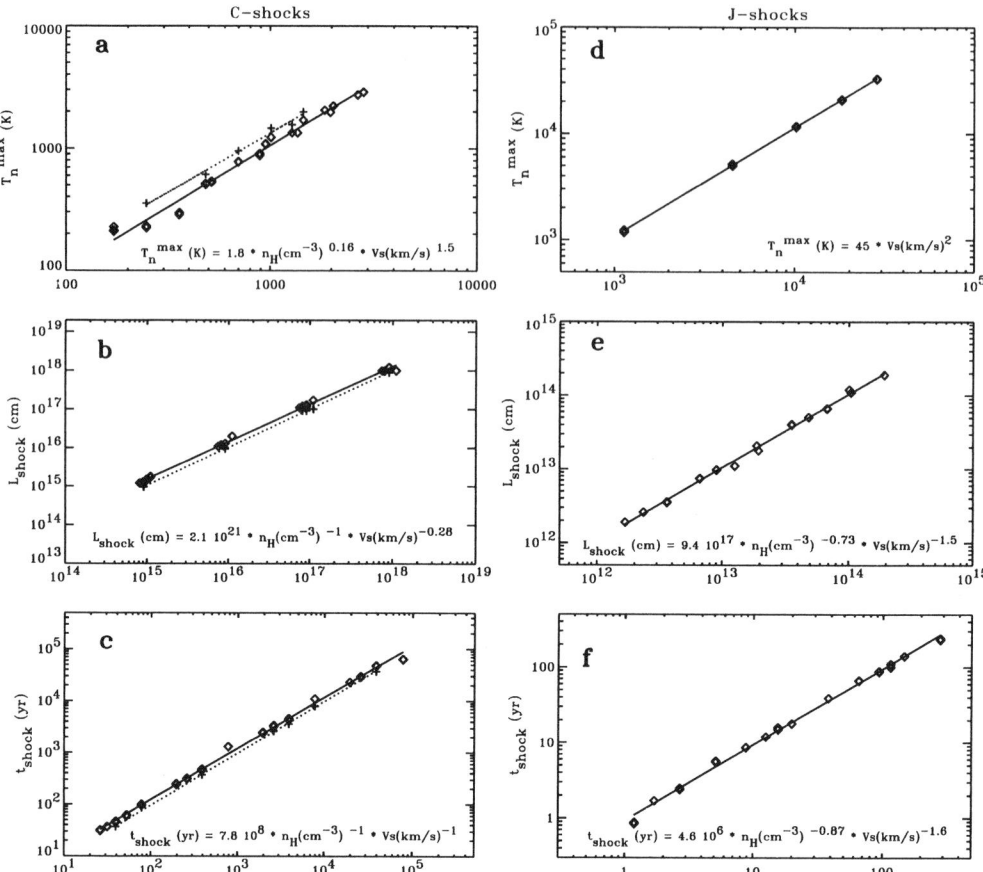

FIGURE 2. Variation of the maximum temperature reached in the shock (T_n^{max}) and of the width and the duration of the shock (L_{shock} and t_{shock}) on n_H and v_s for C-type (a, b, c) and J-type (d, e, f) shocks. Both L_{shock} and t_{shock} are evaluated over the region where $T_n > 50$ K. For C-type shocks, dotted lines refer to models with a lower $B_0(\mu G) = \frac{1}{3}\sqrt{n_H \ (cm^{-3})}$. In each Figure, the ordinate is the value of the parameter which is computed by the model, the abscissa is the value which derives from the given fit.

- v_s = 5, 10, 15, 20, 25 km s^{-1} (except for n_H = 10^3 cm^{-3}, when v_s = 10, 15, 20, 25 km s^{-1}),
- $B_0(\mu G) = 0.1\sqrt{n_H \ (cm^{-3})}$.

Extensive electronic tables are available for the entire grids of C- and J-type shock models (Wilgenbus et al. 2000). Figure 2 presents T_n^{max}, L_{shock} (defined as the width of the region where $T_n > 50$ K) and t_{shock} (time to cross the region with $T_n > 50$ K) as functions of n_H and v_s. In C-type shocks, T_n^{max} increases with n_H and v_s. In J-type shocks, we recover the Rankine-Hugoniot relation, $T_n^{max} \propto v_s^2$. The variations of T_n^{max}, L_{shock}, and t_{shock} are difficult to predict *a priori*, because they result from the balance between the energy input rate ($\propto n_H v_s^3$) and the cooling rate, which depends on collisional processes.

We see in Fig. 2b that L_{shock} is $\propto 1/n_H$ and almost independent of v_s. As a consequence, the total H$_2$ column density is constant, to within 50%, for all the C-type shock models, independent of n_H and v_s, with a typical value $\approx 1.5 \times 10^{21}$ cm^{-2}. As shown

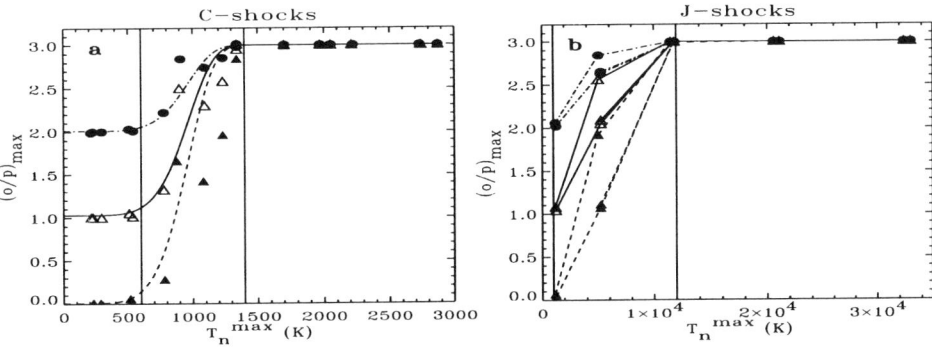

FIGURE 3. The post-shock ortho:para-H_2 ratio, $(o/p)_{max}$, as a function of the maximum temperature T_n^{max} reached by the neutrals in the shock: (a) C-type and (b) J-type shocks.

by the dotted lines in Figure 2, the magnetic field strength has no strong effects on the scaling laws described here.

As we have seen in the Introduction, para-ortho conversion becomes significant when t_{shock} is of the order of τ_{conv}. Therefore, as may be seen in Fig. 3, the post-shock OPR, $(o/p)_{max}$, is a monotonic increasing function of T_n^{max}. There is a threshold value of T_n^{max} for conversion to occur; this value is greater in J-type shocks (≈ 1000 K) than in C-type shocks (≈ 600 K) because the duration of a J-type shock is much smaller. In J-type shocks with $T_n^{max} \gtrsim 1.2 \times 10^4$ K (i.e. $v_s \geq 15$ km s^{-1}), collisional dissociation of H_2 occurs and the increase in $n(H)$ results in τ_{conv} becoming much smaller than t_{shock}. In these models, the OPR attains the statistical value of 3.

3. Comparison with observations

3.1. Empirical determination of the OPR

Observers cannot determine the local densities of the ortho and para forms of H_2 and do not have access to the post-shock OPR, $(o/p)_{max}$, plotted in Fig. 3. They measure column densities, integrated over the entire shock width, and, as we shall see below, deduce an OPR which is always lower than $(o/p)_{max}$.

We now briefly describe the method used to deduce the OPR from H_2 column densities. This method is illustrated in Fig. 4 and is explained more fully by Wilgenbus et al. (2000). The method makes use of an excitation diagram, which is a plot of $\ln[N_J/(g_J g_I)]$ against E_J/k_B, where N_J is the column density measured in rotational level J, E_J is the excitation energy of this level relative to $J = 0$, $g_J = 2J+1$, and $g_I = 1$ (J even) or 3 (J odd). If the levels are thermalized, the plot is a straight line of slope $-1/T_{rot}$, according to the Boltzmann law. However, if the OPR is not at its LTE value, $(o/p)_{LTE}$, the points representing ortho levels are displaced downwards relative to para levels. This displacement allows the OPR to be estimated. For a rotational level J, we have

$$\frac{(o/p)_J}{(o/p)_{LTE}(T_J)} = \left[\frac{N_J}{N_J(LTE, T_J)}\right]^\varepsilon \quad (3.4)$$

where T_J is the excitation temperature derived from levels $J-1$ and $J+1$ (assuming a Boltzmann law) and $N_J(LTE, T_J)$ is the column density of level J, interpolated between levels $J-1$ and $J+1$; ε is equal to 1 for ortho levels (odd J) and -1 for para levels (even J).

As shown in Fig. 4b, the OPR deduced from observations is always lower than the maximum post-shock value. Furthermore, the empirically determined OPR depends on

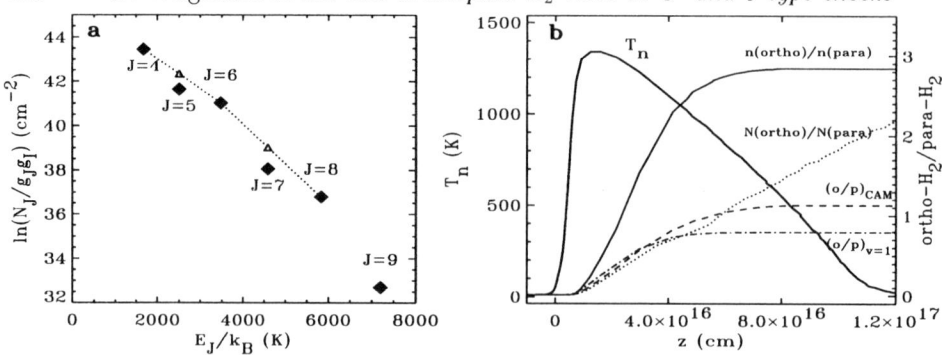

FIGURE 4. (a) Empirical determination of the OPR: diamonds denote observed column densities from lines in the ISOCAM range, and triangles the LTE interpolated values (for ortho-levels). (b) Comparison of various determinations of the OPR, at each position in a model of a C-type shock in which $n_H = 10^4$ cm^{-3}, $v_s = 30$ km s^{-1} and (o/p)$_{\rm init}$=0.01. n(ortho)/n(para) is the local OPR, N(ortho)/N(para) is the ratio of the ortho-and para- column densities, (o/p)$_{\rm CAM}$ is the ratio inferred from the 0-0 S(2) to S(7) lines, and (o/p)$_{v=1}$ is estimated from the 1-0 O(3) and 1-0 S(0) to S(11) lines.

the lines that are observed: (o/p)$_{\rm CAM}$ (inferred from 0-0 S(2) to S(7) lines) is greater than (o/p)$_{v=1}$ (estimated from 1-0 O(3) and 1-0 S(0) to S(11) lines), because levels with different excitation energy are populated in different parts of the shock, where the local OPR's may also differ. As the temperature falls, the value of the OPR determined from levels of highest excitation energy freezes first (see Wilgenbus et al. 2000). Consequently, we find

$$(o/p)_{\rm max} \geq N({\rm ortho})/N({\rm para}) \geq (o/p)_{\rm CAM} \geq (o/p)_{v=1} \geq (o/p)_{\rm init}. \quad (3.5)$$

We conclude that, if an object exhibits an (o/p)$_{v=1}$ greater than (o/p)$_{\rm CAM}$, then the rovibrational and the pure rotational lines cannot come from the same stationary shock wave.

3.2. Diagnostic Diagrams

Neufeld et al. (1998) observed the E, K knots in HH 54 using ISO SWS and showed that a useful diagnostic diagram is obtained by plotting the line intensity ratio S(3)/S(1), which measures the excitation temperature, against S(2)/S(1), which measures both the OPR and the excitation temperature. They extrapolated the results of the C-type models of Timmermann (1998) in order to constrain the initial ortho:para-H$_2$ ratio, the pre-shock density, and the shock speed. In view of the improvements we have made in the treatment of H$_2$ excitation and dissociation (see section 2.1), we present here our equivalent diagnostic diagrams.

We plot in Fig. 5 the ortho:para-H$_2$ ratio derived from observations ((o/p)$_{\rm CAM}$ or (o/p)$_{v=1}$) versus the corresponding average excitation temperature ($\langle T_J \rangle$ and $\langle T_J \rangle_{v=1}$ respectively) for both C- and J-type shocks. Note that the OPR deduced from observations is an average over the shock thickness and never attains 3.0 (see section 3.1).

Because the increase in the observed OPR occurs at lower excitation temperature as n_H decreases, there is no unique pair of values of ((o/p)$_{\rm init}$, v_s) which corresponds to a given observational point in Fig. 5. Thus, there is an intrinsic uncertainty in the pre-shock parameters. Moreover, the curve corresponding to a very low (o/p)$_{\rm init}$ (0.01) defines for each n_H the minimum ortho:para-H$_2$ ratio at a given excitation temperature. An observational point located below this curve is incompatible with the models at the

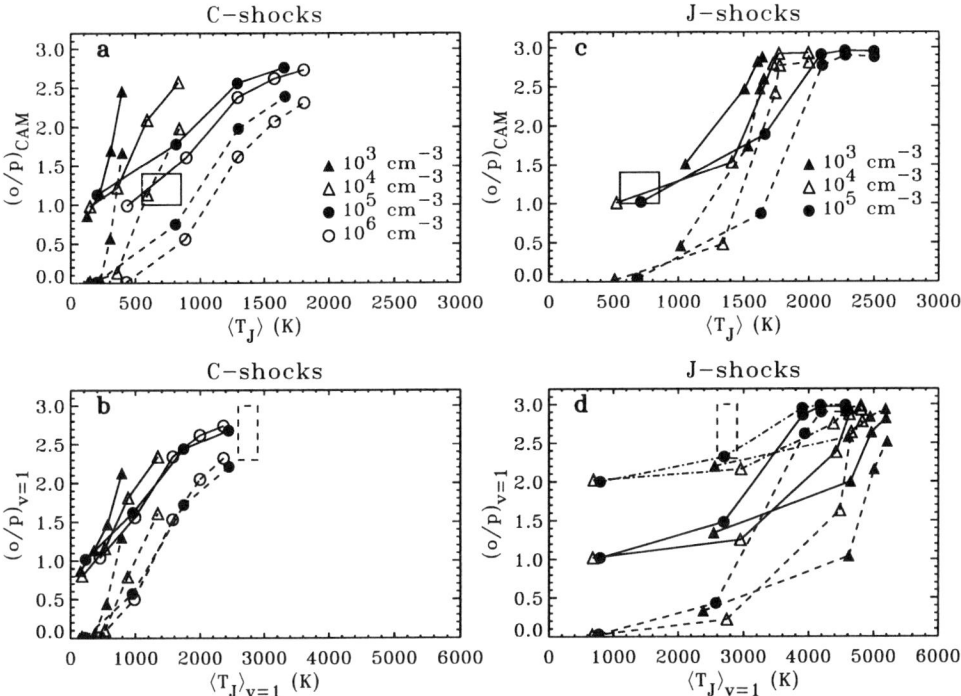

FIGURE 5. Observed OPR as a function of the corresponding excitation temperature for C- and J-type shocks. (a) and (c) show predictions for pure rotational lines observable with ISOCAM; (b) and (d) for rovibrational lines in the $2\mu m$ range. Dashed curves correspond to $(o/p)_{init}=0.01$, solid curves to $(o/p)_{init}=1$, and dot-dashed curves (in (d) only) to $(o/p)_{init}=2$. Each curve corresponds to a given pre-shock density, and each point on a curve to a shock speed. The solid box derives from ISO-SWS observations of HH 54 E,K (Neufeld et al. 1998), while the dashed box derives from H_2 rovibrational line observations at 2 μm of the same object (Gredel 1994).

corresponding value of n_H and implies a higher pre-shock density. In order to determine the initial OPR, we would require an independent constraint on n_H.

By way of illustration, we have plotted in Figs. 5a and 5c the OPR and excitation temperature derived from ISOSWS observations of HH 54 E,K by Neufeld et al. (1998). We conclude that $n_H < 10^4$ cm^{-3} is excluded for C-type shocks. For $n_H = 10^4$, 10^5, and 10^6 cm^{-3}, the inferred pairs of values of v_s and $(o/p)_{init}$ are, respectively, (30 km s^{-1}, ≤ 0.01), (20 km s^{-1}, 0.5), and (15 km s^{-1}, 0.8). For J-type shocks, we conclude to $(o/p)_{init} \approx 1.0$ and $v_s \approx 5$ km s^{-1}, but n_H remains undetermined.

In Figs. 5b and 5d, we have plotted the OPR and excitation temperature obtained from rovibrational spectroscopic observations of the same object (Gredel, 1994). From these observations, we find that a C-type shock with $v_s \geq 35$ km s^{-1} is required to reach such a large value of $(o/p)_{v=1}$; but, at this high speed, conversion is fast enough to yield an OPR close to 3.0, regardless of the value of $(o/p)_{init}$. For J-type shocks, we infer $v_s = 10$ km s^{-1} and $(o/p)_{init} \geq 2.0$, higher than deduced from pure rotational lines. Unless there are spatial variations of the initial ortho:para-H_2 ratio within the beam, a single stationary, planar J-type or C-type shock is unable to account for both the rovibrational and pure rotational lines. Possible explanations are:

- The shock front may be curved within the observing beam (a bow shock). Rovibrational emission occurs closer to the bow apex, where the shock speed is effectively higher, and pure rotational emission further out in the bow shock wings. In the case of HH 54, a C-type bow shock with an initial OPR < 0.8 and apex shock speed $v_s \geq 35\,\mathrm{km\,s^{-1}}$ is a possibility.

- The shock wave may not have reached steady state (Chièze et al. 1998 ; Pineau des Forêts & Flower, these Proceedings). Under these circumstances, components with C- and J-type characteristics are simultaneously present. The rotational lines arise mainly in the C-type component, whilst the vibrational lines come from the J-type component; see Flower & Pineau des Forêts (1999) for a study of Cep A. The initial ortho:para ratio, ahead of the shock wave, might be low (< 0.8), as required by the pure rotational lines, with the ratio increasing in the magnetic precursor to a value of 2.0 just ahead of the J-discontinuity, as required by the vibrational lines. A more detailed investigation of these possibilities, in relation to HH 54 and similar objects, is currently underway.

4. Conclusion

We have studied para to ortho conversion in C- and J- type shocks and showed that its degree is determined mainly by the maximum temperature attained by the neutrals, T_n^{max}. In C-type shocks, conversion starts when $T_n^{max} \approx 600$ K and is complete when $T_n^{max} \geq 1400$ K. J-type shocks cool much faster and thus require higher temperatures: 1000 K to start conversion and 1.2×10^4 K to complete it.

The OPR deduced from observations is always lower than the maximum post-shock OPR. Moreover, if the OPR inferred from rovibrational lines is greater than from pure rotational lines, the two observations must probe different shocks, or different regions of a non-stationary or a bow shock wave.

We have considered the constraints on the initial OPR and the shock speed that are imposed by observations. We find that $(o/p)_{init}$ can be strongly dependent on the assumed n_H; in C-type shocks in particular, a lower n_H implies a lower $(o/p)_{init}$ and a higher v_s.

We are grateful to the Royal Society and the CNRS for financial support under the European Science Exchange Programme. One of the authors (DRF) gratefully acknowledges the award of a research fellowship by the University of Durham.

REFERENCES

CABRIT S. ET AL., 1999 in "The Universe as seen by ISO" ESA SP, **427**, 449
CHIÈZE J.-P., PINEAU DES FORÊTS G., FLOWER D. R., 1998 MNRAS, **295**, 672
DRAINE B. T., 1980 ApJ, **241**, 1021
DRAINE B. T., ROBERGE W. G., DALGARNO A., 1983 ApJ, **264**, 485
FLOWER D. R., PINEAU DES FORÊTS G., 1999 MNRAS, **308**, 271
GERLICH D., 1990 J. Chem. Phys., **92**, 2377
GREDEL R., 1994 A&A, **292**, 580
LACY J. H., KNACKE R., GEBALLE T. R., TOKUNAGA A. T., 1994 ApJ, **428**, L69
LE BOURLOT J., PINEAU DES FORÊTS G., FLOWER D. R., 1999 MNRAS, **305**, 802
MULLAN, D. J., 1971 MNRAS, **153**, 145
NEUFELD D. A., MELNICK G. J., HARWIT M., 1998 ApJ, **506**, L75
SCHOFIELD K., 1967 Planet. Space Sci., **15**, 643
TIMMERMANN R., 1998 ApJ, **498**, 246
WILGENBUS D., CABRIT S., PINEAU DES FORÊTS G., FLOWER D. R., 2000, submitted to A&A

Theoretical Models of Photodissociation Fronts

By B. T. Draine[1] AND Frank Bertoldi[2]

[1] Princeton University Observatory, Princeton, NJ 08544-1001, USA

[2] Max-Planck-Institut für Radioastronomie, D-53121 Bonn, Germany

Observations of H_2 line emission have revealed higher-than-expected gas temperatures in a number of photodissociation fronts. We discuss the heating and cooling processes in photodissociation regions. Observations of NGC 2023 are compared to a theoretical model in which there is substantial gas at temperatures $T = 500 - 1000K$ heated by photoelectric emission and collisional de-excitation of H_2. In general the model successfully reproduces the observed H_2 line emission from a wide range of energy levels. The observed [SiII]34.8μm emission appears to indicate substantial depletion of Si in NGC 2023.

1. Introduction

A significant fraction of the ultraviolet radiation emitted by massive stars impinges on the molecular gas associated with star formation. The resulting photodissociation regions (PDRs) therefore play an important role in re-processing the energy flow in star-forming galaxies. Modelling these PDRs is therefore an important theoretical challenge, both to test our understanding of the physical processes in interstellar gas, and to interpret observations of star-forming galaxies.

It is frequently the case that the illuminating star is hot enough to produce an H II region, in which case the photodissociation region is bounded on one side by an ionization front, and on the other by cold molecular gas which has not yet been appreciably affected by ultraviolet radiation. The $h\nu < 13.6$eV photons propagating beyond the ionization front raise the fractional ionization, photo-excite and photo-dissociate the H_2, and heat the gas via photoemission from dust and collisional de-excitation of vibrationally-excited H_2. Figure 1 shows the different layers within a PDR.

2. PDR Thermometry

The lower rotational levels of H_2 tend to be in LTE (except perhaps for the ortho/para ratio, which adjusts relatively slowly), so we can use the relative level populations of H_2 as a PDR thermometer.

We can also use the absolute column densities $N(v, J)$ to tell us how much molecular gas is at different temperatures. Because the H_2 quadrupole vibration-rotation lines are generally optically thin, only dust extinction affects the radiative transfer.

The level populations $N(v, J)$ can be studied using vibrational transitions (ground-based K,H,J,I,R band spectroscopy) or v=0–0 pure rotation lines (some from the ground, but mainly from space, with ISO or SIRTF, or from the stratosphere, with SOFIA).

3. Modeling PDRs

The central physical process in a PDR is the photo-excitation and photodissociation of H_2 through the Lyman and Werner band lines of H_2, and it is therefore important

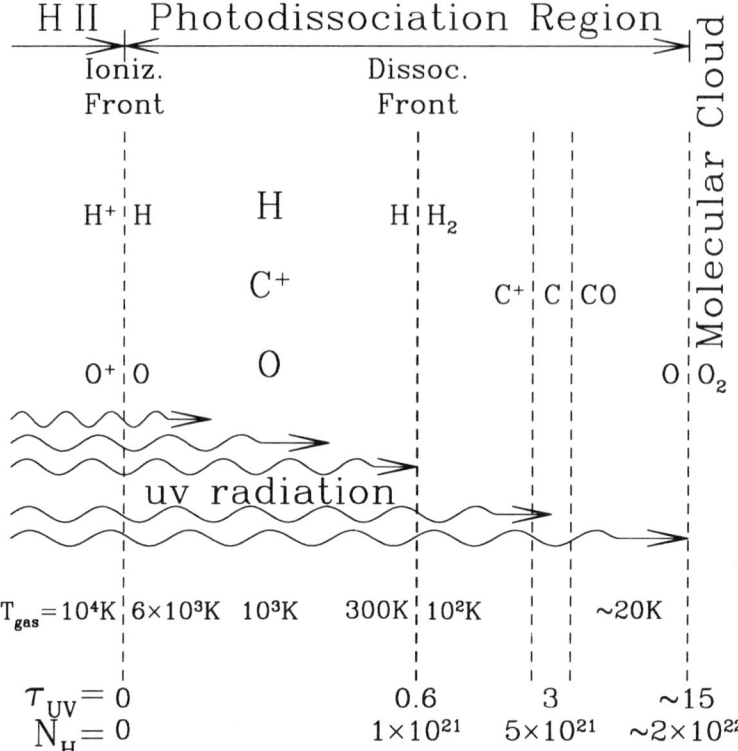

FIGURE 1. Schematic diagram showing the different zones in a photodissociation region.

to model this process accurately. In our models (Draine & Bertoldi 1996) we explicitly include radiative transfer in the 28765 permitted Lyman and Werner band lines between H_2 levels with $J \leq 29$, using wavelengths and oscillator strengths from Abgrall & Roueff (1989) and Abgrall et al. (1993a,b). The lines are treated independently assuming Voigt line profiles and attenuation by dust, but we include a a statistical correction for the effects of line overlap of the H_2 lines (Draine & Bertoldi 1996).

Lyman and Werner band photons are absorbed by H_2 in the ground electronic state $X^1\Sigma_g^+$, resulting in excitation to vibration-rotation levels of either the $B^1\Sigma_u^+$ or $C^1\Pi_u^\pm$ electronic states. The electronically-excited state will decay by spontaneous emission of an ultraviolet photon. About 85% of the downward transitions will be to a bound (v, J) level of the ground electronic state, but about 15% of the transitions will be to the vibrational continuum of the ground electronic state, producing two H atoms, typically moving apart with a kinetic energy of a fraction of ~ 1 eV (Stephens & Dalgarno 1973).

Bertoldi & Draine (1996) discussed the propagation of coupled ionization-dissociation fronts, and showed that except when the PDR is driven by radiation from a very hot star, it is usually the case that the PDR moves into the molecular cloud slowly enough that thermal and chemical conditions are close to being in steady-state balance: most importantly, H_2 destruction is nearly balanced by H_2 formation, and heating is nearly balanced by cooling. It is then convenient to approximate the structure of the PDR by requiring precise thermochemical steady state at each point. The H_2 abundance is therefore determined by a balance between destruction by photodissociation and formation

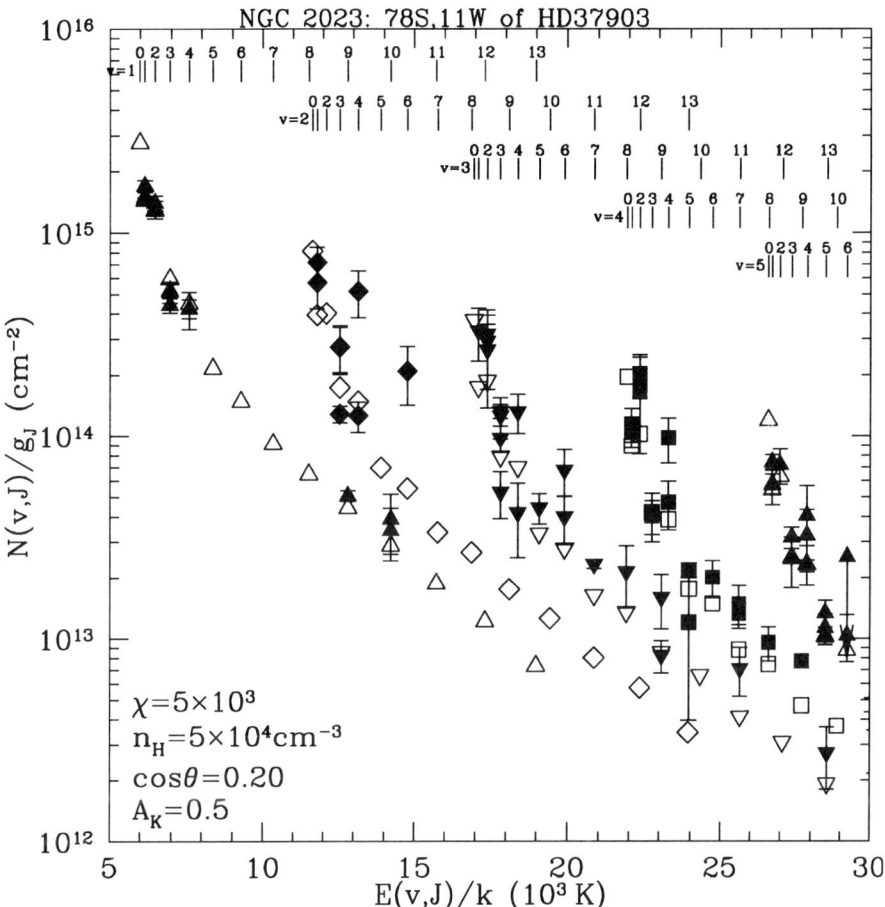

FIGURE 2. $N(v, J)/g_J$ for vibrationally-excited levels of H_2 toward the southern bar in NGC 2023. Open symbols: plane-parallel model with $n_H = 5 \times 10^4 \mathrm{cm}^{-3}$ and $\chi = 5000$ observed from angle $\theta = 78°$ rel. to surface normal. Filled symbols: level populations obtained from observed line intensities after correction for foreground extinction with $A_K = 0.5$mag. Data from Burton et al. (1998) and McCartney et al. (1999).

on grains,

$$R_{gr} n_H n(H) = \zeta_{pd} n(H_2) \qquad (3.1)$$

where ζ_{pd} is the local rate for photodissociation of H_2, and R_{gr} is the rate coefficient for H_2 formation on grains, with the grain abundance assumed proportional to $n_H \equiv 2n(H_2) + n(H) + n(H^+)$. We solve for the steady-state populations of the 299 bound (v, J) levels with $J \leq 29$, including UV pumping, spontaneous decay, and collisional transitions. The distribution over (v, J) of newly-formed H_2 produced on grain surfaces is uncertain; our models allow us to consider various possibilities.

The gas is heated primarily by photoelectrons ejected from dust grains, and (for densities $n_H \gtrsim 10^4 \mathrm{cm}^{-3}$) collisional de-excitation of vibrationally-excited H_2 resulting from UV pumping. In our models we use photoelectric heating rates estimated by Weingartner & Draine (2000a); we assume dust with $R_V = 5.5$ with 5% of the solar carbon in ultra-small carbonaceous grains.

The cooling of the gas is dominated by [CII]158μm, [SiII]35μm, [OI]63μm, and [FeII]26μm. The [OI]63μm line is often optically thick; we use an escape probability treatment.

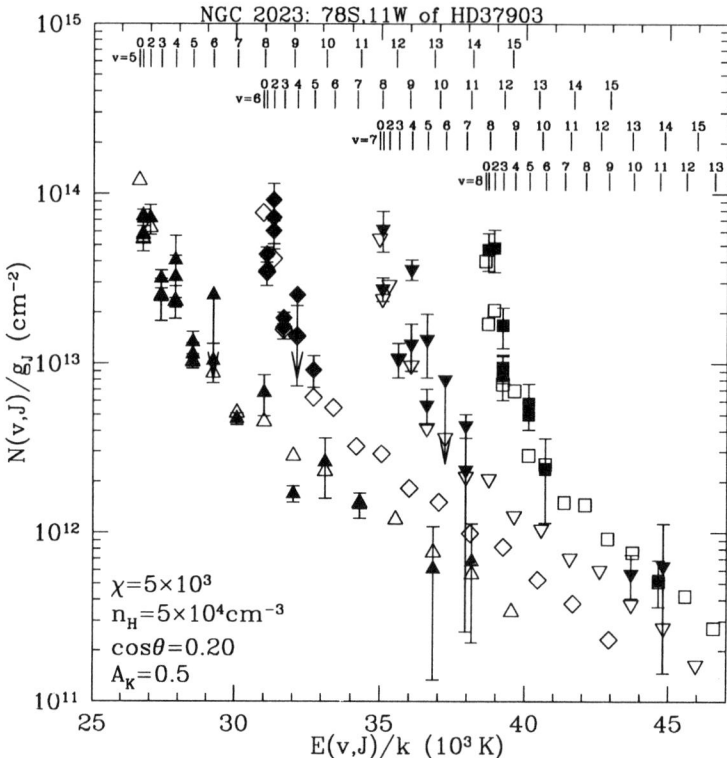

FIGURE 3. Same as Fig. 2, but for $v = 5 - 8$.

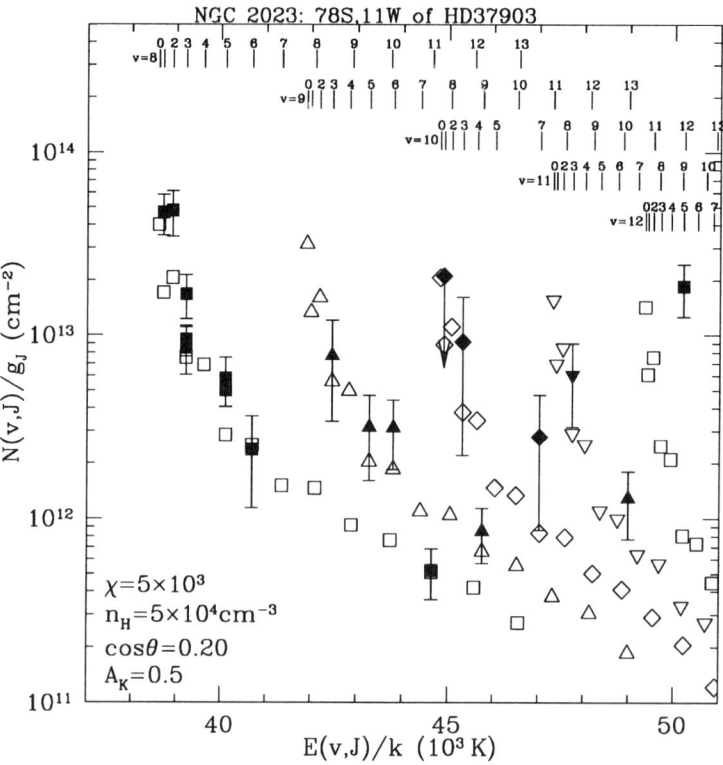

FIGURE 4. Same as Fig. 2, but for $v = 8 - 12$.

4. NGC 2023: A Test Case

The B1.5V star HD 37903 is situated just outside the surface of the L1630 molecular cloud, resulting in the famous reflection nebula NGC 2023. Many emission lines of H_2 have now been measured for NGC 2023, making it an ideal test for PDR models. High resolution images of NGC 2023 in the 1-0S(1) line have recently been published (Rouan et al. 1997; Field et al. 1998; McCartney et al. 1999), showing pronounced filamentary structure, notably strong emission along the "southern bar" 78″ south of HD 37903.

Black & van Dishoeck (1987) produced the first models to explain the H_2 fluorescence from NGC 2023. Their favored model assumed cold H_2 with a density $n_H \approx 10^4 cm^{-3}$ and an ultraviolet radiation field enhanced by a factor $\chi \approx 700$ relative to the Habing (1968) intensity. Draine & Bertoldi (1996) argued that some warm gas with $T \approx 500 - 1000K$ with density $n_H \approx 10^5 cm^{-3}$ was required to explain the observed line ratios.

New observations have become available both from the ground (Burton et al. 1998; McCartney et al. 1999) and from ISO (Bertoldi et al. 2000a). From ground-based spectrophotometry of the bright "southern emission bar" it is now clear that the foreground extinction at K is $A_K \approx 0.5$mag, implying a far-red extinction $A_{0.8\mu m} \approx 2.7$mag. For each observed H_2 emission line we compute the column density $N(v, J)$ in the excited state required to reproduce the observed surface brightness, for an assumed foreground extinction $A_K = 0.5$mag.

A detailed discussion of our modelling of NGC 2023 will be reported elsewhere, but here we show one example of a model developed to reproduce the observations of the southern emission bar. We assume the incident radiation field to be enhanced by a factor $\chi \approx 5000$ relative to the Habing (1968) flux, and a gas density $n_H = 5 \times 10^4 cm^{-3}$. The PDR is assumed to be plane-parallel, inclined relative to the line of sight with $\cos\theta = 0.2$, where θ is the angle between the normal to the PDR (= the direction from the PDR to the illuminating star) and the direction to the Earth.

In Figures 2–4 we show $N(v, J)/g_J$ for the model together with the observed level populations, versus the energy $E(v, J)$ of the excited state. Open symbols are model results; filled symbols are observations. The model is seen to be in quite good agreement with the populations of the vibrationally-excited levels all the way up to $v = 12$! The generally excellent agreement between model and observations indicates that the description of the ultraviolet pumping process is basically sound. The fact that for a given vibrational level the rotational level populations appear to be in good agreement with observations indicates that the model has approximately the correct temperature and density. Only for a few weak lines are the reported fluxes very different from the model predictions: 12-6 Q(5) $\lambda = 0.8225\mu m$, where the reported flux is a factor of 20 stronger than expected, and 9-4 S(11) $\lambda = 0.7663\mu m$, where the observed flux is a factor ~ 7 stronger than expected. With so many other lines in good agreement, perhaps these lines have been misidentified. Note, however, that these are the only two lines from levels with $E(v, J)/k > 48 \times 10^3 K$, so perhaps there are processes populating these very high energy levels which have not been included in the model (see Bertoldi et al. 2000b).

Figure 5 shows the $v = 0$ level populations for our model, together with the level populations inferred from the ISO observations toward NGC 2023 (Bertoldi et al. 2000a), except that the surface brightnesses measured by the relatively large ISO beam have been arbitrarily increased by a factor 1.8 – the reasoning here is that the bright southern bar represents a region with unusually high limb-brightening, probably about a factor of 2 higher than the average over the ISO beam. With this adjustment, we obtain quite good agreement for most of the rotational levels; the largest discrepancy is for $J - 3$, where the model surface brightness is about twice the observed value. Note that the observed

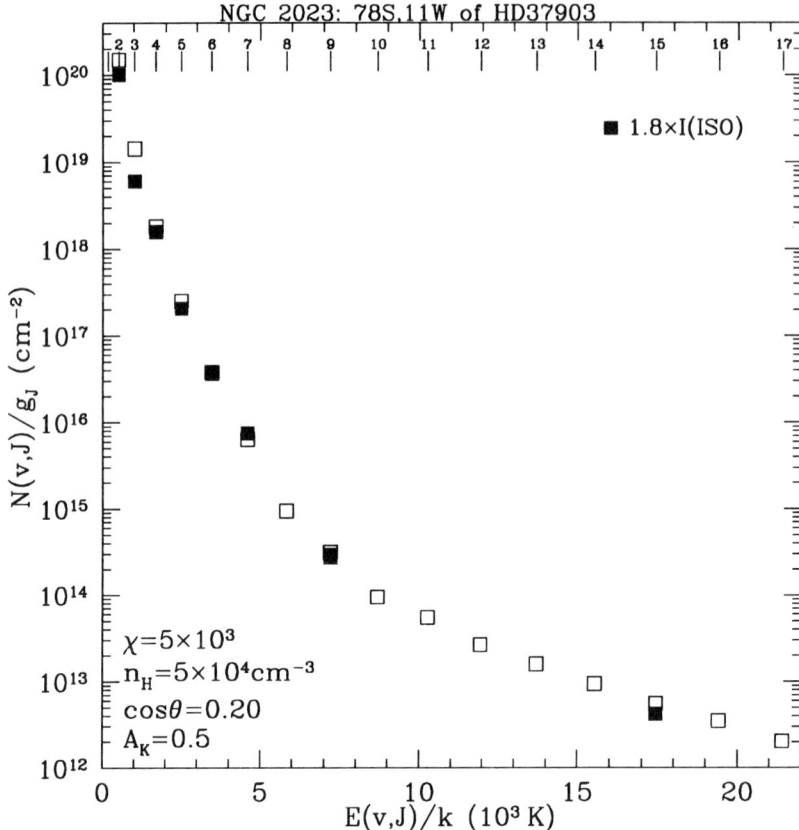

FIGURE 5. Level populations for rotationally-excited levels of the ground vibrational state in NGC 2023. Beam-averaged column densities observed by ISO have been multiplied by 1.8 to allow for likely enhancement of the surface brightness on the southern bar.

emission from the $J = 15$ level is in excellent agreement with the model! The level populations obviously depart strongly from a single-temperature fit; this is mainly due to the range of gas temperatures present (see Figure 6) and the excitation of newly-formed H_2 leaving grain surfaces.

The model is shown in Figure 6. Photoelectric emission from dust grains dominates the heating over much of the PDR, but for $10^{19} \lesssim N_H \lesssim 10^{21} cm^{-2}$ the dominant heat source is collisional de-excitation of H_2 which has been vibrationally excited by UV pumping. The combination of photoelectric heating and heating by de-excitation of H_2 manages to maintain a gas temperature of $\sim 10^3$K over a substantial zone, with the gas temperature still as high as ~ 700K when the H_2 fraction has risen to 10%.

The model assumes gas phase abundances C/H=1.4×10^{-4} (1/3 solar), O/H= 3.2×10^{-4} (1/2 solar), Si/H=9.0×10^{-7} (1/40 solar), Fe/H=1.3×10^{-7} (1/250 solar). Predicted fine structure line intensities for the bright southern bar (for limb-brightening $1/\cos\theta = 5$) are given in Table 1. Most of the lines are not yet observed, but the intensity measured by ISO for [SiII]34.8μm is only 1/25 of the model prediction! Part of the discrepancy can be attributed to beam dilution: the ISO beam is 660 arcsec^2, whereas the bright bar is only $\sim$50 arcsec^2. However, there presumably is [Si II] emission away from the bar, so it appears likely that the Si is depleted by $\sim$400 relative to solar. It is interesting to compare this with the S140 PDR, where the inferred Si abundance was 1/40 solar

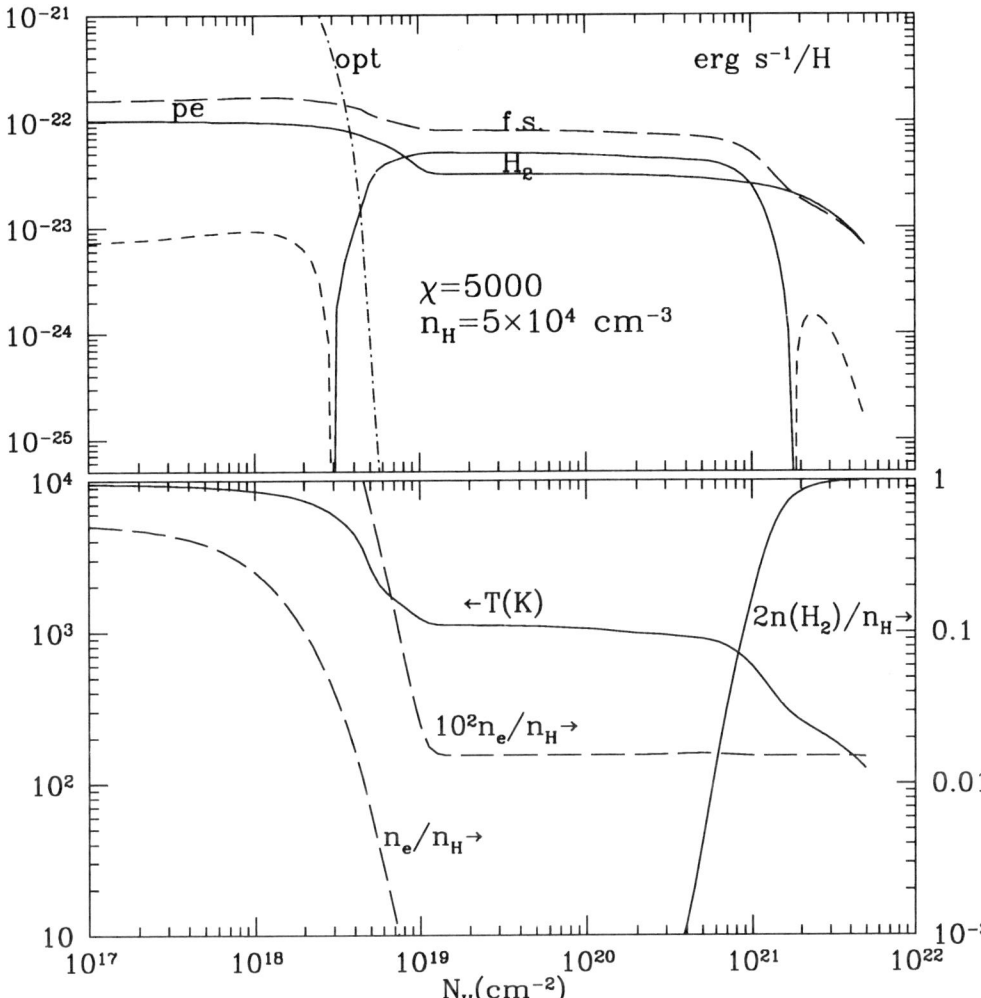

FIGURE 6. Lower panel: temperature T, ionization fraction n_e/n_H, and H_2 fraction as function of column density N_H measured from the ionization front. Upper panel: Heating or cooling contributions due to photoelectric emission from dust (p.e.), fine structure emission (f.s.), optical line emission (opt), and collisional excitation/de-excitation of H_2 (H_2). Solid lines indicate heating; broken lines indicate cooling. H_2 heats the gas over the region $5000 \gtrsim T \gtrsim 300K$.

(Timmermann et al. 1996). Walmsley et al. (1999) have recently discussed the puzzling variations in Si abundance observed in PDRs.

5. Discussion

There are a number of areas for improvement in our modelling of PDRs:
- We need accurate cross sections for collisional excitation and de-excitation of H_2.
- We need to understand the details of the H_2 formation process in PDRs – the overall rate and the (v, J) distribution of newly-formed H_2.
- The rate of photoelectric heating by dust is critical, particularly since drift of the dust grains can alter the dust/gas ratio in the PDR (Weingartner & Draine 1999, 2000b).
- Finally, we need more accurate spectro-photometry, with refined angular resolution,

line	predicted	observed	
[SiII]34.81μm	1.5×10^{-3}	5.7×10^{-5}	erg cm^{-2} s^{-1} sr^{-1}
[FeII]25.99μm	1.8×10^{-4}	-	
[OI]63.18μm	4.5×10^{-2}	-	
[OI]145.5μm	2.9×10^{-3}	-	
[CII]157.7μm	4.6×10^{-3}	-	

TABLE 1. NGC 2023 Fine Structure Line Emission

to allow us to understand the geometry of the filaments and sheets which are apparent in high-resolution images of NGC 2023 and other PDRs.

This research was supported in part by NSF grant AST96-19429.

REFERENCES

ABGRALL, H., & ROUEFF, E. 1989, *A&A Suppl.*, **79**, 313-328.

ABGRALL, H. ET AL. 1993a, *A&A Suppl.*, **101**, 273-321.

ABGRALL, H. ET AL. 1993b, *A&A Suppl.*, **101**, 323-362.

BERTOLDI, F., ET AL. 2000a, in preparation

BERTOLDI, F., & DRAINE, B.T. 1996, *ApJ*, **458**, 222-232.

BERTOLDI, F., ET AL. 2000b, in Astrochemistry, IAU Symposium 197, ed. Y.C. Minh & E.F. van Dishoeck (San Francisco: Astr. Soc. Pac.) in press.

BLACK, J.H., & VAN DISHOECK, E.F. 1987, *ApJ*, **322**, 412-449.

BURTON, M.G., ET AL. 1998, *PASA*, **15**, 194-201

DRAINE, B.T., & BERTOLDI, F. 1996, *ApJ*, **468**, 269-289.

FIELD ET AL. 1998, *A&A*, **333**, 280-286

HABING, H.J. 1968, *Bull. Astr. Soc. Neth.*, **19**, 421-431.

MCCARTNEY, ET AL. 1999, *MNRAS*, **307**, 315-327

ROUAN ET AL. 1997, *MNRAS*, **284**, 395-400

STEPHENS, T.L., & DALGARNO, A. 1973, *ApJ*, **186**, 165-167

TIMMERMANN, R., ET AL. 1996, *Astr.Ap.*, **315**, L281-284.

WALMSLEY, C.M., PINEAU DES FORETS, G., & FLOWER, D.R. 1999, *A&A*, **342**, 542-550.

WEINGARTNER, J.C., & DRAINE, B.T. 1999, in *The Universe as seen by ISO*, ed. P. Cox & M.F. Kessler, (Nordwijk: ESA), 783-786.

WEINGARTNER, J.C., & DRAINE, B.T. 2000a, *ApJ*, submitted

WEINGARTNER, J.C., & DRAINE, B.T. 2000b, in preparation

ISO Spectroscopy of H_2 in Star Forming Regions

By M. E. van den Ancker[1,2], P. R. Wesselius[3]
AND A. G. G. M. Tielens[3,4]

[1]Harvard-Smithsonian Center for Astrophysics, Cambridge, MA 02138, USA

[2]University of Amsterdam, Kruislaan 403, 1098 SJ Amsterdam, The Netherlands

[3]SRON, P.O. Box 800, 9700 AV Groningen, The Netherlands

[4]Kapteyn Astronomical Institute, P.O. Box 800, 9700 AV Groningen, The Netherlands

We have studied molecular hydrogen emission in a sample of 21 YSOs using spectra obtained with the *Infrared Space Observatory* (ISO). H_2 emission was detected in 12 sources and can be explained as arising in either a shock, caused by the interaction of an outflow from an embedded YSO with the surrounding molecular cloud, or in a PDR surrounding an exposed young early-type star. The distinction between these two mechanisms can not always be made from the pure rotational H_2 lines alone. Other tracers, such as PAH emission or [S I] 25.25 μm emission, are needed to identify the H_2 heating mechanism. No deviations from a 3:1 ortho/para ratio of H_2 were found. Both shocks and PDRs show a warm and a hot component in H_2, which we explain as thermal emission from warm molecular gas (warm component), or UV-pumped infrared fluorescence in the case of PDRs and the re-formation of H_2 for shocks (hot component).

1. Introduction

Molecular hydrogen is expected to be ubiquitous in the circumstellar environment of Young Stellar Objects (YSOs). It is the main constituent of the molecular cloud from which the young star has formed and is also expected to be the main component of the circumstellar disk. Most of this material will be at temperatures of 20–30 K and difficult to observe. However, some regions may be heated to temperatures of a few hundred K and produce observable H_2 emission. The intense UV radiation generated by accretion as well as by the central star itself will create a photodissociation region (PDR), of which the surface layer is heated by collisions with photoelectrically ejected electrons from grain surfaces, in any surrounding neutral material. Another possibility to produce warm H_2 is in shocks caused by the interaction of an outflow with the surrounding molecular cloud. Shocks are usually divided into J- or Jump-shocks, and C- or Continuous-shocks. In J-shocks the molecular material is dissociated in the shock front, where the gas is heated to several times 10^4 degrees. Behind the shock front, molecular material will re-form, and warm H_2 may be observed in the post-shock gas. C-shocks, in contrast, are not sufficiently powerful to dissociate molecular material, but may produce observable amounts of H_2 within the shock front itself.

Until recently the study of H_2 in star forming regions has mainly concentrated on the study of the near-infrared ro-vibrational lines observable from the ground. However, the launch of the *Infrared Space Observatory* (ISO) has opened up the possibility to also study the mid-infrared pure rotational lines of H_2, with much lower upper energy levels, and directly detect the thermal emission of warm H_2 in a wide variety of sources. In these proceedings we report on our study of H_2 lines in the ISO spectra of a sample of 21 YSOs. We will show that emission of warm H_2 is common in the environments of intermediate- and high-mass YSOs and can be explained by the phenomena of shocks and PDRs outlined above.

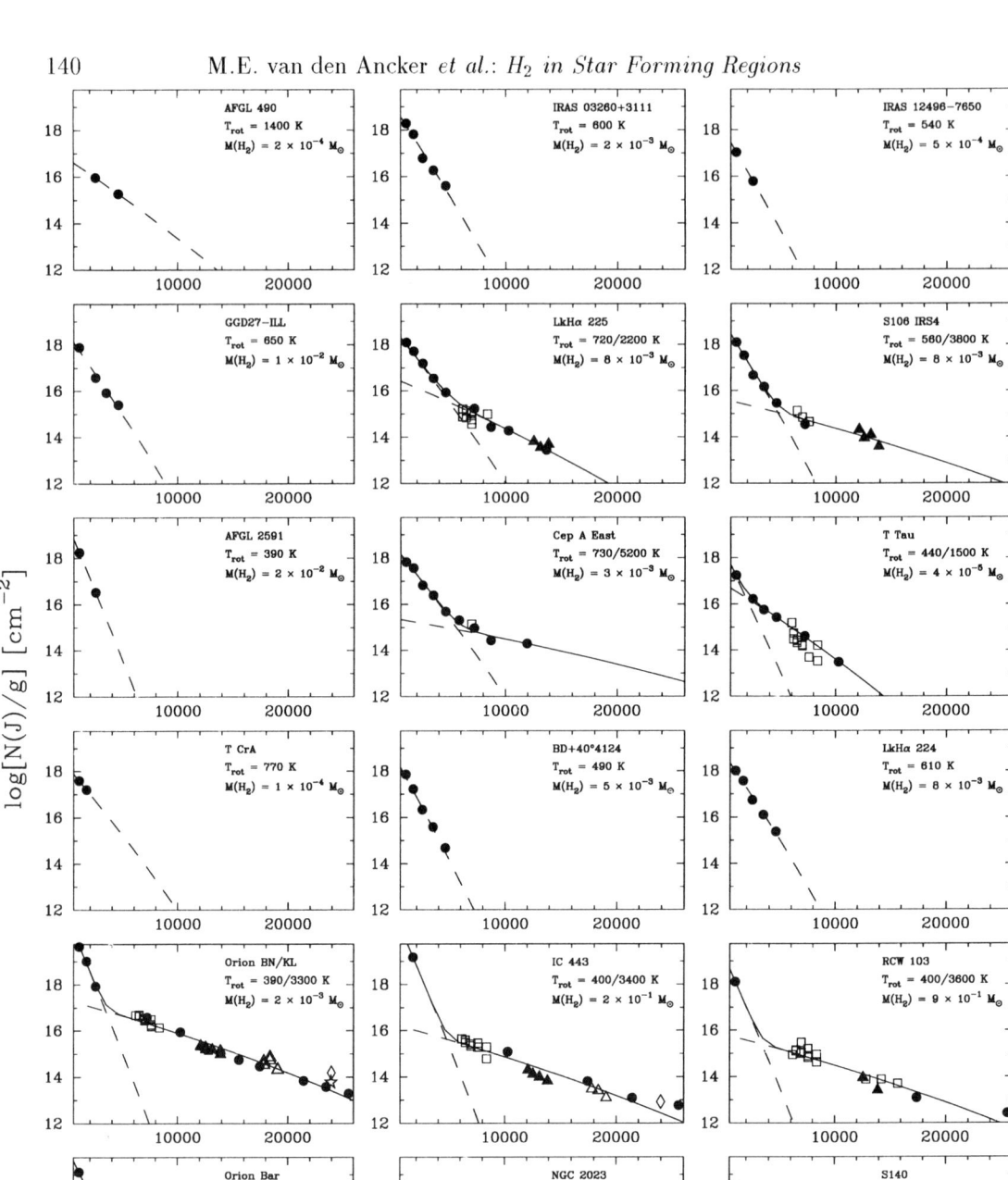

FIGURE 1. H_2 excitation diagrams for programme stars (top four rows) and comparison shocks and PDRs (bottom two rows). Shown are apparent columns of H_2 in the pure-rotational (0–0; filled dots), 1–0 (open squares), 2–1 (filled triangles), 2–0 (open stars), 3–2 (open triangles), 3–0 (filled squares), 4–3 (open diamonds), and 4–1 transitions (open crosses). Observational errors are smaller than the size of the plot symbol. The Boltzmann distribution fits are plotted as dashed lines. The solid lines show the sum of both thermal components for each source.

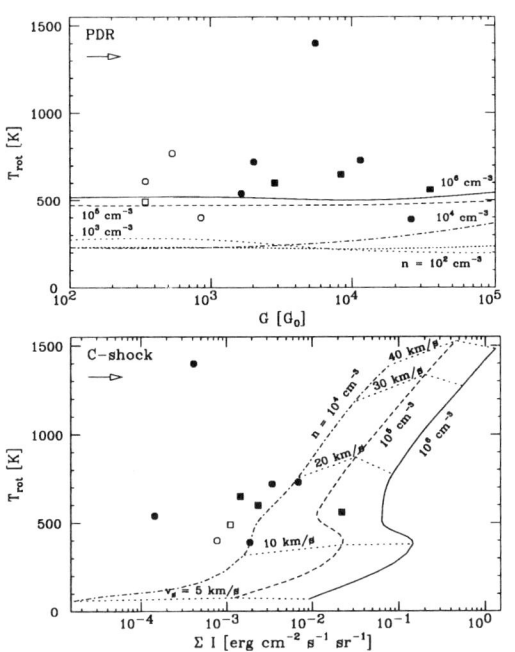
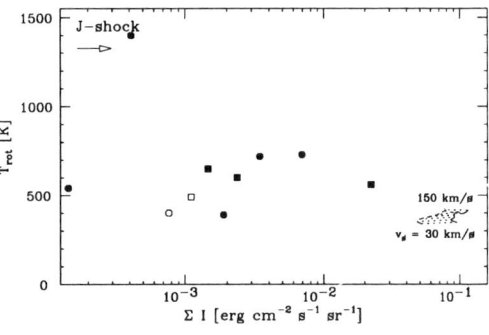

FIGURE 2. Comparison of observed H_2 rotational temperatures to theoretical relation between continuum fluxes and T_{rot} (PDR models) or summed intensity in all observed lines and T_{rot} (shocks). Sources which show PAH emission are plotted as squares. Plot symbols are filled for Class I sources. The arrows show the direction of beam dilution.

2. Observations and Analysis

ISO Short Wavelength Spectrometer (SWS; 2.4 – 45 μm) spectra were obtained for a sample of 21 YSOs, mostly of intermediate and high mass. Data were reduced in a standard fashion using calibration files corresponding to OLP version 7.0. In each object, molecular hydrogen line fluxes or upper limits (total flux for line with peak flux 3σ) of 0–0 S(0) to S(11), 1–0 Q(1) to Q(6) and 1–0 O(2) to O(7) were determined.

Pure rotational (0–0) H_2 emission was detected in 12 out of our 21 sources. Rovibrational (1–0) H_2 emission was detected in 4 sources, all of which were also detected in the pure rotational lines. A first inspection of our data shows that H_2 emission was only found in the vicinity of early-type (< B4) stars or near embedded sources. Qualitatively this is in agreement with one would expect: the strong UV fluxes of early-type stars are expected to produce extended PDRs, whereas embedded YSOs are expected to drive strong outflows, causing a shock as the outflow hits the surrounding molecular cloud.

The 28.2188 μm 0–0 S(0) line was not detected. This shows directly that we did not detect the cool quiescent H_2 in the molecular cloud. A more qualitative analysis of our data can be made by plotting the log of $N(J)/g$, the apparent column density for a given J upper level divided by the statistical weight, versus the energy of the upper level. For the statistical weight we have assumed the high temperature equilibrium relative abundances of 3:1 for the ortho and para forms of H_2. The resulting excitation diagrams are shown in Fig. 1. For comparison we also show excitation diagrams of three sources known to be dominated by shocks (Orion BN/KL peak 1, IC 443 and RCW 103) and three well-known PDRs (the Orion Bar, NGC 2023 and S140), created using data from literature. In Fig. 1 we also show Boltzmann distribution fits to the low-lying pure rotational lines. The fact that for most sources the the points for ortho and para H_2 lie are both well fitted by this nearly straight line proves that our assumption on their relative abundances is correct. For a number of sources, the lines at higher energy levels can be seen to deviate strongly from the Boltzmann fit. In these cases, we have attempted to characterize this behaviour, which may reflect the combined effects of UV-pumped infrared fluorescence and the presence of a very warm, but thin, surface layer in a PDR and may be due to the effect of re-formation of H_2 in J-shocks, by fitting a second Boltzmann distribution to the higher energy level populations. The resulting excitation temperatures and derived mass of molecular hydrogen are also indicated in Fig. 1.

Employing predictions of H_2 emission from PDR, J-shock and C-shock models by Burton et al. (1992), Hollenbach & McKee (1989) and Kaufman & Neufeld (1996), we determined the excitation temperature $T_{\rm rot}$ from the low-lying pure rotational levels as a function of density n and either incident FUV flux G (in units of the average interstellar FUV field G_0) or shock velocity v_s in an identical way as was done for the observations. The results of this procedure, plotted against G (PDRs) or the total flux observed in all lines (shocks) are shown in Fig. 2. Note that there is considerable overlap between the $T_{\rm rot}$ predicted by PDR, J-shock and C-shock models. This means that pure rotational H_2 emission alone cannot distinguish between these mechanisms in all cases and additional information will be needed. However, our ISO spectra provide just such information. Detectable [S I] 25.25 μm emission means that a shock must be present, whereas PAH emission is indicative of the presence of a PDR. The presence or absence of ionic lines such as [Si II] 34.82 μm can further distinguish between a C-shock and a J-shock. The information from these lines is used in Fig. 2 to distinguish between likely shocks (circles) and PDRs (squares).

In general, the PDR sources fall, within errors, in the parameter space outlined by the PDR models. Since the J-shocks only predict a very narrow range of $T_{\rm rot}$, only one source, AFGL 2591, is compatible with the observed H_2 emission arising in such a dissociative shock. All shock sources are compatible with the range in $T_{\rm rot}$ predicted by C-shock models. However, the detection of ionic lines in most of these sources shows that a J-shock component must be present. Most likely real astrophysical shocks are never as simple as the purely dissociative or non-dissociative shocks in the employed models, but are made up of a combination of the two, with the non-dissociative component dominating the H_2 spectrum.

3. Conclusions

We have shown that pure-rotational emission from warm H_2 is readily detectable in the vicinity of intermediate- and high-mass YSOs and can be used to gain insight in the physical conditions in the circumstellar material. The main mechanisms that produce warm H_2 in these types of environments are shocks and PDRs. No deviations from the 3:1 ortho/para ratio of H_2 were found for either type of heating mechanism. Both shocks and PDRs show a warm and a hot component in H_2. The warm component probes the thermal emission from warm gas. For PDRs the hot component may reflect the combined effects of UV-pumped infrared fluorescence and the presence of a thin, very warm surface layer. In shocks the hot H_2 component may be due to the re-formation of H_2 with non-zero formation energy. The warm H_2 component in shocks appears to be dominated by the non-dissociative part of the shock. The evolution of YSOs is expected to be from shock-dominated to PDR-dominated and H_2 may be one of the best tracers of the end of the outflow phase of a young star.

REFERENCES

Burton, M.G., Hollenbach, D.J., Tielens, A.G.G.M. 1992, ApJ 399, 563

Hollenbach, D.J., McKee, C.F. 1989, ApJ 342, 306

Hollenbach, D.J., Tielens, A.G.G.M. 1999, Rev. Mod Phys. 71, 173

Kaufman, M.J., Neufeld, D.A. 1996, ApJ 456, 611

Observations of the H_2 Ortho-Para Ratio in Photodissociation Regions

By Suzanne RAMSAY HOWAT[1], Antonio CHRYSOSTOMOU[2], Peter BRAND, Michael BURTON[4] AND Phil PUXLEY[5]

[1] UK Astronomy Technology Centre, Royal Observatory, Blackford Hill, Edinburgh, EH9 3HJ, UK.

[2] Department of Physical Sciences, University of Hertfordshire, College Lane, Hatfield, Herts. AL10 9AB, UK.

[3] Institute for Astronomy, University of Edinburgh, Blackford Hill, Edinburgh EH9 3HJ, UK.

[4] School of Physics, UNSW, Sydney, New South Wales 2052, Australia.

[5] Gemini 8m Telescopes, 670 N. A'ohoku Pl., Hilo HI 96720, USA

Observations of the near-infrared spectrum of molecular hydrogen in photo-dissociation regions has become a standard tool for revealing the detailed physical conditions and complex density structures of molecular clouds. Most recently, consideration has been give to the detailed behaviour of the ratio of ortho-to-para excited states, and the information that this ratio may contain regarding the history of the molecular cloud (Draine & Bertoldi 1996, Sternberg & Neufeld 1999). This paper will review NIR observations of the H_2 spectrum with particular reference to the ortho-para ratios observed. Recent spectroscopy of both galactic and extragalactic sources provide some interesting constraints on the models.

---

1. Introduction

Modelling of the H_2 emission from photodissociation regions (PDRs) has reached a very high level of sophistication a decade after the first observations of H_2 fluorescent emission, from the planetary nebula NGC2023. The earliest models, which predicted the response of low density H_2 gas to a moderate intensity UV field (Black & van Dishoeck 1987, Sternberg & Dalgarno 1998) have been expanded to include the effects of collisional excitation of the lowest H_2 energy levels (Burton, Hollenbach & Tielens 1990, Sternberg 1991) and of self-shielding of dense H_2 (Draine & Bertoldi 1996). Observations of the H_2 far-red and near-infrared spectrum confirm the model results for emission arising in energy levels as high as $E_k > 40,000K$ (Draine 2000). Recently, theoretical attention has turned to the observed ortho-para ratio of H_2 and the potential that this measure may hold for furthering our understanding of the past and present physical conditions in the PDR. In Section 1, previous observations of the ortho-para ratio are discussed and the theoretical results briefly summarised. In Section 2 we present new ISO spectroscopy of the M17 Northern Bar. In Section 3 a detailed near-infrared spectrum of the proto-planetary nebula Hubble 12 is discussed. In Section 4, early results in a programme to measure the H_2 spectrum in extragalactic sources are shown.

2. Previous observations and theoretical discussions

The ortho-para ratio measured from vibrationally excited H_2 has been examined in the literature ever since the first fluorescent H_2 spectrum was observed from NGC2023 by Gatley *et al.* (1987). Hasegawa *et al.* (1987) discussed the systematic differences in the vibrational and rotational excitation temperatures and the ortho-para ratio between the fluorescent spectrum from NGC2023 and the shock excited spectrum of Orion PK1.

Using ten ro-vibrational lines measured using a CVF, Hasegawa et al. (1987) obtained the ortho-para ratio from the $v = 1$, $v = 2$ and $v = 3$ vibrational levels, finding it to be significantly different from the canonical value of 3 for a gas in local thermal equilibrium. Tanaka et al.(1989) expanded on this work, discussing the use of the rotational temperature and the ortho-para ratio measured from vibrationally excited states as potential discriminants of the excitation mechanism. Since then, numerous observational papers have discussed the ortho-para ratio obtained from the NIR spectrum of the vibrationally excited H_2 lines, amongst them Chrysostomou et al. (1993), Ramsay et al. (1993), Hora & Latter (1996), Harrison et al. (1999). The range of ortho-para ratios typically found in studies of radiatively excited H_2 is from ~1.7–2.5. For example, in a detailed study of the kinematics and excitation of the planetary nebula BD+303639, Shupe et al. (1998) measured an ortho-para ratio of 2.51±0.22, averaged over the $v = 1 - 3$ levels and over a 2arcsec beam. At higher angular resolution, they observed the ortho-para ratio of the $v = 1$ level to vary from a value of 3 closest to the exciting star, to 1.7 deeper into the PDR.

Some of the first theoretical discussion of the variation of the ortho-para ratio and the implications for the physical conditions of the gas are given by
Burton, Hollenbach & Tielens (1992). In anticipation of the observations of the pure rotational emission lines of H_2 which were made possible by the capabilities of the ISO spectrometers, Burton, Hollenbach & Tielens modelled the emission from these lines as a function of density and incident UV field. Burton, Hollenbach & Tielens (1992) pointed out the temperature dependence of the ortho-para ratio. At the edge of a PDR, where the gas temperature is greater than ~200K and there is a population of atomic hydrogen, from the photo-dissociation of the H_2, spin-exchange collisions between H_2 and H will drive the gas into local thermal equilibrium and the ortho-para ratio is ~ 3. Deeper into the cloud, where the temperature drops to $< 200K$, there is still a significant contribution to the H_2 emission from the 0-0 S(0) and 0-0 S(1) lines ($E_k = 510K, 1015K$ respectively). The apparent ortho-para ratio can then fall below 3. Draine & Bertoldi (1996) expanded upon this, showing explicitly the predicted ortho-para ratio measured from vibrationally excited levels for H_2 for the range of densities and UV field strengths covered by their models of PDRs. They noted that it is possible for the apparent ortho-para ratio to rise above the value of three for clouds in which the self-shielding is important. Ortho-H_2 self-shields more effectively allowing ortho-para ratios greater than three. Sternberg & Neufield (1999) have presented a detailed discussions of the effects of self-shielding on the ortho-para ratio. They emphasise the important distinction between the *true*, or total, ortho-para ratio of the gas, and the measured ortho-para ratio (i.e. the ratio between the observed column densities for emission from vibrationally excited ortho- and para-levels). The observed ortho-para ratio will be set by either the temperature of the gas, or the temperature of formation, depending on conditions. In the first case, if LTE is maintained through collisions,

$$\frac{N_o^*}{N_p^*} = \sqrt{(\frac{N_o}{N_p})} = \sqrt{\alpha(T_{\text{gas}})}, \qquad (2.1)$$

where N_o and N_p are the total column densities of the ortho and para molecules respectively, N_o^* and N_p^* are the column densities of vibrationally excited H_2 and $\alpha(T_{\text{gas}})$ is the ortho-para ratio for gas of temperature T_{gas}. For the second case, where collisions are unimportant relative to other process, such as UV pumping, then

$$\frac{N_o}{N_p} = (\frac{N_o^*}{N_p^*})^2 = \alpha^2(T_{\text{form}}). \qquad (2.2)$$

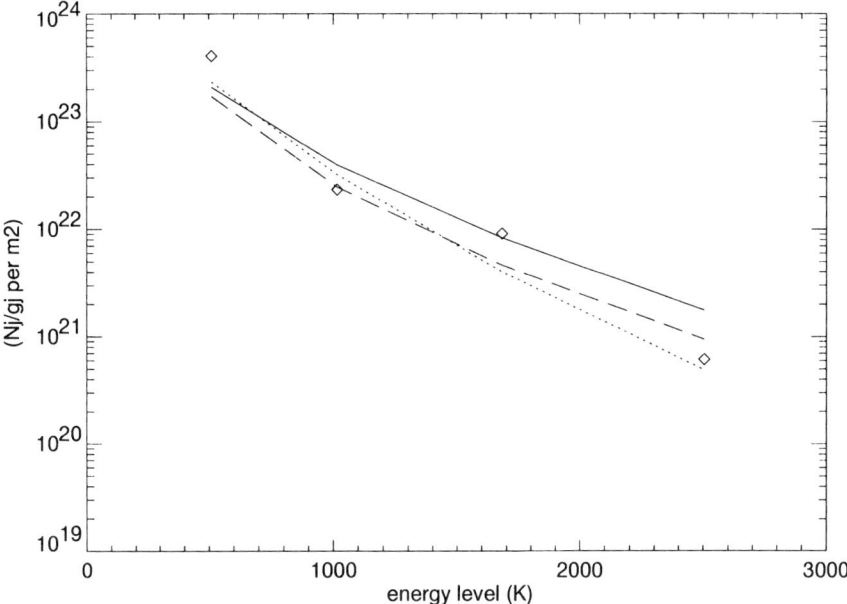

FIGURE 1. The excitation diagram for pure rotational lines of M17 with models from Burton, Hollenbach and Tielens shown for comparison.

where $\alpha(T_{\rm form})$ is the ortho-para abundance ratio set on formation of the gas at temperature $T_{\rm form}$.

Sternberg & Neufeld discuss three models which characterise the behaviour of the ortho-para under different conditions. The most directly applicable of these is described by Equation 1. If the $v = 0$ level is assumed in LTE, with o-p=3, then the vibrationally excited levels will tend to show an ortho-para ratio $\sqrt{(3)} = 1.7$ for $A_v \sim 1$, but will still have o-p=3 at the edge of the cloud. This is as seen by Shupe et al. 1998. Sternberg & Neufeld model the observations of S140 by Timmermann et al., which show that an o-p ratio of 3 for the $v = 0$ levels and an o-p$\sim$ 1.7 are completely consistent with radiative excitation of the gas. The ortho-para ratios from previously observed spectra can largely be explained by this reasoning. However, examples of sources with non-equilibrium ortho-para ratios measured from the ground vibrational state are appearing in the literature. Neufield, Melnick & Harwit (1999) measured a value of $\sim$1.2 for the outflow source HH54. Fuente et al. (1999) determine a value in the range 1.5-2.0 for NGC7023. Other examples may be found in these proceedings (Herpin et al., Boulanger et al., van der Werf et al.).

3. ISO Observations of M17

Chrysostomou et al. (1993) observed the NIR spectrum of the star-forming region in the north of M17 and concluded that the H_2 emission spectrum was that of a moderately dense gas ($n > 10^5 {\rm cm}^{-3}$) with a higher density component ($n > 10^6 {\rm cm}^{-3}$). The $v = 1$ vibrational levels are thermalized by collisions between H_2 and H. ISO observations towards the peak of the H_2 emission observed by Chrysostomou et al. (1993) were made using the SWS AOT2 observing mode. Spectra of the 0-0 S(1) (28μm), 0-0 S(1) (17μm), 0-0 S(2) (12μm) and 0-0 S(3) (9μm) lines were obtained. All of the lines

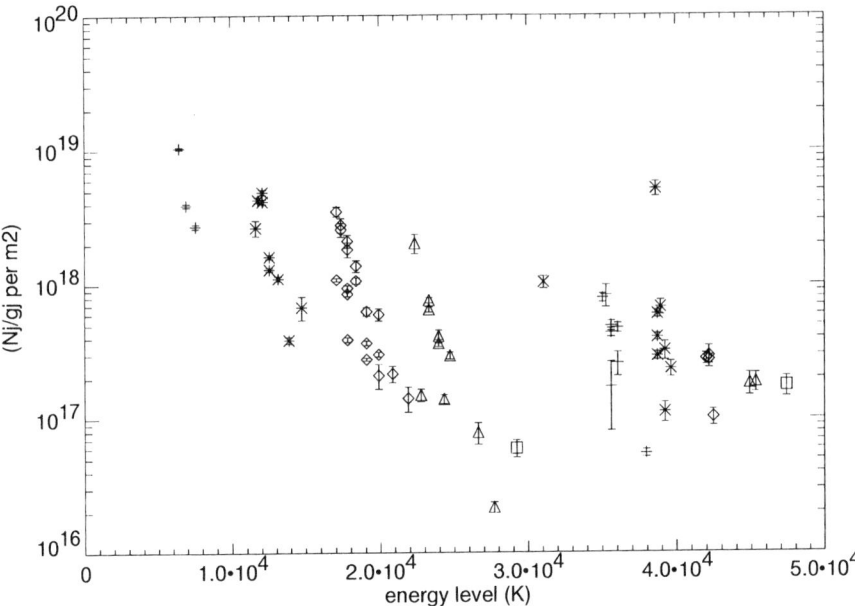

FIGURE 2. The excitation diagram for Hubble 12. Vibrational energy levels from v=1 to v=11 are represented.

were detected, with only the 0-0 S(0) line being significantly affected by fringing. The observed line strengths are shown in Figure 3 along with three example models from Burton, Hollenbach & Tielens (1992). These models have not been normalised to the observed column densities, nor have the observed column densities been corrected for extinction to M17. All of the pure rotational H_2 lines observed will be affected by extinction due to the silicate feature which peaks at 9.7μm. The shape of the silicate feature can be considered constant towards a variety of galactic sources, though it's depth, and hence the absolute extinction, changes. The extinction towards the H_2 emission in M17 Northern Bar has not been measured, though Chrysostomou et al. (1993) did obtained a value of A_k=0.05 from the 1-0 S(1) and 1-0 Q(3) lines in their spectrum. Since the 0-0 S(0) column density may be enhanced by a significant column of cold gas, the ortho-para ratio measured from the 0-0 S(1), S(2), S(3) lines. Taking the Chrysostomou et al. (1993) value for the extinction, a non-equilibrium value of 1.51±0.3 is obtained. This extinction value may not truly represent the extinction in the larger ISO beam, and therefore we consider how much extinction would be required to "correct" the ratio to an equilibrium value of 3. Such a value may be obtained if A_v=40. However, the column densities for the 1-0 S(1) and S(3) line are then an order of magnitude lower than the model column density, with the model normalised to the 0-0 S(2) line. We tentatively conclude that M17 does indeed show a non-equilibrium o-p ratio, but that further observations are required to determine the extinction with a comparable beam size before this result can be confirmed.

4. Red and Infrared Spectroscopy of Hubble 12

Hubble 12 is a young bright proto-planetary nebula, first observed to have a fluorescent H_2 spectrum by Dinnerstein et al. (1988). Since then, a number of studies of the infrared

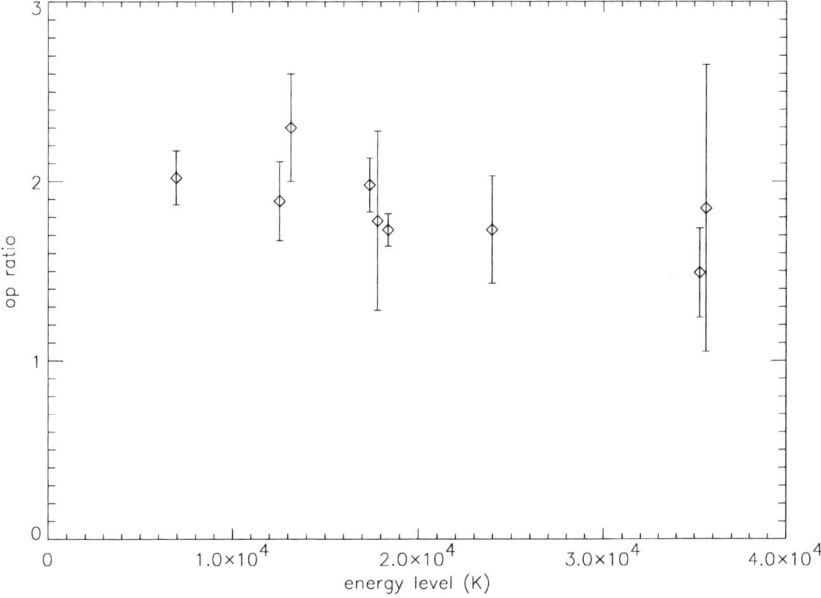

FIGURE 3. The Hubble 12 ortho-para ratio as a function of energy level.

spectrum of Hubble 12 have been carried out (Ramsay et al. 1993, Hora & Latter 1996, Luhmann & Rieke 1996). The K-band spectrum of H_2 emission from Hubble 12 showed it to be a source of pure fluorescent emission, with no evidence for a thermally excited component. Since Hubble 12 is such an excellent laboratory for verifying theories of radiative excitation of H_2, we carried out detailed spectroscopy from the far-red (0.7μm) to the infrared (2.5μm) using the ISIS spectrometer on the WHT and CGS4 on UKIRT. We obtained full coverage across this wavelength range, with the exception of the H-window. We also obtained ISO SWS spectra of the wavelength ranges containing the 0-0 S(1), S(2), S(3) and S(4) lines, but no detections were made. In all, ~70 H_2 lines were detected, from 1-0 S(0) at E_k=6473K to 11-7 O(3) at E_k=47,397K.

The excitation diagram for Hubble 12 is shown in Figure 3. The excitation diagram shows the classic behaviour for a fluorescent spectrum. Lines from each vibrational level lie along a distinct curve. Comparison of the line strengths predicted by Draine & Bertoldi (1996) for low density gas ($10^3 cm^{-3}$) illuminated by a UV field of 100G_o shows no evidence of thermalization of the lower levels. We also determined the ortho-para ratio for each of the vibrational levels for which sufficient lines existed. The ortho-para ratio plot is shown in Figure 4. The ortho-para ratio for the vibrationally excited gas is constant across the energy levels from E_j=6953K to 42192K, with a value of 1.8±0.1. This value is entirely consistent with a total ortho-para abundance ratio of 3.

5. The extragalactic H<small>II</small> region NGC5461

H_2 emission from galaxies is commonly attribute to shock excited emission (e.g. Rieke et al. 1988, Moorwood & Oliva 1990). Since these data often rely on measurements of lines from the $v = 1$ vibrational level and the v=2-1 S(1) line, which are thermalized at high densities, the importance of radiative excitation in star-forming galaxies is not well

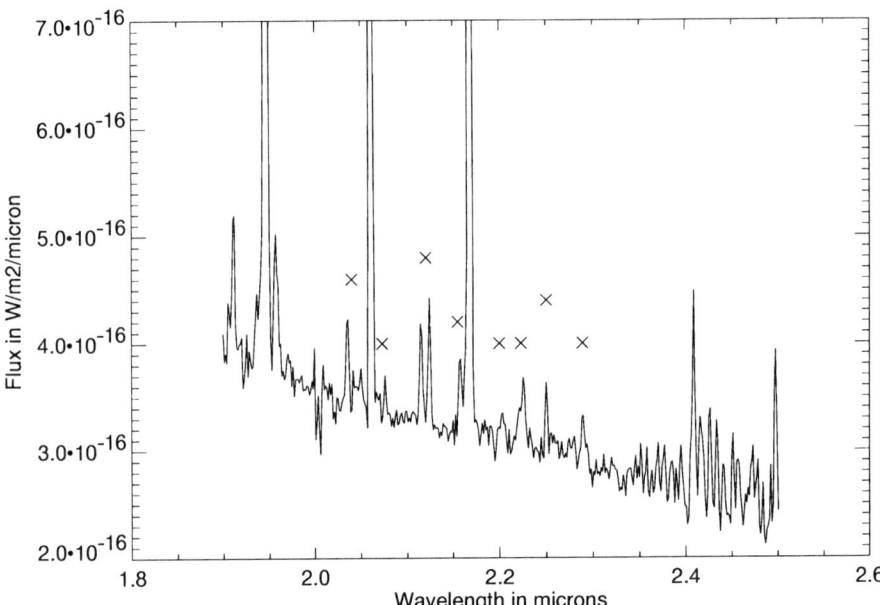

FIGURE 4. The K-band spectrum of NGC5461. The measured H_2 emission lines are indicated with a cross.

understood. NGC5461 is a giant HII region in the galaxy M101. It is one object in a sample of HII regions and blue compact dwarf galaxies selected for observation because of their low continuum. Many star-forming galaxies have bright continuum emission with copious features from late-type stars which obscure the faint, $v > 2$ H_2 lines. The K-band spectrum of NGC5461 is shown in Figure 5, scaled to show the H_2 lines. Emission lines from $v = 1$, $v = 2$ and $v = 3$ were detected, making this the most detailed spectrum yet published of H_2 from an extragalactic source (Puxley, Ramsay Howat & Mountain 1999). The highest excitation line observed is the 3-2 S(3) line at 19089K. The extinction corrected column densities lines have been compared with Black & van Dishoeck's pure fluorescent model 14 and with the equivalent model from Draine & Bertoldi (1996). These comparisons show that the H_2 emission from NGC5461 is from a radiatively excited gas of density less than the critical density at which collisions become important ($\leq 10^5 cm^{-3}$).

The rotational excitation temperature and ortho-para ratio were obtained from both the $v = 1$ and $v = 2$ levels. For the $v = 1(2)$ level the rotational temperature is 1330K±311K (1180K±396K) and the ortho-para ratio is 1.15±0.11 (0.69±0.27). To the best of our knowledge, this ortho-para ratio is the lowest measured from vibrationally excited lines. The ortho-para ratio from the pure rotational lines measured in HH54 was ~1.2. Following the argument of Sternberg & Neufield (1999), that the ortho-para ratio measured from the vibrationally excited lines is the square-root of the total ortho-para ratio for the gas, we obtain a prediction of o-p=1.3 for the total ortho-para ratio. This corresponds to a gas temperature of ~100K. Neufield, Melnick & Harwit (1999) concluded that the 90K temperature implied for the H_2 in HH54 was a remnant from the early history of the gas. In the case of NGC5461 it is unlikely that the low temperature of 100K could be maintained across the 60pc of NGC5461 over which we measured the H_2 line strengths.

The observations contained in this paper were made at the UK Infrared Telescope and the William Herschel telescope. Both are owned by the UK Particle Physics and Astronomy Research Council. The UK Infrared Telescope is operated by the Joint Astronomy Centre. The authors wish to acknowledge the assistance of staff at both telescopes in obtaining these data. Other observations were made with the ISO SWS and reduced using the ISAP reduction package, with the assistance of the ISO support staff.

REFERENCES

BURTON, M.G., HOLLENBACH, D.J. & TIELENS, A.G.G.M. 1990 *Ap. J.* **365**, 620–639.

BLACK, J. & VAN DISHOECK, E. 1987 *Ap. J.* **322**, 412-449.

BURTON, M.G., HOLLENBACH, D.J. & TIELENS, A.G.G.M. 1992 *Ap. J.* **399**, 563–572.

CHRYSOSTOMOU, A., BRAND, P.W.J.L., BURTON, M.G. & MOORHOUSE, A. 1993 *MNRAS* **265**, 329–page.

DINNERSTEIN, H.L., LESTER, D.F., CARR, J.S. & HARVEY, P.M. 1988 *Ap.J.* **327**, L27–pages.

DRAINE, B.T. & BERTOLDI, F. 1996 *Ap.J.* **468**, 269–289.

DRAINE, B.T. 2000 *these proceedings*.

FUENTE, A., MARTÍN-PINTADO, J., RODRÍGUEZ-FERNÁNDEZ, N.J., RODRÍGUEZ-FRANCO, A., DE VICENTE, P. 1999 *Ap.J.* **518**, L45–48.

GATLEY, I., HASEGAWA, T., SUZUKI, H., GARDEN, R., BRAND, P., LIGHTFOOT, J., GLENCROSS, W., OKUDA, H. % NAGATA, T. 1987 *Ap.J.* **318**, L73–76.

HASEGAWA, T., GATLEY, I., GARDEN, R.P., BRAND, P.W.J.L., OHISHI, M., HAYASHI, M. & KAIFU, N. 1987 *Ap.J.* **318**, L77–L80.

HARRISON, A.P., PUXLEY, P.J., RUSSELL, A.P.G. & BRAND, P.W.J.L. 1999 *MNRAS* **297**, 624–pages.

HORA, J.L. & LATTER, W.B. 1996 *Ap.J.* **461**, 288–pages.

LUHMAN, K.L. & RIEKE, G.H. 1996 *Ap.J.* **461**, 298–306.

MOORWOOD, A.F.M. & OLIVA, E. 1990 *A.A.* **239**, 78–84.

NEUFELD, D.A., MELNICK, G.J. & HARWIT, M. 1999 *Ap.J.* **506**, L75–L78.

PUXLEY, P.J., RAMSAY HOWAT, S.K., & MOUNTAIN, C.M. 1999 *Ap.J.*, in press.

RAMSAY, S.K., CHRYSOSTOMOU, A., GEBALLE, T.R., BRAND, P.W.J.L. & MOUNTAIN, C.M. 1993 *MNRAS* **263**, 695–700.

RIEKE, G., LEBOFSKY, M., & WALKER, C. 1988 *Ap. J.* **325** 679–686.

SHUPE, D.L., LARKIN, J.E., KNOP, R.A., ARMUS, L., MATTHEWS, K. & SOIFER, B.T. 1998 *Ap.J.* **498**, 267–277.

STERNBERG, A. & DALGARNO, A. 1988 *Ap.J.* **338**, 197.

STERNBERG, A. & NEUFELD, D.A. 1999 *Ap.J.* **516**, 371–380.

TAKAMI, M., USUDA, T., SUGAI, H., KAWABATA, H., SUTO, H. & TANAKA, M. 1999 *Ap.J.*, in press.

TANAKA, M., HASEGAWA, T., HAYASHI, S., BRAND, P.W.J.L. & GATLEY, I. 1989 *Ap.J.* **336**, 207–211.

TIMMERMANN, R., BERTOLDI, F., WRIGHT, C.M., DRAPATZ, S., DRAINE, B.T., HASER, L. & STERNBERG, A. 1996 *Astron. Astrophys.*, **315**, L281–L284.

Maryvonne Gerin on her bicycle in molecular discussion with Pepe Cernicharo.

Bill Reach and Takahiro Oosato arguing about H_2 cooling, during the refreshing pause.

H_2 emission from CRL 618

By F. Herpin[1], J. Cernicharo[1] AND A. Heras[2]

[1]Dept Física Molecular, I.E.M., C.S.I.C, Serrano 121, E-28006 Madrid, Spain

[2]Space Science Dept. of ESA, ESTEC, P.O. Box 299, 2200 AG Noordwijk, Netherlands

We present a complete study of the H_2 infrared emission, including the pure rotational lines, of the proto Planetary Nebulae CRL 618 with the ISO SWS. A large number of lines are detected. The analysis of our observations shows: (i) an OTP ratio very different from the classical value of 3, probably around 1.76-1.87; (ii) a stratification of the emitting region, and more precisely different regions of emission, plausibly located in the lobes, in an intermediate zone, and close to the torus; (iii) different excitation mechanisms, collisions and fluorescence.

1. Introduction

CRL 618 is one of the few clear examples of an AGB star in the transition phase to the Planetary Nebula stage: a Proto Planetary Nebula (*PPN*). It has a compact HII region created by a hot central C-rich star, and is observed as a bipolar nebula at optical, radio and infrared wavelengths. The expansion velocity of the envelope is around 20 $km\,s^{-1}$, but CO observations show the presence of a high-velocity outflow with velocities up to 300 $km\,s^{-1}$ (Cernicharo et al.1989). High-velocity emission in H_2 is also detected (Burton & Geballe 1986). The high velocity wind and the UV photons from the star perturb the circumstellar envelope producing shocks and photodissociation regions (PDRs) which modify the physical and chemical conditions of the gas (Cernicharo et al.1989, and Neri et al.1992). Clumpiness within the visible lobes and low-velocity shocks being the remnant of the AGB circumstellar envelope are also proposed by Latter et al.(1992). H_2 excitation mechanisms can be investigated through our IR observations. We also propose to study the possible variations of the OTP ratio from the classical value of 3. More details are given in Herpin et al.(2000).

2. Observations

Many H_2 lines were detected (see Fig. 1) with these ISO Short-Wavelength Spectrometer (SWS, de Grauuw et al.1996) observations: 0-0 S(J=0-15), 1-0 Q(J=1-7), 2-1 Q(J=1-7,9,10,13,14,15), 3-2 Q(J=1,3,4,5,7), 4-3 Q(J=2,3,4,7,9,10), and 1-0 O(J=2-7). The fluxes were obtained assuming the source punctual relative to the SWS beam. A SWS spectral resolution of $\lambda/\Delta\lambda = 2000$ was taken.

3. Data analysis

3.1. *Location of the emission*

Because of the large number of data and thus the large energy covering, we based our global data analysis on the $0-0$ S transitions. Moreover, 0-0S(0) to 0-0S(6) emissions are not very sensitive to the extinction and thus are very reliable. A brief look on the plot of the column densities (derived from the line intensities) versus the energies of the upper levels for the uncorrected data (no absorption correction, and OTP ratio of 3 (Fig.2, left) shows first that an OTP ratio (χ) of 3 seems inappropriate as the ortho observations are systematically underestimated. Moreover, correction of the entire set by

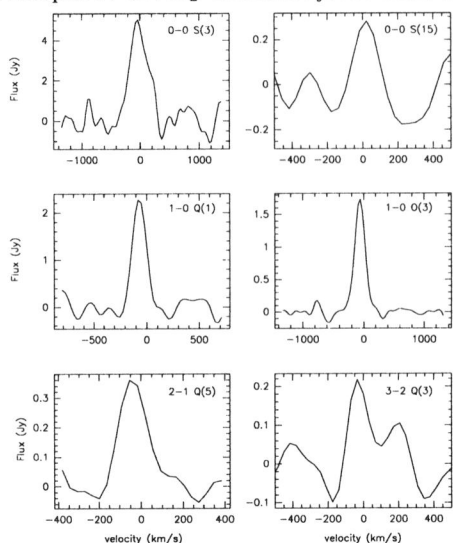

FIGURE 1. Examples of observed H_2 lines centered on the theoretical wavelength. Fluxes are in Jy. In this Fig., only the $0-0$ S(3) line is resolved (velocity resolution $\simeq 200$ km s^{-1}).

the same value of the OTP ratio seems not possible. Thus we introduced different values for different parts of the plot, i.e., different energies, and thus probably different regions of excitation in the PPN environment: $\chi = 1., 1.76$ and 1.87. But some discontinuities are still present. This is not an effect of variations in T_{ex} because in this case slopes would vary. This can only be explained by different regions of emission, and consequently can be corrected by different absorptions. We thus applied 4 different corrections of the absorption: $A = 3.5, 10, 15$ and 20 mag (Latter et al.1992, Thronson 1981). The result is given in the Fig. 2 (right captions).

This study of the resolution of the observed lines (see Fig. 1) shows three different parts of emission. First of all, the $0-0$ S(0) to S(6) lines are totally resolved (with a SWS resolution of 200 km s^{-1}). A broad line, thus resolved, means that the emission arises from a quite large velocity region, this velocity field being then sufficiently important to broaden the line. This suggests that this part of the H_2 emission may arise from the lobes, where the velocity is larger. An intermediate zone of excitation, between lobes and torus, appears for the $0-0$ S(7), S(8) and S(9) emissions, whose half power line widths are around 200 km s^{-1} (third zone on the plot) on the limit of resolution. Finally, all the other lines have small widths (50-160 km s^{-1}), and then probably arise from a low-velocity region, certainly the edge of the torus. A similar study is done for the emissions involving excited vibrational levels. The $1-0$ Q lines are unresolved. We think that these emissions stem from the same intermediate zone as are the $0-0$ S(7)-S(9) emissions. Moreover the "intermediate" slope derived from the column densities is another indication. The same conclusion applies for the $1-0$ O lines, even if their origin is not exactly the same; this emission is probably closer to the lobes, as the derived temperature (cf. Fig.2) is similar as what is found for the $0-0$ S emission in this area. All the lines involving vibrational levels with $v \geq 2$ are completely unresolved and thus arise from a region close to the torus.

3.2. Excitation mechanisms

We attempted to fit the data with a very simple model based on a Boltzmann distribution of the populations (cf. Fig. 2). For the $0-0$ S emissions, the fitted tempera-

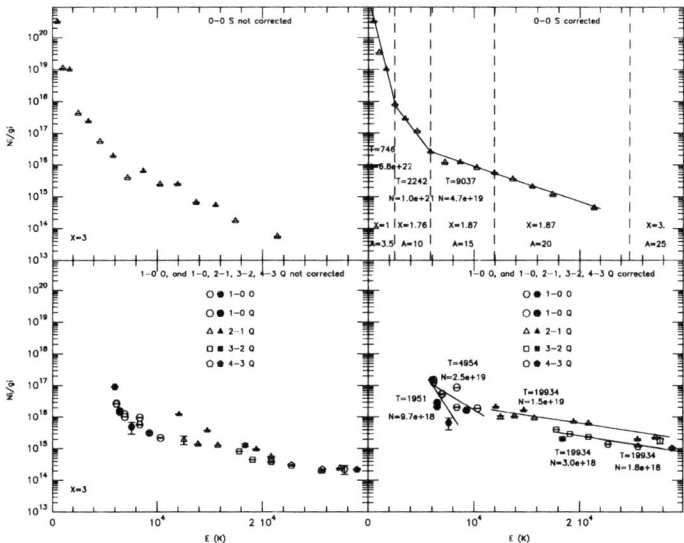

FIGURE 2. Plot of the column densities (in cm^{-2}) versus the energies of the upper levels for uncorrected data (OTP ratio=3, left captions) and corrected data (right captions). Black and white symbols are respectively for para and ortho transitions.

tures and column densities are in agreement with the hypothesis of low-energy emission from shocks in the lobes (low temperatures, and high column densities, corresponding to $n(H_2) \simeq 10^6 - 10^7$ cm^{-3}), and high-energy emissions coming from the torus with higher temperature and weaker density. For the transitions from excited vibrational levels, we admitted that 0-0 S(J), 1-0 Q(J+2) and O(J+4) emissions are produced in the same region, and have the same extinction and OTP ratio. The same argument was used for the other excited vibrational emissions. We obtained different fits with slow slopes (temperatures are larger) for different Δv emissions, which may suggest fluorescent emission.

Concerning the $1 - 0$ emissions, we compared our intensities to theoretical results of Sternberg (1989) and Black (1987). Moreover the temperatures are moderate, as the rot-vibrational column densities. This suggests the thermal contribution in the excitation. A clumpy structure, or one which partly shields the emitting region from the stellar UV flux may explain a low level of fluorescent emission. In thermal sources these rotational and vibrational temperatures are similar, while fluorescent sources are characterized by a much higher vibrational temperature and a much lower rotational temperature (Tanaka et al.1989). That is why emissions 1-0 O and Q are probably more or less of thermal nature ($T_{rot} \simeq 2000-5000$K and $T_{vib} \simeq 1000$ K for $v = 1$), but a component of fluorescent emission is probably also present. The situation seems more complicated for the other vibrational excited transitions. The only case in Sternberg (1989) and also in Black (1987) with all these emissions is the radiative fluorescent model. This strongly suggests that the emissions involving $v \geq 2$ levels are fluorescent. We resume on Fig.3 what we deduced about the excitation of the H$_2$ lines.

3.3. The OTP ratio

We attempt to explain the low OTP ratio by different mechanisms. Shocks breaking molecules and photodissociation, ionization from the central star can produce H, H$^+$ and H$_3^+$...who will react with H$_2$ and alter the OTP ratio. On the other hand, in a

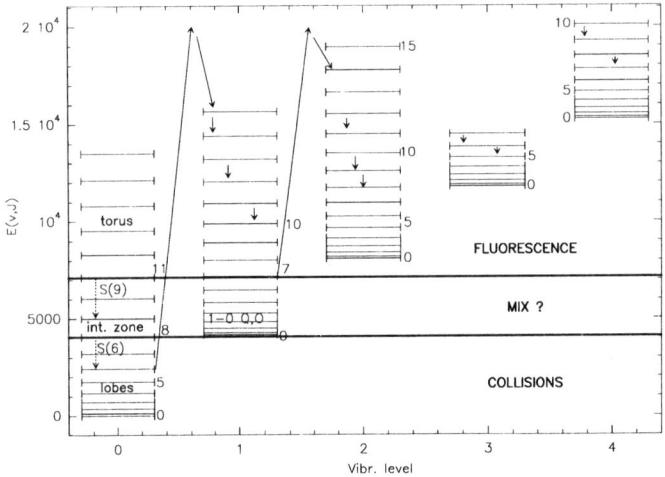

FIGURE 3. Energy levels (v,J) of the H_2 molecule and mechanisms of excitation.

shocked region, the shock velocity may be important in the OTP conversion process. We think that there is in fact no OTP ratio gradient in our data, but large uncertainties, and consequently that the OTP ratio is around 1.76-1.87, the first value of 1 remaining without explanation. We thus think that the OTP ratio is fixed near the central star where low-velocity shocks occur and where atoms and ions can spoil this ratio. Part of this material is swept out into the lobes with this low OTP ratio.

4. Conclusion

Shocks in the lobes produce the low-energy 0-0 S emission, while shocks and perhaps fluorescence induce emission in a transition area. The high-energy emission may be produced by fluorescence in the torus. A mix of fluorescence and collisions imay be on the origin of the 1-0 O and Q lines, while all the other emissions are dominated by fluorescence. Concerning the low OTP ratio, probably around 1.76-1.87, we think that it is fixed near the central star, the material being after swept out into the lobes, with a very non-classical value.

REFERENCES

BLACK, J.H. 1987 *Astrophys. Journal* **322**, 412.
BURTON, M.G. & GEBALLE, T.R. 1986 *Monthly Notice of Royal Astron. Soc.* **223**, 13p.
CERNICHARO, J. ET AL. 1989 *Astron. & Astrophys.* **222**, L1.
DE GRAUUW ET AL. 1996 *Astron. & Astrophys.* **315**, L49.
HERPIN, F. ET AL. 2000 *to be submitted to Astrophys. Journal*
LATTER, W.B. ET AL. 1992 *Astrophys. Journal* **389**, 347.
NERI, R. ET AL. 1992 *Astron. & Astrophys.* **262**, 544.
STERNBERG, A. 1989 *Astrophys. Journal* **347**, 863.
TANAKA, M. ET AL. 1989 *Astrophys. Journal* **336**, 207.
THRONSON, H.A.JR 1981 *Astrophys. Journal* **248**, 984.

Hydrogen in Photodissociation Regions: NGC2023 and NGC7023

By D.Field[1], J.L.Lemaire[2,3], J.P.Maillard[4], S.Leach[2], G. Pineau des Forêts[2], E.Falgarone[5], F.P.Pijpers[6], M.Gerin[5], F.Rostas[2], D. Rouan[2] AND L. Vannier[2]

[1]Institute of Physics and Astronomy, University of Aarhus, DK-8000 Aarhus C, Denmark and Observatoire de Paris-Meudon

[2]Observatoire de Paris-Meudon, F-92195 Meudon Principal Cedex, France

[3]Université de Cergy-Pontoise, F-95806 Cergy Cedex, France

[4]Institut d'Astrophysique, 98bis, Boulevard Arago, F-75014 Paris, France

[5]Ecole Normale Supérieure, 24 Rue Lhomond, 75231 Paris Cedex 05, France

[6]Theoretical Astrophysics Centre, University of Aarhus, DK-8000 Aarhus C, Denmark

High spatial and spectral resolution observations are reported of H_2 infrared emission from the reflection nebulæ NGC2023 and NGC7023. The local molecular gas is strongly perturbed by the presence of the massive stars which power these nebulae. Data yield information on the small-scale structure, the temperature and density and the dynamics of the excited gas. Excited material is found to be hot (400-500K), dense (10^5-10^6 cm^{-3}) and clumped containing substantial flows and velocity fields.

1. Introduction

The two reflection nebulæ NGC2023 and NGC7023 are prototypes of regions in which recently formed massive stars are interacting strongly with their parent gas. The outcome of these interactions is important in understanding the cycle of star formation in which massive stars are created and, by perturbing their surroundings, influence the nature of the gas in which future stars may form. The goal of our work is to examine in detail the perturbed gas around massive young stars. Some of the observations of infrared (IR) emission of molecular hydrogen in NGC2023 and NGC7023, performed in recent years in our group, are described below.

Nebulosity in NGC2023 and NGC7023 is excited by B-stars of temperatures respectively 22,000K and 20,400K. The distance between the star and the illuminated surrounding gas is $\sim$ 0.1 pc in both nebulæ. NGC2023 shows a strong IR excess with emission from small dust particles plus extended red emission, and has an associated molecular cloud with OH, HCHO, HCN, CO, CH, CH^+ and other detections (see Field et al. 1994). NGC7023 is a similar and perhaps still more active object (see Lemaire et al. 1996).

Since H_2 is the principal molecular component of the interstellar medium, interpretation of observations of H_2 emission is relatively free from the detailed chemical complications that beset the interpretation of data of other molecular species. Moreover H_2 IR emission is optically thin, allowing for simple conversion of signal intensity into column density. Emission from H_2 may act as an indicator of structure, of temperature and density and of dynamics. Examples are given below of each of these features of the data.

2. Hydrogen as an indicator of structure in NGC2023

With the advent of high spatial resolution detectors, it is now possible to obtain IR images of spectrally resolved lines of H_2 with a resolution approaching 0.1 arcsec, for

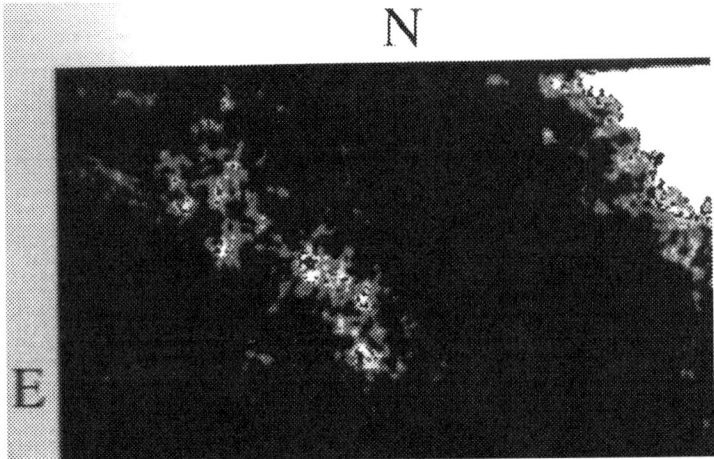

FIGURE 1. Data obtained using the ComeOn Plus adaptive optics system on the ESO3.6m Telescope: a 12.8x7.5 arcsec^2 field showing emission in the v=1-0 S(1) line at 2.121μm in NGC2023 in a region lying 12"E, 3"S of the exciting star HD37903. Saturation in the north-west corner of the image is due to HD37903 which lies 5"W of the edge of the image. For an overview of the H$_2$ emitting structure in NGC2023, the reader is referred to Field et al. 1998.

example for the very bright shocked source OMC1 (e.g. Allen and Burton 1993; Schild et al. 1996; Stolovy et al. 1998). The first seeing-limited ($\sim$ 1" resolution) image of H$_2$ in a reflection nebula, for NGC2023 in the v=1-0 S(1) line at 2.121 μm, was reported in Field et al. 1994, representing an improvement in spatial resolution of over one order of magnitude compared with earlier data. These data revealed remarkably complex small-scale structure.

Observations of NGC2023, using the ESO 3.6m with the Come-On Plus adaptive optics system (Rouan et al. 1997), yielded an image in the v=1-0 S(1) line at a resolution of $\sim$ 0.15", equivalent to $\sim$ 3x10^{-4} pc at the distance of NGC2023. These data were the first to combine adaptive optics and H$_2$ spectral line isolation. The image obtained may be found in Figure 1, showing a cloud of gas, $\sim$ 0.027 pc from the exciting B-star (HD37903). The image in Figure 1 illustrates that the gas close to the star is composed of a number of clumps. Further analysis showed that brightest and darkest regions within the cloud represent a density contrast of at least one order of magnitude, with a number density $(n_H + 2n_{H_2})$ of 10^5cm^{-3} in the brightest regions.

A Fourier analysis of the structures in Figure 1 shows that the number of structures increases as the scale size of structures drops, following a power law. This is consistent with a turbulent cascade, although one which was shown (Rouan et al. 1997) to have a smaller contribution from small-scale structures than non-dissipative turbulence predicts. This result is in direct contrast to the distribution of scales in the shocked star-forming region OMC1, in which data (obtained with the ADONIS adaptive optics system on the ESO 3.6m) show that there is a preferred scale size of $\sim$ 0.003 pc and that the distribution of scale sizes does not follow a power law.

New high signal-to-noise data for both NGC2023 and NGC7023 have recently been obtained using the Canada-France-Hawaii Telescope (CFHT: November 1999) with the KIR camera and the PUEO adaptive optics system. These data are presently being analyzed.

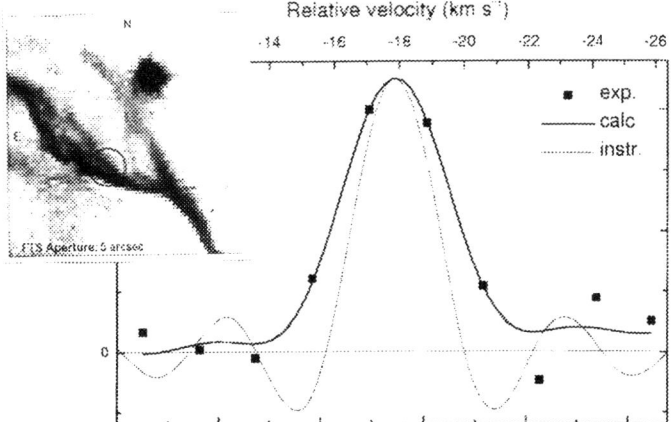

FIGURE 2. Emission in the H_2 v=1-0 S(1) line recorded at a resolution of 125,000 using the FTS spectrograph on the Canada-France-Hawaii Telescope. The spectral range shown is $0.27 cm^{-1}$. The scale of velocity is relative to the laboratory rest frequency for the S(1) line. Black squares are experimental points, the dotted line is the lineshape function associated with the spectrometer and the solid line is a fit to observations, yielding a linewidth (see Sect.3) and a line centre (see Sect.4). The inset shows emission in the same line in a 32×32 arcsec^2 field, as recorded at low spectral resolution in Lemaire et al. 1996. The circle of diameter 5 arcsec shows the position and extent of the region observed by the FTS. The illuminating star, HD200775, lies 48 arcsec to the SE of the centre of the circle marked on the image.

3. Hydrogen as a thermometer in NGC7023

Measurement of the width of a resolved H_2 IR emission line is described below. This represents a relatively model-independent manner in which to measure a temperature in the interstellar gas. The only assumption is that that the linewidth in dominated by thermal motions rather than bulk gas motions. This is a good assumption for H_2 emission in reflection nebulae, in which excitation is via fluorescence (Black and Dalgarno 1976) and shocks are believed absent.

Using the CFHT with the high resolution Fourier transform spectrograph (FTS) to observe NGC7023, we have obtained a spectrum of the H_2 v=1-0 S(1) emission line at a resolution of $\sim 125,000$ (2.4 kms^{-1}) (Lemaire et al. 1999). These data are shown in Figure 2. The insert shows the emission in the v=1-0 S(1) line of H_2, where the 5" circle defines the position and extent of the region from which the spectrum of H_2 was derived. The exciting star, HD200775, lies 48" to the SE of the centre of this circle. The linewidth derived from data in Figure 2 is 3.4 ± 0.8 kms^{-1}. Analysis of observations of CO (Gerin et al. 1998) suggest that turbulent motions may be of the order of 1 kms^{-1}. The corresponding kinetic temperature of the gas is $450 \pm 60K$.

The independent determination of temperature in NGC7023 provides a useful constraint for the determination of the physical conditions in the photodissociation region (PDR) in this object, in particular the number density. The radiation field to which the gas is subjected is constrained by the known properties of the illuminating star to be 10^4 times that of the average interstellar field. There exist numerous theoretical models of steady-state PDRs, e.g. Draine and Bertoldi 1996; Abgrall et al. 1992: Sternberg and Dalgarno 1989, for which a major goal is to reproduce the observed H_2 surface brightness in v=1-0 S(1) and other lines. Using the model in Abgrall et al. 1992, which includes calculation of the thermal balance within the PDR, a number density of $\sim 10^6 cm^{-3}$ reproduces the observed brightness of the v=1-0 S(1) and S(2) lines, within a factor 1.5

Species observed	Velocity /kms^{-1}
HI	4 to 5
H$_2$	3.75± 0.25
CII	2.8± 0.1
CI	2.25± 0.3
^{12}CO(6-5)	2.6± 0.3
^{13}CO(3-2)	2.5± 0.3
^{13}CO(1-0)	1.9± 0.5

TABLE 1. Velocities expressed in the local standard of rest reference frame (v_{lsr}) for HI (Fuente et al. 1996), H$_2$ (Lemaire et al. 1999 and Figure 2), CII, CI and ^{12,13}CO (6-5, 3-2) (Gerin et al. 1998) and ^{13}CO (1-0) (Fuente et al. 1998).

to 2, with a kinetic temperature in the emitting zone of $\sim$ 550K (Lemaire et al. 1999). The success of this and other models is restricted by inaccuracies in the microphysics describing reactive and inelastic collisions involving H and H$_2$, the use of a one-dimensional slab geometry for the PDR and other considerations discussed below in Sect.4.

4. Hydrogen as an indicator of dynamics in NGC7023

The high resolution FTS data shown in Figure 2 provide, in addition to a linewidth, a measurement of the shift of the observed H$_2$ line from the laboratory rest value. After various corrections, described in Lemaire et al. 1999, the line may be shown to be (blue) shifted by -18.59±0.25kms^{-1}. This is equivalent to a velocity expressed in the local standard of rest reference frame (v_{lsr}) of 3.75±0.25kms^{-1}. This value may be compared with LSR velocities for other species obtained from radio data, for H, CII, CI and CO. These velocities are given in Table 1. Data have been derived from spectra taken as far as possible from the same location in the plane of the sky (see Lemaire et al. 1999).

The data in Table 1 show that there is a change of gas velocity as we penetrate further into the molecular cloud, passing from the zone of H$_2$ fluorescence into progressively more shielded zones. Lemaire et al. 1999 therefore demonstrated motion of the photodissociation front relative to the molecular cloud. This motion implies advective mixing of layers of the PDR, causing the temperature structure, for example, to be different from that estimated purely in terms of the atomic and molecular microphysics. Such flows are not included in present PDR models.

Our observations further suggest that PDRs may not be able to achieve a steady-state. The width of the PDR derived in the calculations mentioned in Sect. 3 is $\sim 10^{-3}$ pc, a figure which agrees closely with the observed width of filaments of H$_2$ emission in NGC7023 recorded in recent unpublished observations using PUEO on the CFHT (November 1999). If the dissociation front moves into the PDR with a velocity of 1 kms^{-1} (see Table 1), then a characteristic timescale is 10^3 years. However a PDR such as that in NGC7023, according to results in Goldshmidt and Sternberg 1995, would require several thousand years to reach a steady state and hence a steady state may not be achieved in this PDR. This influences H$_2$ surface brightness, showing enhancement over steady state models.

The high resolution FTS data for NGC7023 provide information on the gas dynamics only in the 5" diameter zone shown in the inset to Figure 2. In order to obtain an overview of the dynamical structure of the H$_2$ emitting region, spectro-imaging observations of NGC7023 have been performed using the BEAR instrument on the CFHT with seeing-

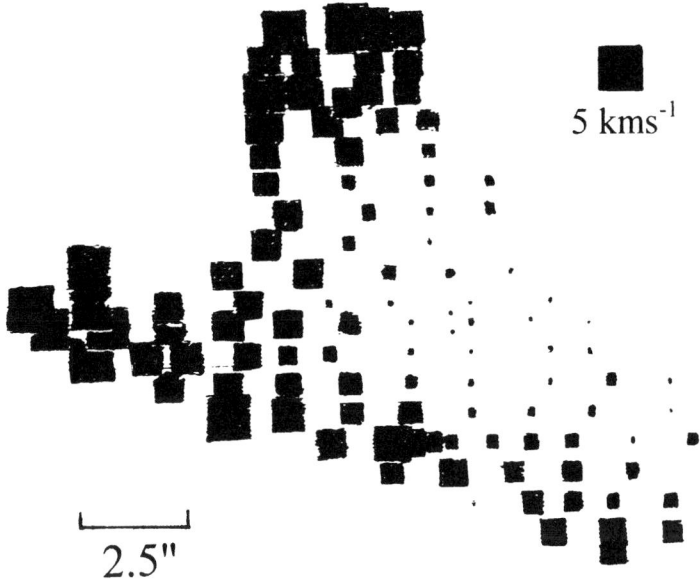

FIGURE 3. Relative velocities of different emitting regions in NGC7023, recorded using the BEAR instrument on the Canada-France-Hawaii Telescope. The size of the square at each position records the velocity of that region, expressed as the shift of the H_2 v=1-0 S(1) line relative to an arbitrary zero located right-centre of the diagram. The square to the top right gives the velocity scale. The centre of this figure is positioned 9"W and 3"S of the centre of the 5" circle shown in the inset to Figure 2.

limited spatial resolution (∼0.7") and spectral resolution of ∼ 23,000. These data allow us to trace the velocity field throughout the region of filamentary H_2 emission shown in the inset to Figure 2. A preliminary analysis of the SW part of this region has been performed. Results are shown in Figure 3. The size of a black square in Figure 3 is proportional to the velocity of that region relative to an arbitrarily chosen position assigned a zero velocity. Velocity differences may be exaggerated by a few tens of per cent due to the present lack of correction for edge effects. Nevertheless, it is clear that that the region is dynamically very active with gas velocities differing by several kms^{-1} over distances of less than 0.01pc (∼4"). Such velocity structure is not found in unperturbed molecular clouds. Its origin has not been studied in detail but evidently arises through the proximity of the massive star. A more general view of the gas motions will be available when BEAR data covering a considerably larger region are fully reduced.

5. Conclusions

Observational data show the remarkable power which H_2 emission has to reveal the properties of molecular clouds in the harsh environment created by the presence of a nearby massive star. The implications for subsequent star formation in such regions have yet to be considered. Indeed it is not clear whether disruption of the molecular material may enhance or reduce the local probability of star formation.

The next generation of observations will combine adaptive optics with spectro-imaging, probing smaller scales of the global velocity field, density and temperature. Combined with improved PDR models, these data will give further insight into this important stage in the cycle of stellar formation and evolution.

Thanks are extended to the Director and Staff of the Canada-France-Hawaii Telescope Corporation and of the European Southern Observatory for making possible the observations recorded in this report. The help of Drs. E. Roueff, J. Le Bourlot, O. Lai, J.-M.Deltorn and D.Simons is also acknowledged. Financial assistance from the Particle Physics and Astronomy Research Council (UK) and the CNRS is acknowledged by DF.

REFERENCES

ABGRALL H., LE BOURLOT J., PINEAU DES FORÊTS G., ROUEFF E., FLOWER D.R.& HECK L. 1992 *Astron.Astrophys.* **253**, 525.

ALLEN D.A. & BURTON M.G. 1993 *Nature* **363**, 54–56.

BLACK J.H. & DALGARNO A. 1976 *Ap.J.* **203**, 132.

DRAINE B.T. & BERTOLDI F. 1996 *Ap.J.* **468**, 269.

FIELD D., GERIN M., LEACH S., LEMAIRE J.L., PINEAU DES FORÊTS G., ROSTAS F., ROUAN D. & SIMONS D. 1994 *Astron.Astrophys.* **286**, 909–914.

FIELD D., LEMAIRE J.L., PINEAU DES FORÊTS, GERIN M., LEACH S., G., ROSTAS F. & ROUAN D. 1998 *Astron.Astrophys.* **333**, 280–286.

FUENTE A., MARTIN-PINTADO J., NERI R., ROGERS C. & MORIARTY-SCHIEVEN G. 1996 *Astron.Astrophys.* **310**, 286.

FUENTE A., MARTIN-PINTADO J., RODRIGUEZ-FRANCO A. & MORIARTY-SCHIEVEN G. 1998 *Astron.Astrophys.* **339**, 575.

GERIN M., PHILLIPS T.G., KEENE J., BETZ A.L. & BOREIKO R.T. 1998 *Ap.J.* **500**, 329.

GOLDSHMIDT O. & STERNBERG A. 1995 *Ap.J.* **439**, 256.

LEMAIRE J.L., FIELD D., GERIN M., LEACH S., PINEAU DES FORÊTS G., ROSTAS F. & ROUAN D. 1996 *Astron.Astrophys.* **308**, 895–907.

LEMAIRE J.L., FIELD D., MAILLARD J.P., PINEAU DES FORÊTS G., FALGARONE E., F.P.PIJPERS, GERIN M. & ROSTAS F. 1999 *Astron.Astrophys.* **349**, 253–258.

ROUAN D., FIELD D., LEMAIRE J.L., LAI O., PINEAU DES FORÊTS G., FALGARONE E. & J.-M. DELTORN 1997 *Mon.Not.R.Astr.Soc.* **284**, 395–400.

SCHILD H., MILLER S. & TENNYSON J. 1997 *Astron.Astrophys.* **318**, 608

STERNBERG A. & DALGARNO A. 1989 *Ap.J.* **338**, 197.

STOLOVY S.R., BURTON M.G., ERICKSON E.F. ET AL. 1998 *Ap.J.* **492**, L151–L155.

A pre-FUSE view of H_2

By M. JURA

Department of Physics and Astronomy, University of California, Los Angles, Los Angeles CA 90095-1562, USA

Observations of the interstellar medium within 1 kpc of the Sun with the *Copernicus* satellite showed a value of the gas to dust ratio that varies by less than a factor of two from its average. The fraction of hydrogen that is molecular is well described by a steady state model that balances formation on grains with photo-destruction. However, in contrast to the local interstellar medium, both in quasar absorption line systems and in circumstellar disks around young stars, there appears to be relatively little H_2. We particularly focus on estimating the amount of H_2 in circumstellar disks around main sequence stars – the environment where planets form.

1. Introduction

One of the main results achieved with the *Copernicus* satellite was the systematic measurement of interstellar H and H_2 within about 1 kpc of the Sun. It was found (Savage et al. 1977, Bohlin, Savage & Drake 1978) that the dust to gas ratio is uniform to within a factor of 2 of its average value with the mass in gas being approximately 100 times larger than the mass in dust. Also, the fraction of hydrogen that is molecular, $[2N(H_2)]/[N(H) + 2N(H_2)]$, is well described by a standard steady state model (Hollenbach, Werner & Salpeter 1971). In this standard model, the H_2 is formed on the surface of grains with a rate of about 3×10^{17} cm^3 s^{-1} (Jura 1975) and destroyed by the absorption of ultraviolet photons with a rate near 5×10^{-11} s^{-1} when the gas is optically thin (Jura 1974). The general applicability of this model to many more regions will be tested with FUSE.

Although the standard model is quite successful at explaining the abundance of H_2 in the local interstellar medium, there is no reason to think that it must apply everywhere. There might be substantial variations in the amount of H_2 for a variety of reasons including: (1) a different abundance of grains; (2) grains that are inefficient at forming H_2 because, for example, they are much warmer or colder than local interstellar grains; (3) a different composition or size distribution of the grains; (4) time-dependent effects; or (5) variations in the mean intensity of the ultraviolet radiation field.

As discussed elsewhere at this conference, quasar absorption line systems can have column densities as large as 10^{21} H cm^{-2} yet no detectable H_2. In this article, we focus on the disks around main sequence stars where planets might form or have already formed. The H_2 in these disks may play a key role in determining the gravitational and chemical evolution of the matter into either Jovian or terrestrial planets.

2. Dust and Gas Around Main Sequence Stars

One of the major discoveries with IRAS was that many nominally "normal" main sequence stars such as Vega or β Pic possess substantial amounts of orbiting circumstellar dust. While each circumstellar system is different, in this paper we adopt as a prototype the well studied system around HR 4796 (Jura 1991). This A0 main sequence star at a distance of 67 pc from the Sun is encircled by dust with a characteristic temperature of 110 K (Jura et al. 1998) which lies in a well defined ring with a radius of ~55 AU (Koerner et al. 1998, Jayawardhana et al. 1998, Schneider et al. 1999, Augereau et al. 1999)

In order to determine the gas to dust ratio, we estimate independently the mass of gas and the mass of dust. The dust mass, M_{du}, is most accurately measured from the submillimeter flux, F_ν. For observations at wavelength λ of a star at a distance D_* from the Sun where the grains have a temperature, T_{gr}, we may write in the Rayleigh-Jeans limit of the Planck function that:

$$M_{du} = (F_\nu \lambda^2 D_*^2)/(2 k T_{gr} \chi_\nu) \tag{2.1}$$

where χ_ν is the opacity (cm^2 g^{-1}) of the ensemble of dust particles. Greaves et al. (1999) reported a flux at 850 μm of 19.1 $\pm$ 3.4 mJy, and they assumed $\chi_\nu(850\ \mu m) = 1.7$ cm^2 g^{-1}. We therefore infer that $M_{dust} = 1.1 \times 10^{27}$ g. Greaves et al. (1999) derived a slightly larger value for M_{dust} since they employed a slightly smaller value of the grain temperature.

Although eventually it may be possible to measure directly the amount of H$_2$ from the emission in the pure rotational lines of this molecule as has been performed for some pre-main sequence stars (van Dishoeck et al. 1998, Thi et al. 1999), currently we have no direct measure of the mass of circumstellar H$_2$ around HR 4796. Instead, we use an indirect method of estimating the amount of H$_2$ by extrapolating from the CO radio emission for which, currently, there are only upper limits (Liseau 1999, Greaves et al. 1999, Zuckerman et al. 1995). The most sensitive bound has been achieved by Liseau (1999) who found for the J = 2-1 CO line that the integrated line intensity, $\int T_{mb}\, dv$, is ≤ 0.031 K km s^{-1}. From Jura et al. (1997), we write that the average column density of CO in the telescope beam, $\overline{N(CO)}$, is:

$$\overline{N(CO)} = (3k^2 T_{ex}) \int T_{mb}\, dv/(4\pi^3 \mu^2 h\nu^2) \tag{2.2}$$

where μ is the dipole moment of the CO molecule, ν is the frequency of the CO (2-1) transition and T_{ex} is the excitation temperature of the gas. The projected area, A_{beam}, in a beam assumed to be Gaussian with half power diameter, θ_{HPBW}, is:

$$A_{beam} = \pi \theta_{HPBW}^2 D_*^2/(16 \ln 2) \tag{2.3}$$

Since Liseau (1999) used the 15m SEST facility, $\theta_{HPBW} = 23"$. Therefore, we derive that $M_{CO} \leq (1.0 \times 10^{22}\, T_{ex})$ g. The excitation temperature of the gas is uncertain and difficult to calculate. In most models, $T_{ex} < T_{gr}$ (Chiang & Goldreich 1997). If we conservatively assume that the gas and dust temperature are equal so that $T_{gas} = 110$ K, then $M_{gas} \leq 1.1 \times 10^{24}$ g.

These results for the mass of CO and dust imply that $M_{CO}/M_{dust} \leq 1.0 \times 10^{-3}$. For most "normal" assumptions of the CO/H$_2$ ratio, this upper limit to M_{CO}/M_{dust} implies that the mass in H$_2$ is less than the mass in dust, and, therefore, the gas to dust ratio around HR 4796 is very much lower than found in the local interstellar medium.

One possible explanation for the absence of CO around HR 4796 is that there is a large amount of gas, but it is mostly atomic because of the ultraviolet radiation emitted by HR 4796 (Liseau 1999). At the moment, we have no independent observational check of this hypothesis. Weak optical absorption lines of Na I and Ca II with equivalent widths of 3 mÅ toward HR 4796 have been reported by Holweger et al. (1999). Currently, however, it is uncertain whether these lines arise from circumstellar gas or whether they have an interstellar origin.

Theoretical calculations also do not give a convincing argument about whether most of the hydrogen is atomic or molecular. Current models (Aikawa et al. 1997) for the gas phase chemistry in circumstellar disks do not pertain directly to the conditions that we infer for the material around HR 4796, and further calculations to model the circumstellar environment in this system are warranted. It is unlikely that the small amount of gas-phase CO is the result of freezing onto grains as has been suggested for pre-main sequence stars (Willacy et al. 1998), since the grain temperature is 110 K while CO freezes onto grains if the temperature is less than about 20 K (Leger 1983).

If the gas to dust ratio has a "normal" value, then the hydrogen density around HR 4796 is likely to be much higher than in diffuse interstellar clouds where a $\sim 10^2$ cm^{-3}. In particular, a very simple picture of the dust distribution is that of a cylinder with radius 55 AU and height of 0.1 of its radius. In this case, the dust density is 6×10^{-18} g cm^{-3}, which, for a "normal" gas to dust ratio, implies a density of hydrogen nuclei of 4×10^8 cm^{-3}. For such a high density in a diffuse interstellar cloud, we would expect most of the hydrogen to be molecular even if there are variations of factors of 100 in the formation rate on grains or the ultraviolet destruction rate. Although uncertain, we suggest that the low value of CO around HR 4796 implies a low gas to dust ratio. Other systems, such as the disk around the pre-main sequence star GG Tau also exhibit unusually low gas to dust ratios when compared to the "standard" interstellar value (Dutrey et al. 1994).

3. Implications

It is not too surprising that the gas to dust ratio around HR 4796 appears to be much lower than found in the local interstellar medium. The grains around HR 4796 are "typically" near 10 μm in diameter (Jura et al. 1997, Koerner et al. 1998) – much larger than interstellar grains which have a characteristic size near 0.1 μm (Kim, Martin & Hendry 1994). Therefore, it is possible that the grains and the gas are not dynamically coupled to each other, and that they have different histories.

One process which might remove gas but not large particles from the circumstellar environment of HR 4796 is a wind from its pre-main sequence companion, the M-type star HR 4796B. This secondary star is an X-ray source (Jura et al. 1998) which suggests that it has an associated wind. It is possible that the wind from the companion has efficiently swept away much of the gas around HR 4796, but the large dust particles remain in orbit because they are not dynamically coupled to the gas.

Finally, 4796 – like other main sequence stars with large amounts of circumstellar dust – is quite young; its inferred age is less than $\sim 10^7$ yr (Jura et al. 1998, Stauffer et al. 1995). Since we infer $\sim 10^{27}$ g of dust which may have an ice composition, it is plausible that snowballs, comets or even planets may eventually form.

4. Conclusions

In the local interstellar medium, the amount of H_2 seems to be well explained by a standard steady state model for formation on grains and destruction by ambient interstellar ultraviolet radiation. However, in other environments – such as the disks around main sequence stars – it may be that the gas to dust ratio is much lower than in the local interstellar medium. If this is the case, these disks around main sequence stars may be regions where the gas and dust have separated from each other with important consequences for the subsequent evolution of the matter.

REFERENCES

AIKAWA, Y., UMEBAYASHI, T., NAKANO, T. & MIYAMA, S. M. 1997 *Ap. J.* **486**, L51-L54.
AUGEREAU, J. C., LAGRANGE, A. M., MOUILLET, D., PAPALOIZOU, J. C. B. & GROROD, P. A. 1999 *Astr. & Ap.* **348**, 557-569.

BOHLIN, R. C., SAVAGE, B. D. & DRAKE, J. F. 1978 *Ap. J.* **224**, 132-142.

CHIANG, E. I. & GOLDREICH, P. 1997 *Ap. J.* **490**, 368-376.

DUTREY, A., GUILLOTEAU, S. & SIMON, M. 1994 *Astr. & Sp.* **286**, 149-159.

GREAVES, J. S., MANNINGS, V. & HOLLAND, W. S. 1999 *Icarus*, in press

HOLLENBACH, D., WERNER, M. W. & SALPETER, E. E. 1971 *Ap. J.* **163**, 165-180.

HOLWEGER, H., HEMPEL, M. & KAMP, I. 1999 *Astr. & Ap.* **350**, 603-611.

JAYAWARDHANA, R., FISHER, S., HARTMANN, L., TELESCO, C., PINA, R. & FAZIO, G. 1998, *Ap. J.* **503**, L79-L82.

JURA, M. 1974 *Ap. J.* **191**, 375-379.

JURA, M. 1975 *Ap. J.* **197**, 575-580.

JURA, M. 1991 *Ap. J.* **383**, L79-L80.

JURA, M. GHEZ, A. M., WHITE, R. J., MCCARTHY, D. W., SMITH, R. C. & MARTIN, P. G. 1995 *Ap. J.* **445**, 451-456.

JURA, M., KAHANE, C., FISCHER, D. & GRADY, C. 1997 *Ap. J.* **485**, 341-349.

JURA, M., MALKAN, M., WHITE, R., TELESCO, C., PINA, R. & FISHER, R. S. 1998 *Ap. J.* **505**, 897-902.

KIM, S.-H., MARTIN, P. G. & HENDRY, P. D. 1996 *Ap. J.* **422**, 164-175.

KOERNER, D. W., RESSLER, M. E., WERNER, M. W. & BACKMAN, D. E. 1998 *Ap. J.* **503**, L83-L86.

LEGER, A. 1983 *Astr.&Ap.* 123, 271-278.

LISEAU, R. 1983 *Astr. & Ap.* **348**, 133-138.

SAVAGE, B. D., DRAKE, J. F., BUDICH, W. & BOHLIN, R. C. 1977 *Ap. J.* **216**, 291-307.

SCHNEIDER, G., SMITH, B. A., BECKLIN, E. E., KOERNER, D. W., MEIER, R., HINES, D. C., LOWRANCE, P. J., TERRILE, R. J., THOMPSON, R. I. & RIEKE, M. 1999, *Ap. J.* **513**, L127-L130.

STAUFFER, J. R., HARTMANN, L. W. & BARRADO Y NAVASCUES, D. 1995 *Ap. J.*, **454**, 910-916.

THI, W.-G., VAN DISHOECK, E. F., BLAKE, G. A., VAN DADELHOFF, G.-J., & HOGERHEIJDE, M. R. *Ap. J.* **521**, L63-L66.

VAN DISHOECK, E. F., THI, W. F., BLAKE, G. A., MANNINGS, V., SARGENT, A. I., KOERNER, D. & MUNDY, L. G. 1998 *Ap. Space Sci.* **255**, 77-82.

WILLACY, K., KLAHR, H. H., MILLAR, T. J. & HENNING, TH. *Astr. & Ap. 338*, 995-1005.

ZUCKERMAN, B., FORVEILLE, T. & KASTNER, J. H. 1995 *Nature* **373**, 494-496.

H_2 absorption line measurements with ORFEUS

By P. RICHTER[1,2], H. BLUHM[1], O. MARGGRAF[1] AND K.S. DE BOER[1]

[1]Sternwarte der Universität Bonn, Auf dem Hügel 71, 53121 Bonn, Germany
[2]Washburn Observatory, University of Wisconsin-Madison, 475 N. Charter Street, Madison, WI 53706, U.S.A

We review recent H_2 absorption line measurements in the diffuse interstellar medium, using FUV spectra from the *Orbiting and Retrievable Far and Extreme Ultraviolet Spectrometer* (ORFEUS). We investigate molecular hydrogen gas along lines of sight toward 5 stars in the Magellanic Clouds and toward 3 stars within the Milky Way. Molecular fractions in gas within the Magellanic Clouds are significantly lower than typically found in gas in the Milky Way, most likely caused by the lower dust content. The finding of H_2 in a Galactic high-velocity cloud led us to speculate that the high-velocity gas in front of the Magellanic Clouds is part of the Galactic fountain. Sight lines toward the Galactic stars show well defined absorption by molecular hydrogen, deuterium and metals, allowing the study of physical and chemical conditions in the local interstellar gas in great detail.

1. Introduction

Molecular hydrogen is by far the most abundant molecule in the interstellar medium. Its measurement, however, is difficult: H_2 has no permanent dipole moment and no radio emission is seen from H_2, in striking contrast to the second most abundant molecule in the ISM, carbon monoxide (CO). For the study of the diffuse interstellar medium the FUV absorption spectroscopy is the only method to obtain information about the molecular hydrogen content, but satellites are required for this method, since the earth's atmosphere is opaque for radiation in the FUV domain. Considerable effort has been invested in the seventies to investigate H_2 absorption lines with the *Copernicus* satellite toward nearby bright stars. Later satellite missions, such as the IUE and HST, were able to observe more distant background sources, but they did not cover the wavelength range below 1150 Å where the H_2 absorption lines from the Lyman and Werner bands are seen. The ORFEUS telescope (Barnstedt et al. 1999), launched for its second mission in 1996, was the first instrument able to measure H_2 absorption lines in the FUV outside the Milky Way and was used to investigate molecular hydrogen absorption in gas of the Magellanic Clouds (de Boer et al. 1998, Richter et al. 1998,1999a), in high-velocity gas in the Galactic halo (Richter et al. 1999b) and in local Galactic gas (Marggraf et al. 2000). Some of these ORFEUS spectra were also used to investigate metal abundances in the local ISM and in gas of the Galactic halo. For the sight line toward BD +39 3226 a D/H ratio for local Galactic gas was determined (Bluhm et al. 1999).

In this article we review the most important results from ORFEUS II data derived at Bonn University and discuss their significance for our understanding of the properties of the ISM.

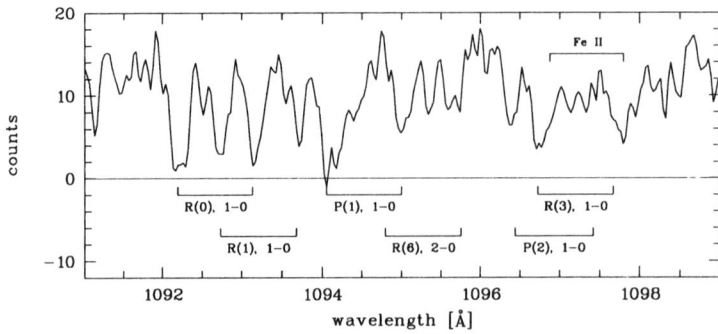

FIGURE 1. A portion of the ORFEUS II spectrum of the LMC star LH 10:3120 is shown. H_2 absorption is seen in local Galactic gas (0 km s^{-1}) and in LMC gas (+270 km s^{-1}). H_2 transitions are labeled below the spectrum

2. The ORFEUS telescope

ORFEUS is equipped with two alternatively operating spectrometers, the echelle spectrometer (Krämer et al. 1990) and the Berkeley spectrometer (Hurwitz & Bowyer 1996). The spectroscopic data presented here were obtained with the Heidelberg/Tübingen echelle spectrometer during the second ORFEUS mission in Nov./Dec. 1996. The echelle has a resolution of somewhat better than $\lambda/\Delta\lambda = 10^4$ (Barnstedt et al. 1999), working in the spectral range between 912 and 1410 Å. A detailed description of the instrument is given in Barnstedt et al. (1999).

3. H_2 measurements in the Magellanic Clouds

A set of 5 stars in the Magellanic Clouds has been observed during the ORFEUS II mission for the investigation of molecular hydrogen in the diffuse ISM. Only one sight line out of four toward the Large Magellanic Cloud (LMC) shows clear H_2 absorption at LMC velocities, the one to LH 10:3120 in the north-western part of the LMC. In the spectrum of LH 10:3120 (Fig. 1) H_2 absorption is seen in local Galactic gas at a radial velocity of 0 km s^{-1} and in LMC gas near +270 km s^{-1} (de Boer et al. 1998). H_2 column densities have been determined by using a curve-of-growth technique. For the LMC gas toward LH 10:3120 we derive a value of $N(H_2)_{LMC} = 6.6 \times 10^{18}$ cm^{-2}. No CO absorption is found in the strongest of the CO absorption bands, the C-X band near 1088 Å. We find an upper limit of 3.3×10^{13} cm^{-2} for the CO column density in the LMC gas, from which we obtain a $N(H_2)/N(CO)$ ratio of $\geq 2.0 \times 10^5$.

The spectrum of the star HD 5980, the only ORFEUS target in the Small Magellanic Cloud (SMC), shows H_2 absorption at SMC velocities near +150 km s^{-1} in two components (Richter et al. 1998). For this gas we derive a total column density of $N(H_2)_{SMC} = 5.1 \times 10^{16}$ cm^{-2}. For all other lines of sight toward the Magellanic Clouds, upper limits for the H_2 column densities were determined by inspecting the strongest of the H_2 transitions in the region near 1000 Å (Richter 2000).

Values of $N(H\,I)$, $N(H_2)$ and $E(B-V)$ for all 5 lines of sight measured with ORFEUS are used to investigate the diffuse molecular ISM of the Magellanic Clouds (MC). In order to extend the dataset, we include recent results from low dispersion spectra with HUT from Gunderson et al. (1998) for two lines of sight toward Sk $-66\,19$ and Sk $-69\,270$

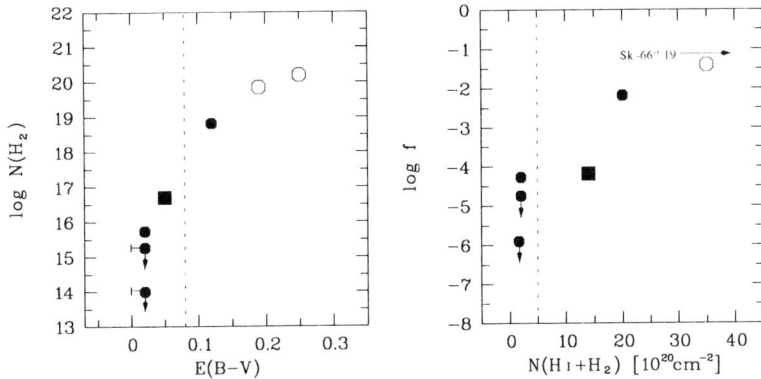

FIGURE 2. Correlations between atomic hydrogen, molecular hydrogen and colour excess for Magellanic Cloud gas along 7 lines of sight. LMC stars measured with ORFEUS are labeled by filled dots, the one SMC target is given by a filled square. Data from Gunderson et al. (1998) for two additional lines of sight to the LMC have been included, here given as open circles. An interpretation for the shown correlations is given in Section 3

in the LMC. They estimate column densities of molecular hydrogen in the LMC gas by fitting H_2 line profiles. Values for $E(B-V)_{MC}$ have been adopted from the literature (see Richter 2000 for details). H I column densities have been taken either from previous publications or have been derived by fitting multi-component Voigt profiles to the damped Lyα absorption line at 1215 Å. Fig. 2 presents correlations between $N(H_2)$, $E(B-V)$, $f = 2N(H_2)/[N(H\,\textsc{i})+2N(H_2)]$, and $N(H\,\textsc{i}+H_2) = N(H\,\textsc{i}) + 2N(H_2)$. The left panel shows $\log N(H_2)$ plotted versus $E(B-V)$. In principle, we find the typical relation known from the *Copernicus* H_2 survey from Savage et al. (1977) for Galactic gas. They found that $\log N(H_2)$ undergoes a transition from low to to high values at $E(B-V) \approx 0.08$ (Fig. 2 left panel, dashed line). For the Galaxy, a similar distribution is seen for f, the fraction of hydrogen in molecular form, when plotted against the total hydrogen column density $N(H\,\textsc{i}+H_2)$. The *Copernicus* sample shows that the transition from low ($f \leq 10^{-2}$) to high ($f > 10^{-2}$) molecular fractions in the local Galactic gas is found at a total hydrogen column density ('transition column density' $N_T(H\,\textsc{i})$) near 5.0×10^{20} cm^{-2} (right panel, dashed line). Inspecting the right panel of Fig. 2, we find for the Magellanic Clouds gas along two of 7 lines of sight high total hydrogen column densities ($\geq 10^{21}$ cm^{-2}) but low molecular hydrogen fractions ($f \leq 10^{-2}$). The data points of these two lines of sight toward LH 10:3120 and HD 5980 indicate that the transition column density N_T in the Magellanic Clouds is significantly larger than in the Milky Way.

Obviously, the fraction of hydrogen in molecular form in the Magellanic Clouds is influenced by the lower dust content. The data suggests that the diffuse ISM in the Magellanic Clouds contains much less molecular material than the diffuse Milky Way gas. With new data from the FUSE satellite, launched in June 1999, one will be able to study the molecular fractions in the Magellanic Clouds in far more detail.

4. H_2 absorption in the Galactic halo

Toward one star in the LMC (HD 269546), weak H_2 absorption has been detected in gas of a high-velocity cloud (HVC) in the Galactic halo (Richter et al. 1999b). This measurement shows for the first time that molecular gas principally exists in some regions

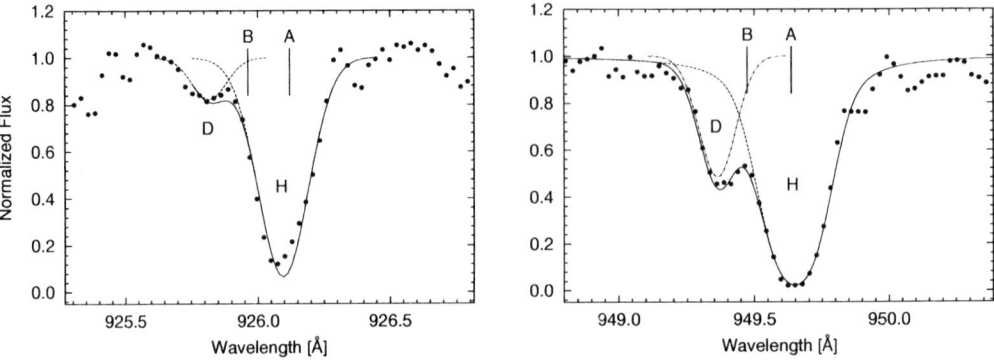

FIGURE 3. For two of the Lyman lines a 500 km s^{-1} wide section of the spectrum is displayed. *Left:* Deuterium and hydrogen Lyman η line in the ORFEUS spectrum (data points) with a two-com ponent Gauss profile. *Right:* Deuterium and hydrogen Lyman δ line (data points) with a two-component fit (Voigt profile convolved with instrumental Gauss). In both plots the positions of the velocity components A and B found in the metal lines are marked f or the H I lines. Component B has only neglegible effect on the absorption profile

of the Milky Way halo. Since the formation of H_2 requires the presence of dust, the detection of H_2 in the HVC in front of the LMC indicates that this HVC contains dust and heavy elements and cannot be part of infalling low-metallicity gas accreted from intergalactic space, as might be the case for other HVC complexes (Wakker et al. 1999). Strong Fe II absorption (Richter et al. 1999b) also points to a Galactic origin of this HVC, but a careful ionization correction is required for the determination of accurate metal abundances. We speculate that this HVC is part of the Galactic fountain (Shapiro & Field 1976).

Further H_2 absorption line measurements are required to further investigate the molecular content in high-velocity clouds. The strong correlation between H_2 and dust might help to discriminate between the different models for the formation of high-velocity clouds.

5. Interstellar deuterium and H_2 toward BD +39 3226

While the primordial deuterium abundance is tightly connected with the baryonic density in the Universe, the interstellar abundance depends also on the evolution of the Milky Way. Therefore, a large number of D/H measurements in different environments is desirable.

A FUV echelle spectrum obtained during the ORFEUS II mission is used to derive H I, D I, H_2, and metal column densities along the line of sight toward BD +39 3226 (Bluhm et al. 1999). This sdO star is located at $l = 65.00$, $b = +28.77$, in a distance of about 270 pc. It has a pure helium line spectrum in the optical and some additional lines of highly ionized metals in the UV, which are well separated from the interstellar lines because of the stellar radial velocity of $\simeq -279$ km s^{-1}. Interstellar absorption by metals is found at heliocentric velocities of $\simeq -24$ km s^{-1} (component A) and, very weak, at $\simeq -75$ km s^{-1} (component B). The latter component is neither resolved in the H I lines nor detectable in the H_2 lines. A theoretical fit to the Ly α line, taking into account the stellar He II line ≈ 1.57 Å bluewards, gives a H I column density of $1.12^{+0.29}_{-0.23} \cdot 10^{20}$ cm^{-2}. This value is consistent with a curve-of-growth analysis of the D and H Lyman series,

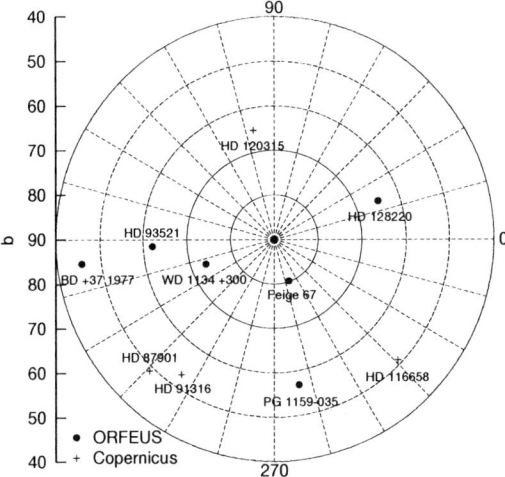

FIGURE 4. Positions of ORFEUS and Copernicus targets in the northern polar region of the Galaxy. Not all spectra of ORFEUS targets can be used for the analysis of interstellar H_2 due to strong stellar absorption lines or low flux in the spectral range of interest

including Ly δ, ϵ, ζ, and κ, which leads to $N(\text{H\,\textsc{i}}) = 1.6^{+0.9}_{-0.6} \cdot 10^{20}$ cm^{-2} and $N(\text{D\,\textsc{i}}) = 1.45^{+0.50}_{-0.38} \cdot 10^{15}$ cm^{-2}. Averaging the H\,\textsc{i} column densities we derive $D/H = 1.2^{+0.5}_{-0.4} \cdot 10^{-5}$. Within its uncertainties our result is in agreement with other measurements of D/H in the local ISM (see e.g. McCullough 1992 and Linsky 1998).

We also investigated metal lines in ORFEUS and IUE spectra. Their column densities indicate only moderate depletion (≤ 1.7 dex), in agreement with the small $E(B-V) = 0.05$ and the low abundance of H_2 ($N(H_2) = 4.8^{+2.0}_{-1.6} \cdot 10^{15}$ cm^{-2}). H_2 is examined for absorption by rotational exitation states up to $J = 7$. The column densities of these exitation states reveal a H_2 ortho-to-para ratio of 2.5, a Boltzmann excitation temperature of 130 K, and UV-pumping with an equivalent temperature of < 1800 K. At a detection limit of $N \approx 10^{13}$cm^{-2}, no absorption by HD is found in the spectrum.

6. H_2 toward the northern Galactic pole

Little is known on the distribution of molecular hydrogen in the local interstellar medium (LISM), especially in directions away from the Galactic disk, where only few luminous targets have been observed by *Copernicus*. The ORFEUS spectrometer now allowed high resolution measurements of targets in these directions in the sky.

We examine the distribution of H_2 in the LISM toward the northern Galactic pole at latitudes $b > 45°$. The high resolution UV echelle spectra of the nearby sdO stars BD +37 1977 ($l = 187.1$, $b = +45.6$) and HD 128220 ($l = 19.9$, $b = +64.9$) observed during the ORFEUS II mission, are analyzed in the following. A detailed analysis for molecular hydrogen toward HD 93521 can be found by Gringel et al. (2000) and Barnstedt et al. (2000). Distances to BD +37 1977 and HD 128220 are calculated to be 1030 pc and 440 pc, respectively, with an error of about 30%.

We determine column densities for H_2 rotational excitation states up to $J = 7$ (Marggraf et al. 2000). For BD +37 1977 we find a low total H_2 column density of $(2.1^{+0.4}_{-0.2}) \times 10^{15}$ cm^{-2}. Probably, the molecular gas is excited by UV pumping to an equivalent excitation temperature of $T = 395 \pm 11$ K. Towards HD 128220 the total H_2 column density is $(6.5^{+3.3}_{-4.1}) \times 10^{17}$ cm^{-2}, in at least two unresolved absorption components. The Boltzmann excitation temperature for the lowermost states, $J = 0$ and 1, has a value of $T = 77 \pm 10$ K. The equivalent temperature of stages excited by UV pumping ($J = 4$ to 7) is $T < 1500$ K. They are observed in only one of the components. We further obtain column densities for H I from Lyman profile fitting and for S II and Si II. Only 10% or less of the H I seen in absorption is found in H I 21 cm spectra from the Leiden-Dwingeloo survey (Hartmann & Burton 1998). This indicates either gas concentrated in a few (or only one) dense, possibly circumstellar clumps, or extended flocculent molecular gas in the northern polar disk region of our Galaxy. Assuming pressure equilibrium, a layer of flocculent molecular gas at a height $z \simeq 80$ pc could be possible. To further explore this issue an analysis of further lines of sight is necessary.

We thank the Tübingen and Heidelberg ORFEUS team for their great support. PR was supported by a grant from the DARA (now DLR) under code 50 QV 9701 3. HB is supported by the GK *The Magellanic Clouds and other dwarf galaxies*. OM is supported by grant Bo 779/24 of the DFG.

REFERENCES

BARNSTEDT J., KAPPELMANN N., APPENZELLER I., ET AL. 1999 *A&AS* **134**, 561–567.

BARNSTEDT J., GRINGEL W., KAPPELMANN N., GREWING M. 2000 *A&A* in press.

BLUHM H., MARGGRAF O., DE BOER K. S., ET AL. 1999 *A&A* **352**, 287–296.

DE BOER K.S., RICHTER P., BOMANS D.J., HEITHAUSEN A., KOORNNEEF J. 1998 *A&A* **338**, L5–L8.

GRINGEL W., BARNSTEDT J., DE BOER K. S., ET AL. 2000 *A&A* submitted.

GUNDERSON K.S., CLAYTON G.C., GREEN J.C. 1998 *PASP* **110**, 60–67.

HARTMANN, D. & BURTON, W. B. 1997 *Atlas of Galactic Neutral Hydrogen* Cambridge Univ. Press.

HURWITZ M., BOWYER S. 1996 *Astrophysics in the extreme ultraviolet* eds. S. Bowyer & R.F. Malina, Kluwer; p. 60.

KRÄMER G., BARNSTEDT J., EBERHARD N., ET AL. 1990 *Observatories in Earth Orbit and Beyond* eds. Y. Kondo, Kluwer, Ap.Sp.Sci.Lib., Vol. **166**, 177.

LINSKY J. L. 1998 *Space Sci. Rev.* **84**, 285–296.

MARGGRAF O., ET AL. 2000 *in preparation.*

MCCULLOUGH P. R. 1992 *ApJ* **390**, 213–218.

RICHTER P. 2000 *A&A* submitted.

RICHTER P., DE BOER K.S., BOMANS D.J., ET AL. 1999a *A&A* **351**, 323–329.

RICHTER P., DE BOER K.S., WIDMANN H., ET AL. 1999b *Nature* **402**, 386–387.

RICHTER P., WIDMANN H., DE BOER K.S., ET AL. 1998 *A&A* **338**, L9–L12.

SAVAGE B.D., BOHLIN R.C., DRAKE J.F., BUDICH W. 1977 *ApJ* **216**, 291–307.

SHAPIRO P.R., FIELD G.B. 1976 *ApJ* **205**, 762–765.

WAKKER B., HOWK J.C., SAVAGE B.D., ET AL. 1999 *Nature* **402**, 388–390.

Ultraviolet Observations of Molecular Hydrogen in Interstellar Space

By Theodore P. SNOW

Center for Astrophysics and Space Astronomy, Campus Box 389, University of Colorado,
Boulder CO 80309. (e-mail: tsnow@casa.colorado.edu)

Molecular hydrogen is the most abundant constituent in interstellar space, but is difficult to observe in its most common form. Because H_2 has no dipole moment, it does not have allowed rotational or vibrational transitions and therefore in its cold, unexcited state it has no radio or infrared spectral features. Therefore the most sensitive method for detecting cold H_2 is through its allowed electronic bands, which lie in the far-ultraviolet portion of the spectrum. Previous UV instruments have provided some information on far-UV spectra of cold H_2, but all were limited in either throughput or spectral resolving power, or both. The current *FUSE* mission has a combination of high throughput and moderately high spectral resolution, and is providing information on molecular hydrogen in interstellar regimes that have been previously unexplored. This review summarizes previous UV observations of H_2 and then gives an overview of early *FUSE* results, with an emphasis on H_2 in translucent clouds.

1. Introduction

The mass in the galactic interstellar medium is dominated by molecular hydrogen, which begins to outweigh atomic hydrogen in diffuse clouds and is expected to become completely dominant for translucent and dense clouds; i.e. clouds having visual extinctions greater than about $A_V = 2$ magnitudes. Even in diffuse clouds, H_2 is important, playing a crucial role in cloud chemical and physical processes.

Despite the importance of understanding H_2 and its distribution, physics, and chemistry, relatively little direct information is available because of the obtuse spectroscopic properties of the molecule. It has no dipole moment, hence no allowed rotational or vibrational transitions, so that cold H_2 can be observed in absorption only through its electronic bands in the UV or over very long path-lengths through the quadrupole transition near 2.2 μm (Lacy *et al.* 1984). Excited H_2 is another story, and has been observed via near-infrared quadrupole emission (first detected by Gautier *et al.* 1976); via UV fluorescence emission (Witt *et al.* 1989); and more recently, by rotational emission in the mid-IR, by the *ISO* mission (Falgarone *et al.*, this volume).

This review provides a summary of earlier far-UV observations of H_2, and then an overview of recent observations with the *FUSE* instrument. This overview will be followed by a description of the H_2 observing programs that are under way using *FUSE*, with emphasis on the translucent cloud program.

2. Previous UV Observations of H_2

The first far-UV detection of interstellar H_2 absorption bands was accomplished by Carruthers (1970), using a simple rocket-borne spectrometer to detect several H_2 bands, including lines from a range of rotationally excited levels, in the line of sight toward the star ξ Persei. Following this first detection, a number of other far-UV experiments also found molecular hydrogen, as it became apparent that H_2 starts to become important for clouds having A_V values of about 0.1 and higher.

Among the earlier instruments capable of observing far-UV H_2 absorption, the *Copernicus* satellite provided by far the most comprehensive information (e.g. Spitzer *et al.* 1973; Spitzer, Cochran, & Hirshfeld 1974; Spitzer & Jenkins 1975; Shull & Beckwith 1982), but was generally limited to stars having extinctions of $A_V \leq 1$ mag. The *Copernicus* results showed that that molecular hydrogen abundance jumps dramatically near a column density of $N(H_2) \approx 10^{19}$ cm^{-2}, where the molecule becomes self-shielding; that the column density of H_2 and of total hydrogen (including H I) correlates well with extinction (Bohlin, Savage, & Drake 1978); that the high-lying rotational states ($J \geq 2$) are excited by some source other than collisions in the gas; and that in diffuse clouds with significant reddening, the ratio of $J = 1$ to $J = 0$ populations (i.e., the ortho- to para- hydrogen ratio) is in thermal equilibrium and thus serves as a kinetic temperature indicator while obscuring any prior history of this ratio as produced during H_2 formation. The excitation of high J-levels was soon attributed to radiative pumping, as hydrogen molecules cascade back down through a range of excited vibrational and rotational levels following the absorption of UV photons (Spitzer & Zweibel 1974; Jura 1975a,b).

Vibrationally excited H_2 was not found by *Copernicus*, but was later detected by the *Hubble Space Telescope* (Federman *et al.* 1995). Many lines arising from excited vibrational levels appear at the longer UV wavelengths where the *HST* is sensitive, but so far no one has pursued the search beyond the original detection (but Snow, Black, & McCray have a pending *HST* Cycle 9 program to do so). The analysis of vibrational excitation can yield important information on excitation conditions in cold gas, particularly the radiation field, in cold gas.

Subsequent to the end of the *Copernicus* mission in 1980, there was a prolonged lapse in UV observations of H_2 absorption, which was finally broken by a series of missions (or at least the papers therefrom) in the late 1980s and 1990s. Among these were the two *Voyager* probes, whose far-UV broad-band spectrometers were capable of at least detecting the presence of H_2 if not providing sufficient spectral resolution for a detailed analysis (Longo *et al.* 1989; Snow, Allen, & Polidan 1990); the *IMAPS* Shuttle-borne instrument, which obtained high-resolution spectra of H_2 in a few lines of sight (Jenkins & Peimbert 1997 and references cited therein); the *Hopkins Ultraviolet Telescope* or *HUT*, which had only moderate spectral resolution but detected H_2 in several lines of sight (Bowers *et al.* 1995; Blair *et al.* 1996); and the *ORFEUS* spectrometer, which obtained quite good spectra of H_2 in several lines of sight, including stars in the Magellanic Clouds (e.g., Richter *et al.* 1998 and references cited therein; Barnstedt *et al.* 2000).

For completeness, it is worth noting here that there have been a few detections of UV H_2 fluorescence emission from excited regions near hot stars (e.g. Witt *et al.* 1989) and from the diffuse interstellar gas (Martin, Hurwitz, & Bowyer 1990), but these detections have more to tell us about local excitation conditions than about the distribution or state of molecular hydrogen in the general cold gas.

Also in the interest of completeness, it should be pointed out that H_2 has been detected at infrared wavelengths, primarily through quadrupole emission near 2.2 μm (e.g. Gautier *et al.* 1976; Shull & Beckwith 1982; and many papers cited in the review of photodissociation regions by Hollenbach & Tielens 1997). But quadrupole *absorption* has also been detected, albeit only in lines of sight with huge extinctions (Lacy *et al.* 1984; Kulesa, this volume). Such detections are very useful, for example in extending the known CO/H_2 correlation to large column densities, something that can only be nibbled at using UV observations, which are constrained to relatively low column densities.

3. A Brief Overview of the *FUSE* Mission

For a detailed reviews of the *FUSE* spacecraft and mission, see the recent papers by Moos *et al.* 2000 and Sahnow *et al.* 2000, and references cited therein, for even more detailed descriptions.

In essence, *FUSE* consists of four parallel telescope and grating systems that share two large-format detectors. The instrument provides complete spectral coverage from the Lyman limit at 912 /AA/ to about 1175 /AA, with overlap regions selected to match the spectral bands expected to have the greatest scientific value. Optical coatings are either LiF (better for wavelengths above 1000 Å) or SiC (for wavelengths below 1000 Å), and the detectors are open-faced micro-channel arrays using delay-line technology for locating photon events. The nominal spectral resolving power ($R = \lambda/\Delta\lambda$) is $R = 30,000$, and the effective area is as high as 70-80 cm^2 in the region of greatest spectral overlap and scientific interest, around 1000-1050 Å.

In practice a number of constraints have so far prevented *FUSE* from achieving all of its performance goals, although many have been met and at this writing fine-tuning is taking place that is expected to continue improving things. In particular, the alignment of the four spectral channels has proven difficult, and the difficulty seems to be exacerbated by flexure in the optical assembly that is induced by thermal effects as the spacecraft transits between day and night. As a conseqence, refining the focus of the optical assembly was delayed for some time after launch, because it was deemed important to get the alignment optimized first. Without optimum focus, the resolving power is not yet at its planned value, and is found to be in the range $R = 12,000\text{-}20,000$. Again, as of this writing the focus is undergoing adjustment and testing, and there are signs of improvement in the spectral resolving power. The throughput of the instrument appears to be exactly as expected, though, and the scattered light is minimal.

Due to the constraints, however, the data to be described in this paper have a spectral resolving power of approximately $R = 14,000$, and include only spectra longward of 1000 /AA, because the observatons were made before all four channels were co-aligned. Nevertheless, these data are extremely useful, in some ways close to optimal, for the analysis of the far-UV H$_2$ bands.

4. Early *FUSE* Results on H$_2$

Among the high-priority programs being carried out by the *FUSE* Principal Investigator team (see http:// violet.pha.jhu.edu/) is the study of molecular hydrogen in various environments. There are two PI team "medium" programs devoted to H$_2$: one on H$_2$ in diffuse cloud lines of sight (headed by J. M. Shull of the University of Colorado); and one on H$_2$ in translucent clouds (headed by the author, also of Colorado). The diffuse H$_2$ program is using spectra obtained for a variety of other PI-team programs (e.g., the surveys of galactic halo abundances, of the hot interstellar medium, and of hot star winds), while the translucent-cloud program is confined to stars specifically observed for this purpose (because no one observes highly-reddened stars in the far-UV unless it is absolutely necessary to the scientific goals of the program!). It is interesting and predictable that virtually every *FUSE* spectrum is contaminated by H$_2$ absorption bands – so part of our contribution at Colorado has been to fit and remove this contamination, as for studies of hot star wind profiles (e.g. Fullerton *et al.* 2000).

The first *FUSE* paper on H$_2$ in diffuse lines of sight (Shull *et al.* 2000) notes that nearly every stellar and extragalactic source observed shows the presence of molecular hydrogen absorption in its spectrum. This initial survey also shows that the kinetic temperature

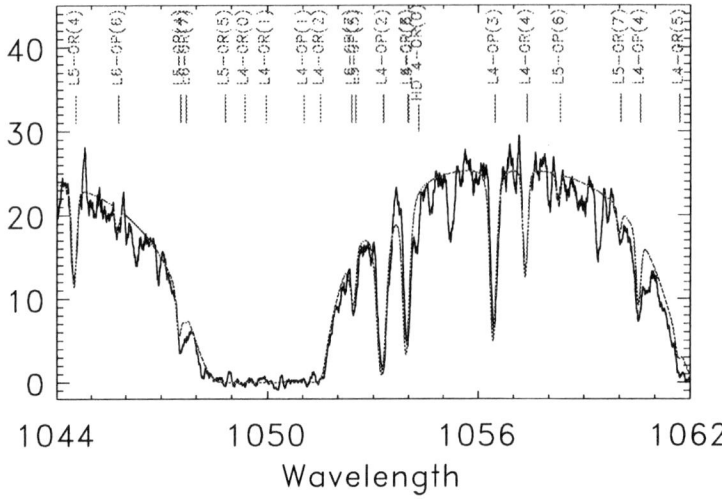

FIGURE 1. A portion of the *FUSE* Far-UV Spectrum of HD 73882. Several very prominent H_2 bands are seen in this spectrum, which lies in the region where the Lyman bands of molecular hydrogen are dominant.

inferred from the $J = 1/J = 0$ ratio is higher for the galactic halo than in the disk, and that the excitation conditions in lines of sight toward the Large and Small Magellanic Clouds are similar to those in our galactic halo. Perhaps the most significant finding is the ubiquity of H_2 in all kinds of environments, indicating that the basic material for star formation may be more widespread than previously thought.

The translucent-cloud program includes some 35 stars having total visual extinctions in the range $A_V = 1.5$ to 5 magnitudes, representing a variety of dust extinction curve types and atomic and molecular abundances as shown by ancillary data on various species observed from the ground. The first of these stars to be observed is HD 73882, which is described in the paper by Snow *et al.* 2000, and the results for which are summarized here.

5. HD 73882: A Case Study of H_2 in a Translucent-Cloud

The star HD 73882 is well known in the translucent cloud community because it is of early spectral type (O8.5V); it is bright enough (V = 7.27) to allow high-S/N observations at optical wavelengths; and it has sufficient foreground dust ($E_{B-V} = 0.72$; $A_V = 2.44$) and gas to have large column densities of many atomic and molecular species. A series of papers by van Dishoeck, Black, Gredel, and colleagues have provided data on CO (from mm-wave observations), on other diatomics such as CH, CN, CH^+, and C_2; and atomic species including Ca II, Na I, and K I (Gredel, van Dishoeck, & Black 1993; van Dishoeck *et al.* 1991).

The dust extinction curve has been derived and characterized by Fitzpatrick & Massa (1986, 1988, 1990), who also computed the atomic hydrogen column density, based on *IUE* spectra. The extinction curve shows a steep far-UV rise with significant curvature, generally similar in nature to other "dense cloud" curves, as characterized by Cardelli, Clayton, & Mathis (1988). But the value of R_V, the ratio of total to selective extinction as derived from infrared photometry, is $R_V = 3.9$, which is a bit unusual for a dark-cloud line of sight, where it is more normal to find values near 3.0. Indeed Cardelli *et al.* describe the extinction curve for HD 73882 as peculiar, because it does not conform strictly to the curves for other molecular clouds. The relatively high value of R_V, along with the high value for the wavelength of maximum polarization (Serkowski, Mathewson, & Ford 1975), suggests that small grains are not as dominant in this line of sight as in other dense clouds having strong molecular lines and steep far-UV extinction rises.

According to Massa, Savage, & Fitzpatrick (1983), the extinction in the line of sight to HD 73882 is probably dominated by one or more dense molecular clouds between Earth and the star, as opposed to foreground diffuse clouds or H II regions. The molecular column densities support this view, with relative abundances (compared with hydrogen) similar to those predicted by translucent cloud models (e.g. van Dishoeck & Black 1988).

Our analysis of the molecular hydrogen toward HD 73882 is based in large part on the line-of-sight velocity structure inferred from high-resolution optical spectra of interstellar lines of atomic species such as Na I and K I, which are thought to form in the same cold clouds where H_2 is formed. (Our high-resolution optical data for this line of sight have been obtained by D. E. Welty, and are described in more detail in Snow *et al.* 2000). The resulting H_2 column density, particularly for the excited rotational levels for which the lines are only moderately damped, depends critically on the assumed velocity structurevelocity dispersion parameter b. For the overall H_2 abundance, which is dominated by the $J = 0$ and $J = 1$ levels whose lines are thoroughly damped, the b-value is unimportant.

The molecular hydrogen column density we have measured toward HD 73882 (cf. Fig.1) is 1.1×10^{21}, a factor of nearly 3 larger than any previously-measured value obtained through far-UV absorption spectra. This demonstrates that *FUSE* will allow us to investigate H_2 abundances and excitations in interstellar regimes previously unexplored – as was expected in view of the factor of 10^4 increase in far-UV sensitivity over previous instruments such as *Copernicus* (primarily owing to the use of an array detector instead of a scanning photomultiplier).

The molecular fraction $f = 2N(H_2)/[2N(H_2) + N(H\ I)]$ is $f = 0.67$, higher than any previously derived hydrogen molecular fraction except for the uncertain value of $f = 0.80$ toward HD 24534 (Snow et al. 1998), for which the H_2 column density is based on poor-quality *Copernicus* data (this star will be observed with *FUSE*, which should provide a precise measurement of the H_2 abundance and a refined value of f).

Despite the enhanced H_2 fraction, however, the abundances of CO, CN, CH, and C_2, relative to the total hydrogen abundance, are more similar to those found in diffuse clouds (e.g., toward ζ Oph) than in dark clouds. Because these molecular abundances represent only a small fraction of the presumed total carbon abundance, we conclude that the clouds toward HD 73882 have not reached the predicted transition point where carbon becomes primarily molecular (e.g. van Dishoeck & Black 1986).

The CO/H_2 and CH/H_2 ratios toward HD 73882 are consistent with linear extrapolations from the values found previously for diffuse clouds (Dickman 1975; Mattila 1986). The CO/H_2 correlation in particular is important for estimating the masses of dark IS clouds (e.g. Frerking, Langer, & Wilson 1982), because CO can be measured via mm-wave emission lines in regions having very large extinctions. Our result suggests that the diffuse-cloud relationship can be extrapolated safely into the translucent cloud regime, something we will explore further as we obtain data on additional *FUSE* targets.

Our derived molecular hydrogen column density, combined with the line-of-sight extinction properties cited above, show that the ratio of hydrogen, both molecular and total, to dust extinction, is also similar to the value found for diffuse clouds (Bohlin et al. 1978).

We were able to deduce the rotational excitation of H_2 toward HD 73882. The ratio of $J = 1$ to $J = 0$ (i.e., the ratio of ortho- to para-hydrogen) is usually assumed to be an indicator of cloud kinetic temperature, because in these clouds the collisional timescale for depopulating these levels is far shorter than the radiative timescale. Our analysis of the $J = 1/J = 0$ ratio yields $T_{kin} = 59$ K. This is consistant with the value $T_{kin} = 77 \pm 17$ K found for diffuse clouds by *Copernicus* (Savage et al. 1977; see also Shull et al. 2000).

The excitation of higher rotational levels clearly shows that the higher J-levels toward HD 73882 are not in thermal equilibrium with the gas kinetic temperature. Similar results found for diffuse clouds have been interpreted in terms of UV pumping – the H_2 molecules cascade down through vibrational and rotational levels following electronic excitation upon the absorption of a UV photon. We have not attempted a detailed model fit to the high-J excitation toward HD 73882 because these models (e.g. Jura 1975b) rely strongly on the ratios of certain J-levels (e.g., $J = 4/J = 2$), and in our analysis these ratios are still very uncertain. We can conclude, however, that the observed excitation probably requires a significant local UV radiation field. The excited H_2 therefore may arise in a cloud boundary zone, rather than in the cloud core where the molecules are well protected from UV radiation. If this is the case, then we might expect to see velocity shifts or different b-values for the high-J lines as compared with the $J = 0$ and $J = 1$ lines – but current data quality precludes a sensitive search for such effects. We may hope eventually to use *FUSE* spectra to search for high-J velocity shifts or enhanced b-values, but not until the instrumental spectral resolution and wavelength calibration are optimized.

It is noteworthy that even in a line of sight thought to be dominated by molecular cloud material within a single cloud complex, the velocity structure is very complicated. Our analysis of the molecular hydrogen absorption bands toward HD 73882, particularly of the excited rotational levels, would have been severely compromised if not for the high-resolution ground-based optical absorption and mm-wave emission line data to which we had access. It is thus imperative for future *FUSE* interstellar-line observations (or those obtained with other low- to moderate-dispersion instruments) to obtain appropriate high-resolution spectra in order to have any hope of unambiguous interpretation of the lower resolution data.

Another very important caution arises from our analysis of the continuum depression caused by overlapping H_2 bands. This effect is important for H_2 column densities above $\sim 10^{20}$ cm^{-2}, as the damping wings of adjacent bands encroach on each other. This affects the calculation of H_2 abundances through profile fitting, and it will have a significant effect on the derivation of far-UV extinction curves. The overlapping of bands and the resultant continuum depression peak near 1000 Å, where the Lyman and Werner series overlap – but the effect is significant throughout the entire spectral region where the H_2 bands arise. Needless to say, the effect will be most severe for the most heavily-reddened stars.

6. Summary and Outlook

These first results from the *FUSE* PI-team survey of molecular hydrogen in translucent interstellar clouds demonstrate that *FUSE* has vast potential for providing data on the most common of all interstellar constituents, and that the data obtained by *FUSE* will provide for many important analyses of the chemical and physical conditions in such clouds. In the first star studied, HD 73882, we found a higher H_2 abundance and molecular fraction than previously observed, while also finding that the analysis depends critically on assumptions made about the line-of-sight cloud velocity structure. Clearly it will be important in future *FUSE* studies of similar clouds to obtain ground-based high-resolution optical spectra to guide the analysis.

Toward that end, the *FUSE* PI group who are pursuing H_2 in translucent clouds are also carrying out ancillary observations, from the ground and from space. In conjunction with the *FUSE* observing program, we are using many other instruments to supplement the far-UV spectroscopy: e.g., we (primarily D. E. Welty, with Snow and D. C. Morton) have observed most of the program stars in optical wavelengths at high or ultra-high spectral resolution, in order to disentangle the cloud velocity structure in each line of sight; we (led by D. G. York) are undertaking moderately high-resolution, very high S/N optical spectroscopy of all of our *FUSE* targets in order to obtain very high S/N data on the diffuse interstellar bands; and we (with participation by F. Chaffee) are conducting near-IR observations in order to compare grain mantle features such as water ice, the 3.4-μm hydrocarbon band, and the silicate feature at 9.6 μm, with the UV data on dust extinction and gas-phase depletions.

The ultimate goals of our program are to fully understand the masses of dense interstellar clouds, to probe the chemistry and physics of translucent clouds, and to probe the transition region between diffuse and dense interstellar clouds.

This work is based on data obtained for the Guaranteed Time Team by the NASA-CNES-CSA *FUSE* mission operations by the Johns Hopkins University. Financial support to U.S. participants has been provided by NASA contract NAS5-32985.

REFERENCES

Barnstedt, J., Gringel, W., Kappelmann, N., & Grewing, M. 2000, A&S, in press
Blair, W. P., Long, K. S., & Raymond, J. C. 1996, ApJ, 468, 871
Bohlin, R. C., Savage, B. D., & Drake, J. F. 1978, ApJ, 224, 132
Bowers, C. W., Blair, W. P., Long, K. S., & Davidsen, A. F. 1995, ApJ, 44, 748
Cardelli, J. A., Clayton, G. C., & Mathis, J. S. 198, ApJL, 329, L33
Carruthers, G. r. 1970, ApJL, 161, L81
de Boer, K. S., Richter, P., Bomans, D. J., Heithausen, A., Koornneef, J. 1998, A&A, 338, L5
Dickman, R. L. 1975, ApJ, 202, 50
Falgarone, E., Verstraete, L., Pineau Des Forêts, G., Flower, D., & Puget, J.-L. 2000, in H_2 in Space, Eds.: F. Combes, G. Pineau des Forêts (Cambridge University Press, Astrophysics Series), in press
Federman, S. R. 1981, A&A, 96, 198
Federman, S. R., Cardelli, J.A., van Dishoeck, E. F., Lambert, D. L., & Black, J. H. 1995, ApJ, 445, 325
Fitzpatrick, E. L. & Massa, D. 1986, ApJ, 307, 286
Fitzpatrick, E. L. & Massa, D. 1986, ApJ, 328, 734
Fitzpatrick, E. L. & Massa, D. 1990, ApJS, 72, 163
Frerking, M. A., Langer, W. D., & Wilson, R. W. 1982, ApJ, 262, 590
Fullerton, A., *et al.* 2000, ApJL, in press
Gautier, T. N., III, Fink, U., Larson, H. P., & Treffers, R. R., 1976, ApJL, 207, L129
Gredel, R., van Dishoeck, E. F., & Black J. H. 1993, A&A, 269, 477
Hollenbach, D. J. & Tielens, A. G. G. M. 1997, ARA&A, 35, 179

Jenkins, E. B. & Peimbert, A. 1997, ApJ, 477, 265
Jura, M. 1975a, ApJ, 197, 575
Jura, M. 1975b, ApJ, 197, 581
Kulesa, C. 1999, private communication
Lacy, J. H., Knacke, R., Geballe, T. R., & Tokunaga, A. T., ApJL, 428, L69
Longo, R., Stalio, R., Polidan, R., & Rossi, L. 1989, ApJ, 339, 474
Massa, D., Savage, B. D., & Fitzpatrick, E. L. 1983, ApJ, 266, 662
Mattila, K. 1986, A&A, 160, 157
Moos, H. W. et al. 2000, ApJL, in press
Morton, D. C. 1974, ApJL, 193, L35
Rachford, B. L., Snow, T. P., et al. 2000, in preparation
Richter, P., et al. 1998, A&A, 338, L9
Savage, B. D., Drake, J. F., Budich, W., & Bohlin, R. C. 1977, ApJ, 216, 291
Sahnow, D. et al. 2000, ApJL, in press
Serkowski, K., Mathewson, D. L., & Ford, V. L. 1975, ApJ, 196, 261
Shull, J. M. & Beckwith, S. 1982, ARA&A, 20, 163
Shull, J. M. *et al.* 2000, ApJL, in press
Snow, T. P., Allen, M. M., & Polidan, R. S. 1990, ApJL, 359, 23L
Snow, T. P., Hanson, M. M., Black, J. H., van Dishoeck, E. F., Crutcher, R. C., & Lutz, B. L. 1998, ApJL, 496, L113
Snow, T. P. & Witt, A. N. 1996, ApJL, 468, L65
Snow, T. P. *et al.* 2000, ApJL, in press
Spitzer, L., Cochran, W. D., & Hirshfeld, A. 1974, ApJ, 28, 373
Spitzer, L, Drake, J. F., Jenkins, E. B., Morton, D. C., Rogerson, J. B., & York, D. G. 1973, ApJL, 181, 116
Spitzer, L. & Jenkins, E. B. 1975, ARA&A, 13, 133
Spitzer, L. & Zweibel, E. G. 1974, ApJL, 191, L127
van Dishoeck, E. F., Phillips, T. G., Gredel, R., & Black, J. H. 1991, ApJ, 366, 141
van Dishoeck, E. F. & Black, J. H. 1986, ApJS, 62, 109
Welty, D. E., & Hobbs, L. M. 2000, in preparation
Welty, D. E., Morton, D. C., & Snow, T. P. 2000, in preparation
Witt, A. N., Stecher, T. P., Boroson, T. A., & Bohlin, R. C. 1989, ApJL, 336, L21

FUSE and deuterated molecular hydrogen

By R. FERLET[1], M. ANDRÉ[2], G. HÉBRARD[1], A. LECAVELIER[1], M. LEMOINE[3], G. PINEAU DES FORÊTS[3], E. ROUEFF[3] AND A. VIDAL-MADJAR[1]

[1]Institut d'Astrophysique de Paris, CNRS, 98 bis Bld Arago, F-75014 Paris, France

[2]Department of Physics and Astronomy, Johns Hopkins University, Baltimore, MD 21218, USA

[3]Observatoire de Paris, 5 place Janssen, F-92195 Meudon cedex, France

The Lyman and Werner band systems of deuterated molecular hydrogen (HD) occur in the far UV range below 120 nm. This spectral window is now open at moderate resolution and high sensitivity with the *FUSE* satellite. *FUSE* spectra of hot stars with high extinction through translucent clouds will give access to the deuterium abundance inside molecular clouds where D is essentially in the form of HD. Measurement of HD/H_2 ratio becomes thus a new powerful method to evaluate the D/H ratio in the interstellar medium.

An example is given with the *FUSE* spectrum of the high extinction O9III star HD 73882 ($E_{B-V} = 0.7$). Very preliminary analysis and an estimate of the HD/H_2 ratio are presented.

1. Introduction

It has long been recognized that the primordial abundance of deuterium represents the most sensitive probe of the baryonic density Ω_b of the Universe (see, e.g., Schramm & Turner 1998; Olive et al. 1999). On the other hand, abundance of deuterium at any epoch is a lower limit to its primordial abundance, since deuterium is destroyed, not created, in stars of any mass. For this reason, deuterium abundance is also an efficient tracer of the universal star formation rate. Unfortunately, the evolution of deuterium abundance from the primordial to the solar metallicity is still unclear.

Measurements of the atomic D/H ratio have been performed in different astrophysical sites, namely in moderate to high redshift quasar absorbers, in the presolar nebula and in the local interstellar medium (for reviews see, e.g., Ferlet & Lemoine 1996; Linsky 1998; Vidal-Madjar et al. 1998a; Lemoine et al. 1999). These studies indicate that the abundance of deuterium may vary in the local interstellar medium by a factor as high as ~ 2 over spatial scales as small as few tens of pc (Vidal-Madjar et al. 1998b; Jenkins et al. 1999). The determination of presolar and quasar absorbers (D/H) abundances is also limited by the existing scatter in the results.

Observation of deuterated molecules is another mean to estimate the deuterium abundance. To date, over 20 single D-bearing species and two doubly deuterated molecules, D_2CO and ND_2H, have been observed at radio frequencies in the interstellar medium both in cold dark clouds and in warmer star forming regions. However, chemical fractionation takes place in cold regions and mantles desorption of grains are often invoked in star forming regions. Moreover, these observations must be combined with a presupposed D/H ratio to provide a wealth of information on interstellar chemistry. For these reasons, it is difficult to derive an accurate estimate of the deuterium fractional abundance from deuterated molecular abundances.

Recently, the pure rotational line at $112\,\mu m$ of HD J=1→0 has been detected with *ISO* in giant planets (Feuchtgruber et al. 1997) and in the Orion bar (Wright et al. 1999). Bertoldi et al. (1999) have also detected the excited rotational line at $19.43\,\mu m$ of HD J=6→5 in Orion KL, a molecular outflow region. Although *ISO* opened the sky to HD emission, the derived column densities depends strongly on the modeling of HD excitation and on the extinction corrections. With *Copernicus*, H_2 and HD have also been detected in absorption in the ultraviolet in diffuse interstellar clouds such as ζ Oph (Wright & Morton 1979). Unfortunately, the observed low HD/H_2 values (few $\times 10^{-7}$ to few $\times 10^{-6}$) reflect the mostly atomic nature of these diffuse clouds: to deduce a D/H ratio from these data requires a detailed model of formation/destruction of HD.

Nevertheless, as we argue in the present contribution, the Far Ultraviolet Spectroscopic Explorer (*FUSE*) should provide in the near future accurate estimates of the deuterium abundance through the measurement of the HD/H_2 ratio in dense molecular clouds, in which D and H are essentially in their molecular form. Such measurements, which do not suffer from chemical fractionation corrections, will shed new light on the problem of the D/H ratio.

2. *FUSE* Observations

FUSE was successfully launched on June 24, 1999 from Cape Canaveral. It observes in the wavelength range 905 Å–1187 Å with a sensitivity of about 10^4 times that of *Copernicus* (Moos 1998; Moos et al. 2000). Many absorption lines from different rotational states of HD are available in this range, although they are often blended with other lines from H_2 toward the most reddened lines of sight. *FUSE* is thus able to probe dense molecular regions in which hydrogen and deuterium are essentially in the form of H_2 and HD.

Such observations were conducted with *FUSE* toward the reddened star HD 73882 ($E_{B-V}=0.72$) for a total of $\simeq 21$ h integration time. Unfortunately, the spectrograph was still unfocused and the spectral resolving power does not exceed $\lambda/\Delta\lambda \simeq 14000$; the signal-to-noise ratio is $\simeq 20$ per resolution element. The broad H_2 bands are clearly seen, as shown in Fig. 1. Several other features are also detected including atomic species such as CI, OI, NI, $FeII$, PII, few lines of HD $J = 0$ and CO bands. Examples of spectral regions with H_2 lines from different $J-$ levels are shown in Figs. 2 and 3.

We have performed a very preliminary analysis of these data using profile fitting of all detected lines of all above species, in the whole spectral range. Table 1 give the resulting molecular column densities. It should be noted that all lines are saturated, so that column densities are highly sensitive to the assumed value of the intrinsic broadening parameter b which combines thermal and turbulent broadening. Uncertainties in the velocity structure of the line of sight and in the line spread function of the instrument make matters worse. In a first step, we assumed a single molecular absorbing component along the line of sight toward HD 73882, as seen in previous observations of CH, C_2, CN and CH^+ (Gredel et al. 1994). Nevertheless, it turns out that the total H_2 column density is about insensitive to these problems because the dominant $J = 0$ and $J = 1$ levels are on the damped part of the curve of growth. The error on $\log_{10}(N(H_2)$ is roughly ± 0.2. The fitting value found for the b parameter is small, combining $\simeq 0.4$km/s for turbulent velocity and $\simeq 20$ K for temperature. If the actual b turns out to be higher, then the present column densities should be considered as rough upper limits.

From this preliminary analysis, achieved for the first time at such a depth within a molecular interstellar cloud, we find: $CO/H_2 \sim 8 \cdot 10^{-5}$ (Fig 4) and $HD/H_2 \sim 3 \cdot 10^{-5}$ (Fig 5). The HD 73882 line of sight seems to be mostly molecular, in agreement with the previous observations of CH, C_2, CN and CH^+ (Gredel et al. 1994). We also note that it has a peculiar extinction curve, with an enhanced far-ultraviolet extinction (Massa et al. 1983) and a weak bump at 2200 Å. Moreover, the $R_V = A_V/E_{B-V}$ parameter is 3.39, larger that the mean galactic value $\simeq 3.1$ (Cardelli et al. 1989).

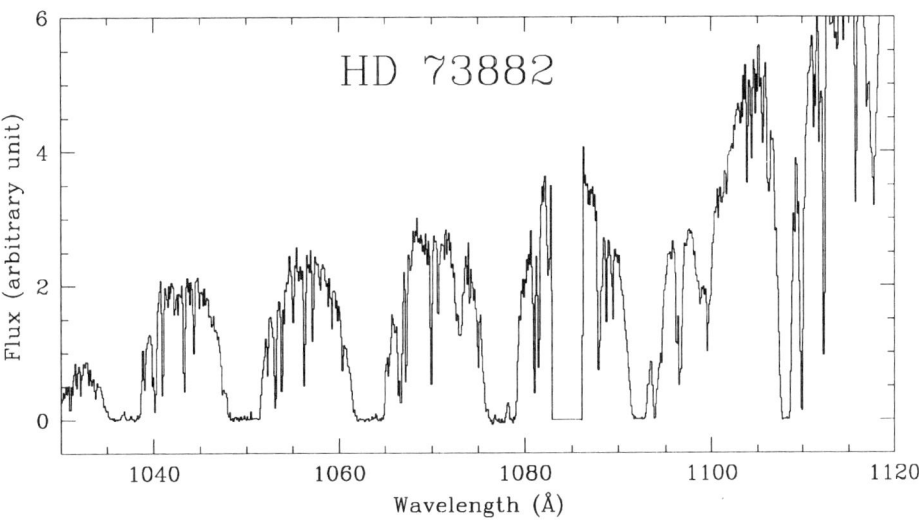

FIGURE 1. *FUSE* spectrum of HD 73882 from 1030 to 1120 Å. Many atomic and molecular lines are visible. In particular, the following absorption lines of H_2 are strongly saturated:
lines from H_2 (J=0) at 1037, 1049, 1063, 1077, 1092 and 1108 Å,
lines from H_2 (J=1) at 1038, 1051, 1065, 1079, 1094 and 1110 Å.

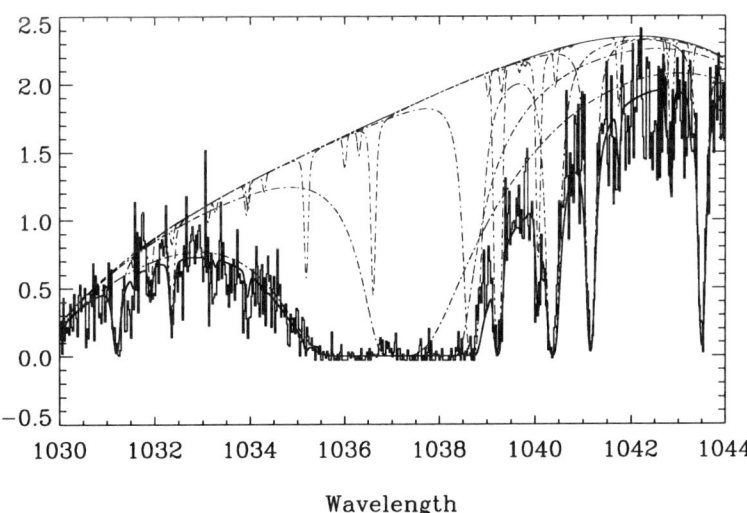

FIGURE 2. Fit of the spectrum from 1030 to 1044 Å. The following lines of H_2 are identified:
J=0 at 1036.6 Å; J=1 at 1037.2 and 1038.2 Å;
J=2 at 1040.4 Å; J=3 at 1031.2 and 1043.5 Å;
J=4 at 1032.4 Å; J=5 at 1040.0 Å.
HD $J=0$ lines are also present at 1031.9 and 1042.8 Å.

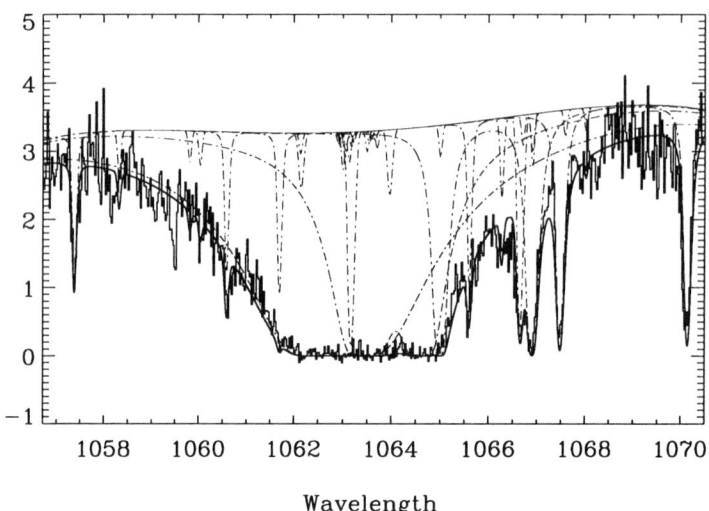

FIGURE 3. Fit of the spectrum from 1057.0 to 1071 Å. The following lines of H_2 are identified: J=0 at 1062.9 Å; J=1 at 1063.4 and 1064.6 Å; J=2 at 1066.9 Å; J=3 at 1067.5 and 1070.1 Å; J=4 at 1057.3 and 1060.6 Å; J=5 at 1065.6 Å

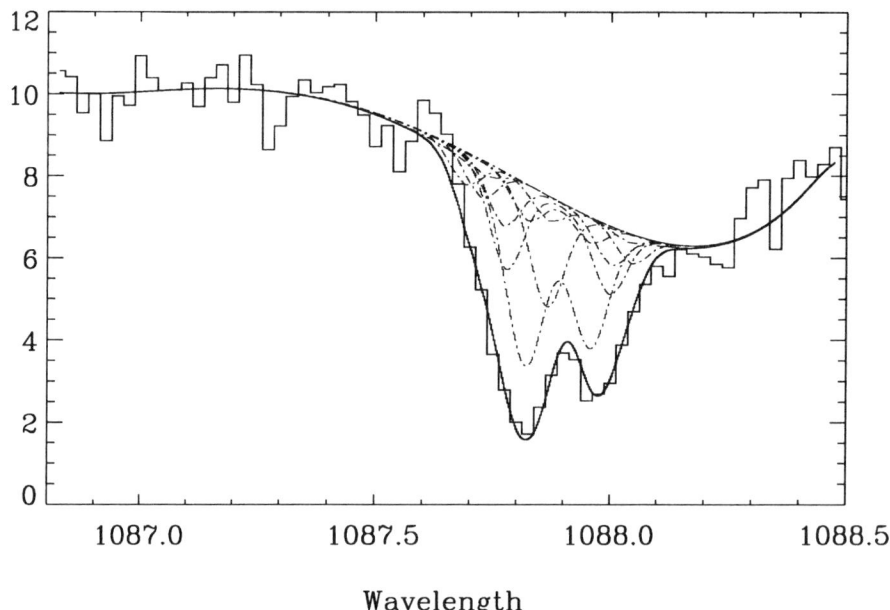

FIGURE 4. Fit of the CO lines from $J = 0$ to $J = 3$. We used the same turbulent velocity as for the H_2 lines.

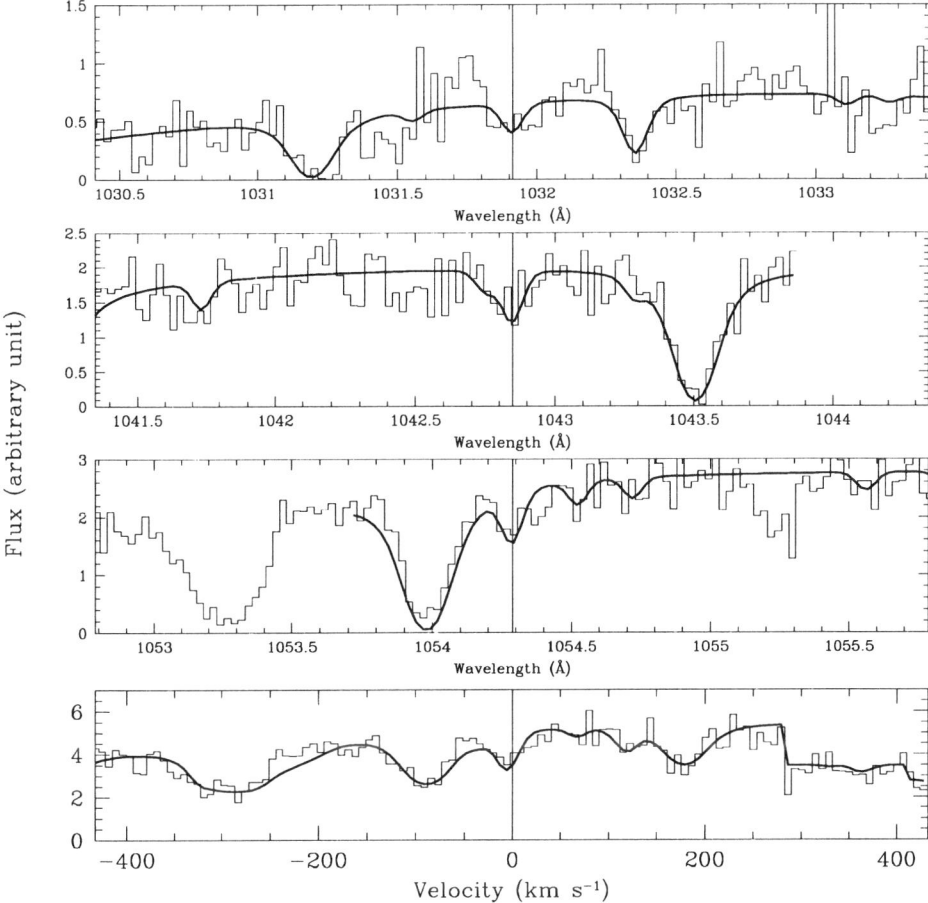

FIGURE 5. HD absorption lines detected toward HD 73882 and the corresponding fits. The three upper panels show the HD lines at 1031.9, 1042.8 and 1054.3 Å. The bottom panel is the sum of the three lines.

3. Discussion

In order to check if the reservoir of deuterium in the HD 73882 sight-line is the deuterated molecular hydrogen, we have modeled the photodissociation region with a density of 500 cm^{-3} as determined from C_2 observations, a standard ultraviolet interstellar radiation field and a cosmic ionization rate $\xi = 5 \times 10^{-17}$ s^{-1}. The model (Roueff & Nodé-Langlois, 1998; Roueff & Le Petit 2000) is an extension of the photodissociation region (PDR) model described in Abgrall et al. (1992) and in Le Bourlot et al. (1993). The atomic to molecular transition is studied in a semi-infinite plane parallel model where the radiation field is impinging on one side of the cloud. The mechanisms involved in the photodissociation of HD are very similar to those related to H_2, apart for self-shielding which is much more efficient for the abundant molecular hydrogen. Similarly to H_2, formation of HD occurs on grains. However, HD is also efficiently formed through the reaction:

TABLE 1. Molecular column densities toward HD 73882.

Molecule	J	$\log_{10} N$ (cm^{-2})
H_2	0	21.1
H_2	1	20.5
H_2	2	19.3
H_2	3	18.7
H_2	4	17.6
H_2	5	17.7
H_2	6	15.7
H_2	7	15.2
CO	*	17.1
HD	0	16.7

* The CO column density is the summation over all detected levels.

$$H_2 + D^+ \to HD + H^+.$$

Therefore, whenever H_2 predominates, this reaction also becomes the dominant process of HD formation.

It remains beyond the scope of this paper to perform detailed comparison between models and observations. Nevertheless, one can summarize the main physical and chemical features of the model in the following figures. Fig. 6 displays the abundances of atomic and molecular hydrogen as a function of the extinction A_V. The H to H_2 transition occurs at an A_V of about 0.02, at much lower extinction than the D to HD transition which takes place at $A_V = 0.13$ (Fig. 7). This reflects the smaller contribution of self-shielding in the photodissociation mechanism for HD. Finally, Fig. 8 shows the photo-destruction probabilities of H_2, HD, CO and C as functions of A_V, again at different extinctions. According to the extinction toward HD 73882, it seems that we are presently dealing with an interstellar region in which the HD molecule could indeed be essentially the reservoir of deuterium.

One should finally note that charge transfer from H^+ to atomic oxygen initiates a chemical scheme capable of producing OH and that the equilibrium abundance of OH is a measure of the density of H^+, hence of the ionizing flux on which the efficiency of the deuterium fractionation (1) depends (Black & Dalgarno 1973). Column densities of OH can be measured from the ground (Felenbok & Roueff 1996); such measurements would improve the accuracy of the deuterium abundance determination through HD, if steady-state conditions apply.

4. Conclusion

We have reported on *FUSE* observations in the direction of the reddened star HD 73882. The HD lines are clearly detected and lead to an estimate of the HD/H_2 ratio of the order of 3×10^{-5}. This result is obtained under the assumption of a single molecular absorbing component and is very sensitive to the intrinsic width of the lines. As a matter of fact, it is favoured by the χ^2 of the fits but should be taken more as an upper limit. The present observations are possibly sampling regions deep within the cloud where the HD molecule represents the reservoir of D atoms. If confirmed with future *FUSE* observations in the

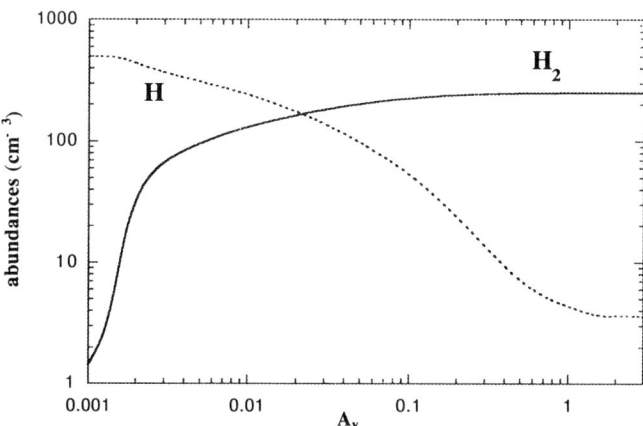

FIGURE 6. Comparison of the relative evolution of abundance of H and H_2 with A_v. We assume that a semi infinite plane-parallel cloud is exposed to an isotropic ultraviolet radiation field. The cloud has a constant density n_H=500 cm^{-3} (deduced from C_2 observations) and the radiation field is taken as the normal interstellar one (scaling factor χ =1). ξ is the cosmic ionization rate.

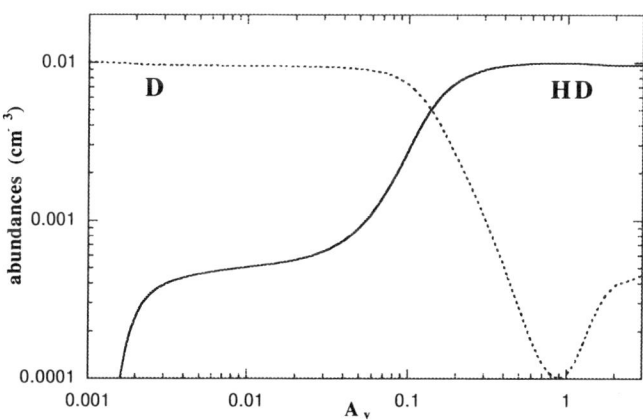

FIGURE 7. Same as Fig. 6 but for D and HD.

direction of other molecular clouds, it will offer a new and reliable tool for probing D/H in dense regions of the interstellar medium, via the simple relation:

$$D/H = 0.5 \times HD/H_2.$$

Acknowledgments. *FUSE* is an *Origins* Mission. It is funded by NASA Explorer Program in cooperation with the Canadian Space Agency and the Centre National d'Études Spatiales of France. *FUSE* was developed and is being operated for NASA by the Johns Hopkins University in collaboration with the University of California, Berkeley

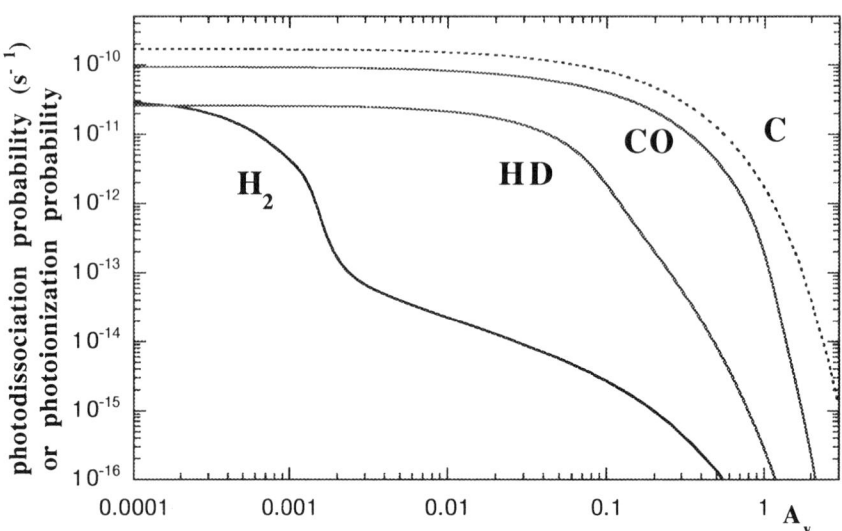

FIGURE 8. Photodissociation probability in s^{-1} for H$_2$, HD and CO, and photoionization probability of C, as function of A$_v$.

and the University of Colorado. Financial support has been provided by NASA contract NAS5-32985.

REFERENCES

ABGRALL, H., LE BOURLOT, J., PINEAU DES FORÊTS, G., ROUEFF, E., FLOWER, D. R. & HECK, L., 1992, *A&A* **253**, 525

BERTOLDI, F., TIMMERMANN, R., ROSENTHAL, D., DRAPATZ, S. & WRIGHT, C., 1999, *A&A* **346**, 267

BLACK, J.H. & DALGARNO, A., 1973, *ApJ* **184**, L101

CARDELLI J.A., CLAYTON G.C. & MATHIS J.S., 1973, *ApJ* **345**, 245

FELENBOK, P. & ROUEFF, E., 1996, *ApJ* **465**, L57

FERLET, R. & LEMOINE, M., 1996, in *Cosmic Abundances*, Astron. Soc. Pacific Conf. Series **99**, 78

FEUCHTGRUBER, H., LELLOUCH, E., BÉZARD, B., ENCRENAZ, TH., DE GRAAUW, TH. & DAVIS, G.R., 1997, *A&A* **341**, L17

GREDEL, R., VAN DISHOECK, E.F. & BLACK, J.H., 1994, *A&A* **285**, 300

JENKINS, E.B., TRIPP, T.M., WOZNIAK, P., SOFIA, U.J. & SONNEBORN, G., 1999, *ApJ* **520**, 182

LE BOURLOT, J., PINEAU DES FORÊTS, G., ROUEFF, E. & FLOWER, D.R., 1993, *A&A* **267**, 233

LEMOINE, M. ET AL., 1999, *New Astronomy* **4**, 231

LINSKY, J.L., 1998, *Space Science Reviews* **84**, 285

MASSA D., SAVAGE, B.D. & FITZPATRICK, E.L., 1983, *ApJ* **266**, 662

Moos, W., 1998, in *Origins, ASP Conference Series*, **148**, 304

Moos, H.W., et al. 2000, *ApJL*, submitted

Olive, K. A., Steigman, G. & Walker, T. P., 1999, Phys. Rep. in press

Roueff, E. & Nodé-Langlois, T., 1998, *Rapport de stage de l'École Polytechnique*

Roueff, E. & Le Petit, 2000, in preparation

Schramm, D. N., Turner, M. S., 1998, Rev. Mod. Phys. **70**, 303

Vidal-Madjar, A., Ferlet, R. & Lemoine, M. 1998a, *Space Science Reviews* **84**, 297

Vidal-Madjar, A., et al. 1998b, *A&A* **338**, 694

Wright, C.M., van Dishoeck, E.F., Cox, P. & Kessler, M.F., 1999, *ApJ* **515**, L29

Wright, E.L. & Morton, D.C., 1979, *ApJ* **227**, 483

Ted Snow retrieving an old student of his, Valery Le Page.

Guillaume Pineau des Forêts basking in the sun, with Pierre Cox.

ISO-SWS observations of H_2 in Galactic sources

By Christopher M. WRIGHT[1]

[1]School of Physics, University College, Australian Defence Force Academy, University of New South Wales, Canberra ACT 2600, Australia

A review is presented of ISO observations of molecular hydrogen, H_2, toward various Galactic source types, such as shocks and photon dominated regions. In so doing I examine the similarities and differences in the H_2 spectrum found under these different excitation conditions and mechanisms, and how the observations impact on some of the latest models.

1. Introduction

Before the launch of the Infrared Space Observatory (ISO, Kessler et al. 1996) observations of H_2 were restricted to a hot component with an excitation temperature of about 2000 K in shock excited sources, and a non-thermally (i.e. fluorescently) excited component in photon dominated regions (PDRs). These components were typically probed with the 1–0 S(1) line at 2.12 μm as well as several other near-infrared ro-vibrational transitions, and in some cases pure rotational transitions from high J levels (e.g. Gredel 1994, Knacke & Young 1981). Only a few observations, principally toward the Orion star forming region, of lower energy pure rotational transitions existed, e.g. the 0–0 S(2) and 0–0 S(1) lines at 12.2786 and 17.0348 μm, but which already pointed to the existence of a lower temperature component in such sources (e.g. Beck, Lacy & Geballe 1979; Parmar, Lacy & Achtermann 1991, 1994; Richter et al. 1995; Burton & Haas 1997).

The main motivation behind ISO H_2 observations is simply that in molecular clouds, where stars are forming, most of the mass resides in H_2. Further, emission from the lowest pure rotational transitions, where most of the H_2 mass is expected to reside, occurs in the mid-infrared, 5–30 μm, where the Earth's atmosphere is opaque. Therefore, by opening up the window beyond 2.4 μm, ISO, and specifically the Short Wavelength Spectrometer (SWS, de Graauw et al. 1996), offered the possibility of a much stronger observational effort to be directed towards H_2 spectroscopy of Galactic shocks, PDRs and x-ray dominated regions (XDRs), providing clues to such physics as the temperature structure, cooling contribution, ortho-to-para abundance ratio and even the deuterium abundance. In the following I review ISO observations of H_2 toward various Galactic source types, and examine how the observations impact on some of these issues.

2. Shocks

Table 1 presents a summary of most of those H_2 observations of shocks that have been published since ISO's launch in November 1995. Apart from T Tau, T_{ex} has quite a narrow range, from $\sim$ 600 to 800 K. The column density of this warm gas is at least an order of magnitude greater than that of the hot $\sim$ 2000 K component. Comparison with the C-type planar shock models of Kaufman & Neufeld (1996, KN96) usually requires at least two components to adequately match the data, which are indicated in Table 1. The first component, of a low density, low velocity and high covering factor, reproduces the warm, $\sim$ 700 K, gas, whilst the second component, of a higher density, higher velocity and low covering factor, reproduces the hot, $\sim$ 2000 K, gas. It is however not clear that the success at matching these shocks with the KN96 models provides unequivocal support for the models, since Rosenthal et al. (in preparation) have shown that a range of shock

Source	T_{ex}^a K	$N_{H_2}^b$ $\times 10^{20}$ cm^{-2}	v_{s1} km s^{-1}	n_{s1} cm^{-3}	Φ_{s1}	v_{s2} km s^{-1}	n_{s2} cm^{-3}	Φ_{s2}	Ref.^{c}
				Shocks					
Cep A West	688	1.8	20	10^4	0.5	35	$10^{6.5}$	0.001	1
Cep A East	740	1.2	20	10^6					2
DR 21 West	628	5.8	20	10^4	1.25	30	10^6	0.005	3,4
DR 21 East	671	4.4	20	10^4	1.25	30	10^6	0.0055	3
Orion IRc2	698	12	20	10^4	3.0	35	$10^{6.5}$	0.0175	5,6
Orion Pk2	819	11	20	10^4	4.0	35	$10^{6.5}$	0.03	6
Orion Pk1	767	16							7
IC 443	822	2.1	20	10^4	0.85	40	10^6	0.005	8
T Tau	440	0.15							9
HH 54	650	0.9							10
			Photon Dominated Regions						
S140	500	2.1							11
IC 1396A pos1	412	0.3							6
IC 1396A pos2	448	0.8							6
S106 H2	705	0.8							6
S106 PDR	778	0.4							6
S106 IRS4	490	9.1							2
Orion Bar	552	7.1							6
NGC 7023	580	0.9							12,13
IC 63	685	0.1							14

^{a}Excitation temperature, T_{ex}, calculated between the extinction corrected $v=0$ and $J=3$ and 7 levels, apart from S106 H2 and S106 PDR ($v=0$ and $J=4$–7).
^{b}Column density, N_{H_2}, calculated assuming the kinetic temperature is equal to T_{ex}.
^{c}(1) Wright et al. (1996), (2) van den Ancker et al. (1998), (3) Wright, Timmermann & Drapatz (1997), (4) Smith et al. (1998), (5) van Dishoeck et al. (1998), (6) Wright et al. (in preparation), (7) Rosenthal et al. (1999), (8) Cesarsky et al. (1999), (9) van den Ancker et al. (1999), (10) Neufeld, Melnick & Harwit (1998), (11) Timmermann et al. (1996), (12) Bertoldi et al. (in preparation), (13) Fuente et al. (1999), (14) Thi et al. (1999)

TABLE 1. List of published shock and PDR sources observed by ISO

models in the literature can adequately match the H$_2$ data toward the Pk 1 position of the Orion shock. Further, the two component model cannot account for cooler ($T_{ex} \sim$ 150 K) gas detected toward Orion Pk2 in the 28.2 μm 0–0 S(0) line, nor hotter ($T_{ex} \geq 3000$ K) gas detected from very high levels ($J_{up} \geq 18$). Figure 1 shows a segment of spectrum toward Orion IRc2, along with the population diagram and KN96 model predictions. One measurement of the HD 19.4 μm $J = 6 - 5$ transition toward Orion Pk1 by Bertoldi et al. (1999) yielded a [D/H] of $(0.76 \pm 0.29) \times 10^{-5}$.

3. Photon Dominated Regions

Table 1 presents a summary of most of those PDR observations that have been published since the launch of ISO. For all the sources the gas density is of order 10^5–10^6 cm^{-3}, whilst the radiation field ranges from 10^3–$10^5 G_0$ (e.g. Burton, Hollenbach & Tielens 1992). There is quite a large range in T_{ex}, from $\sim$400 K up to 700 K and even 800 K. As noted by Draine & Bertoldi (1999) such high temperatures were unexpected based on pre-ISO PDR models, and suggested that perhaps a higher photoelectric heating efficiency was required. Indeed, Weingartner & Draine (1999) have calculated a new efficiency, based on an enhanced dust-to-gas ratio in the PDR due to gas-grain drift, and there is now good agreement between the models and data (e.g. Draine 2000). Figure 1

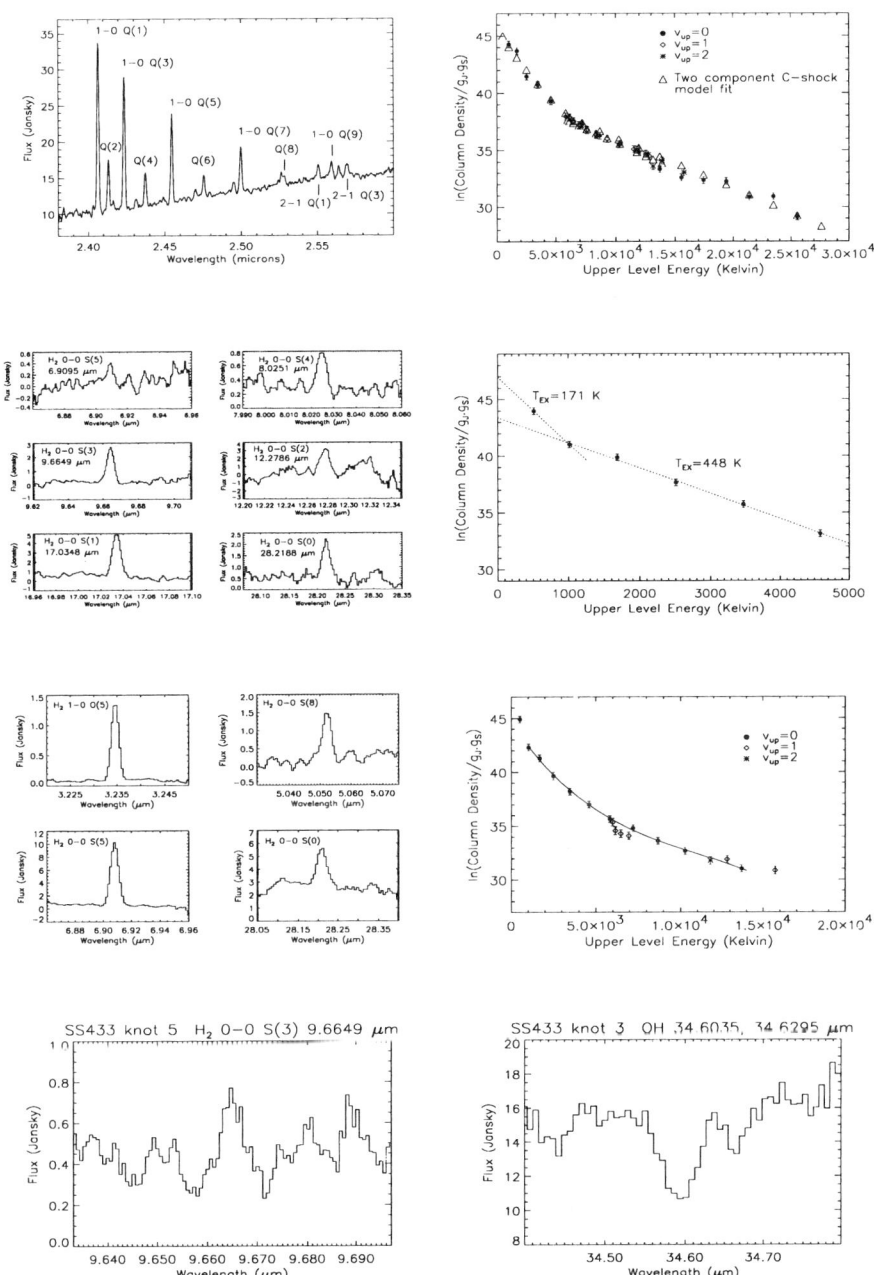

FIGURE 1. Top row: A segment of the H_2 ro-vibrational spectrum toward the Orion shock, and the resulting extinction corrected population diagram. All H_2 lines are marked, whilst all other emission lines are due to hydrogen recombination transitions from n→5. 2nd row: Spectra of several H_2 pure rotational lines observed toward a position in the reflection nebula IC 1396A, and the resulting extinction corrected population diagram. 3rd row: A selection of H_2 spectra toward a position in the supernova remnant and putative XDR RCW 103, along with the resulting extinction corrected population diagram with a cubic fit to the v=0 levels. Last row: Spectra of the H_2 0–0 S(3) and unresolved OH doublet lines toward the infra-red knots 5 and 3 respectively of the x-ray binary SS 433.

shows a sample of spectra and resulting population diagram from the reflection nebula IC1396 A. One measurement of the HD 112 μm $J = 1 - 0$ line toward the Orion PDR by Wright et al. (1999) yielded an inferred [D/H] of $(1.0 \pm 0.3) \times 10^{-5}$.

I am extremely grateful to my many ISO collaborators, and to the Australian Research Council for funding my travel to wonderful Paris.

REFERENCES

BECK, S.C., LACY, J.H. & GEBALLE, T.R., 1979, *Astrophys. J. Letters*, 234, L213

BERTOLDI, F., TIMMERMANN, R., ROSENTHAL, D., DRAPATZ, S. & WRIGHT, C.M., 1999, *Astron. Astrophys*, 346, 267

BURTON, M.G., HOLLENBACH, D.J. & TIELENS, A.G.G.M., 1992, *Astrophys. J*, 399, 563

BURTON, M.G. & HAAS, M.R., 1997, *Astron. Astrophys.*, 327, 309

CESARSKY, D., COX, P., PINEAU DES FORÊTS, G. ET AL., 1999, *Astron. Astrophys.*, 348, 945

DE GRAAUW, TH., HASER, L.N., BEINTEMA, D.A. ET AL., 1996, *Astron. Astrophys.*, 315, L49

DRAINE, B.T., 2000, to be published in the proceedings of the conference H_2 in Space, eds. F. Combes & G. Pineau des Forêts, Cambridge University Press

DRAINE, B.T. & BERTOLDI, F., 1999, in *The Universe as seen by ISO*, eds P. Cox & M.F. Kessler, ESA Publications Division SP-427, Noordwijk, p553-559

FUENTE, A., MARTIN-PINTADO, J., RODRIGUEZ-FERNANDEZ, N.J. ET AL., 1999, *Astrophys. J. Letters*, 518, L45

GREDEL, R., 1994, *Astron. Astrophys.*, 292, 580

KAUFMAN, M.J. & NEUFELD, D.A., 1996, *Astrophys. J.*, 456, 611

KESSLER, M.F., STEINZ, J.A., ANDEREGG, M.E. ET AL., 1996, *Astron. Astrophys.*, 315, L27

KNACKE, R.F. & YOUNG, E.T., 1981, *Astrophys. J. Letters*, 249, L65

NEUFELD, D.A., MELNICK, G.J. & HARWIT, M., 1998, *Astrophys. J. Letters*, 506, L75

PARMAR, P.S., LACY, J.H. & ACHTERMANN, J.M., 1991, *Astrophys. J. Letters*, 372, L25

PARMAR, P.S., LACY, J.H. & ACHTERMANN, J.M., 1994, *Astrophys. J.*, 430, 786

RICHTER, M.J., GRAHAM, J.R., WRIGHT, G.S., KELLY, D.M. & LACY, J.H., 1995, *Astrophys. J. Letters*, 449, L83

ROSENTHAL, D., BERTOLDI, F., DRAPATZ, S. & TIMMERMANN, R., 1999, in *The Universe as seen by ISO*, eds P. Cox & M.F. Kessler, ESA Publications Division SP-427, Noordwijk, p561-564

SMITH, M., EISLÖFFEL, J. & DAVIS, C.J., 1998, *Mon. Not. R. Astr. Soc.*, 297, 687

THI, W.F., VAN DISHOECK, E.F., BLACK, J.H. ET AL., 1999, in *The Universe as seen by ISO*, eds P. Cox & M.F. Kessler, ESA Publications Division SP-427, Noordwijk, p767-770

TIMMERMANN, R., BERTOLDI, F., WRIGHT, C.M. ET AL., 1996, *Astron. Astrophys.*, 315, L281

VAN DEN ANCKER, M.E., WESSELIUS, P.R., TIELENS, A.G.G.M. & WATERS, L.B.F.M., 1998, in *ISO's view on Stellar Evolution*, eds. L.B.F.M. Waters et al., *Astrophys. & Sp. Sci.*, 255, 69-75

VAN DEN ANCKER, M.E., WESSELIUS, P.R. ET AL., 1999, *Astron. Astrophys.*, 348, 877

VAN DISHOECK, E.F., WRIGHT, C.M., CERNICHARO, J. ET AL., 1998, *Astrophys. J. Letters*, 502, L173

WEINGARTNER, J.C. & DRAINE, B.T., 1999, in *The Universe as seen by ISO*, eds P. Cox & M.F. Kessler, ESA Publications Division SP-427, Noordwijk, p783-786

WRIGHT, C.M., DRAPATZ, S., TIMMERMANN, R. ET AL., 1996, *Astron. Astrophys.*, 315, L301

WRIGHT, C.M., TIMMERMANN, R. & DRAPATZ, S., 1997, in *First ISO Workshop on Analytical Spectroscopy*, eds. A.M. Heras et al., ESA Publications Division SP-419, Noordwijk, p311-312

WRIGHT, C.M., VAN DISHOECK, E.F., COX, P., SIDHER, S. & KESSLER, M.F., 1999, *Astrophys. J. Letters*, 515, L29

H2 in Molecular Supernova Remnants

By William T. REACH AND Jeonghee RHO

California Institute of Technology, Infrared Processing and Analysis Center, MS 100-22, Pasadena, CA 91125, USA

We discuss *ISO* observations of infrared ionic and H_2 lines toward molecular shocks in the supernova remnants 3C 391, W 28, and W 44. The total surface brightness of the H_2 lines toward these lines of sight exceeds that of atomic fine structure lines, showing that these lines of sight are dominated by dense molecular shocks. The H_2 excitation and the presence of bright ionic lines require that there are multiple shocks into gas with a range of pre-shock densities from 10–10^3 cm^{-3}.

1. Introduction

Massive stars end their lives in supernova explosions, and they do not live long enough to travel far from their parent molecular clouds. Therefore, supernovae frequently occur inside molecular clouds, providing compression, turbulence, cosmic rays, radiation, and heat. Using the *Infrared Space Observatory*, we performed a set of observations designed to search for infrared emission from the gas and dust that gets excited in molecular shock fronts. When the shock front passes through a molecular cloud, the gas cools via the most 'convenient' transitions available to it: low-density gas cools via atomic fine structure lines from the abundant ions, while molecular gas cools via the large number of rotational and/or vibrational transitions available. The first results of our project were the detection of bright [O I] 63 μm lines (Reach & Rho 1996), proving that abundant energy was being pumped into the gas by the shock fronts. The second results were the detection of H_2O and OH lines due to formation of molecules in the dense, warm, post-shock gas (Reach & Rho 1998). In this contribution, we describe the results of observing the H_2 rotational lines. The H_2 molecular is easily excited in warm molecular gas, making molecular supernova remnants some of the brightest sources of extended H_2 emission in the infrared.

Although most of the energy of molecular supernova remnants is expected to be emitted in the infrared, supernova remnants were difficult to detect with *IRAS* (Arendt 1996). The targets we observed were chosen from the new class of 'mixed-morphology' supernova remnants, which have shell-like radio emission and thermal, centrally-filled X-ray emission (Rho & Petre 1998). The specific locations within the remnants were chosen based on the locations of OH masers (Frail et al. 1996).

2. Far-infrared spectra

Figure 1 shows the complete spectra from 40–190 μm for our three target remnants; the spectra were made with the *ISO* Long-Wavelength Spectrometer (Clegg et al. 1996). The long-wavelength (> 120 μm) continuum is mostly due to unrelated line-of-sight emission (note that these remnants are located in the Galactic plane); the shorter wavelength continuum is at least partially due to the remnants (cf. Reach & Rho 1996). The brightest spectral lines are clearly evident even on this large scale and at this moderate spectral resolution. They are due to the most abundant interstellar elements with ground states of non-zero spin (such that low-energy ground-state fine-structure transitions exist). By comparing the list of observed lines to our 'periodic table for atomic fine structure lines'

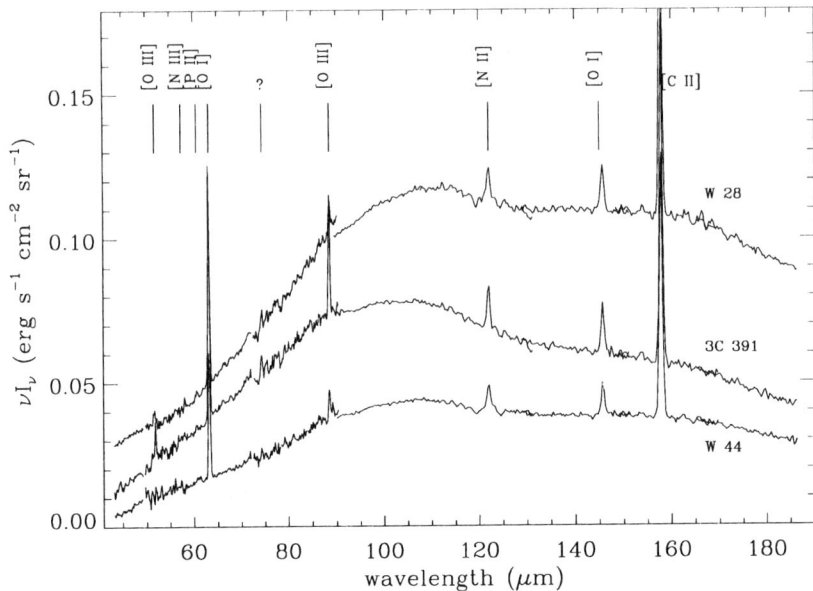

FIGURE 1. Far-infrared spectra of 3 molecular supernova shocks, made with the *ISO* Long-Wavelength Spectrometer. The prominent fine structure lines are labeled.

(Reach & Rho, in preparation), we found that all of the lines we observed can be explained on basic principles (including the first detection of the [P II] 60.6 μm line), except for some lines of OH and H_2O (Reach & Rho 1998) and one unidentified line at 74.3 μm. It is highly unlikely that the unidentified line is an atomic fine structure line, because we have checked that none of the abundant elements have appropriate transitions at this wavelength. Therefore, we surmise that the 74.3 μm line is probably a molecular feature. So far, we have not found a molecule (or ion) that could be responsible for this line.

3. H_2 line spectra

For the rest of this contribution, we concentrate on the H_2 lines. We used the *ISO* Short Wavelength Spectrometer (de Graauw et al. 1996) to observe 2 H_2 lines for each remnant. We chose the S(3) and S(9) pure rotational lines, because the C-shock model predicted much more emission in the S(9) line (Draine et al. 1983), and the J-chock model predicted much more emission in the S(3) line (Hollenbach & McKee 1989), such that a detection of one or the other line should be a discriminant between these two models. At the same time as the two H_2 lines were observed, we took spectra of the [Si II] 35 μm and [Fe II] 26 μm lines with other detectors in the instrument. Figure 2 shows the spectra of the H_2 lines. For all 3 remnants, both lines were detected, with more energy emitted from the S(3) transition.

We estimated the total brightness of H_2 line emission as follows; the results of each step in the calculation are shown in Table 1.

(i) The observed brightness of the lines is shown in the first two rows; in both cases, the observed flux in the beam was divided by the beam solid angle to get the surface brightness.

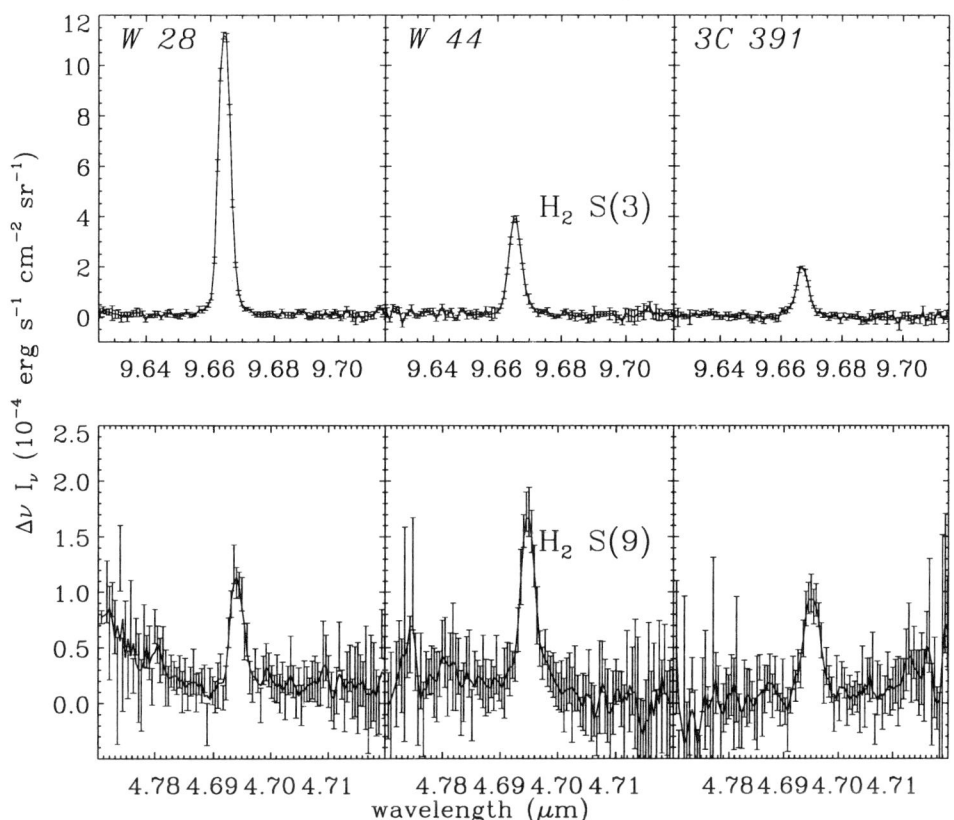

FIGURE 2. *ISO* SWS spectra of two H_2 lines observed toward three supernova remnants. The top row shows the S(3) lines, and the bottom row shows the S(9) lines for each remnant. The flux densities were multiplied by the spectral resolution $\Delta\nu$ and divided by the beam solid angle.

step	quantity	3C 391	W 28	W 44	units
(i)	$I[S(9)]$	1.9×10^{-4}	1.2×10^{-4}	2.4×10^{-4}	erg cm^{-2} s^{-1} sr^{-1}
	$I[S(3)]$	4.7×10^{-4}	12.7×10^{-4}	5.8×10^{-4}	erg cm^{-2} s^{-1} sr^{-1}
(ii)	A_V	13	4	7	mag
(iii)	T	1230	1004	1240	K
(iv)	I_{tot} H_2	1.2×10^{-2}	1.6×10^{-2}	1.5×10^{-2}	erg cm^{-2} s^{-1} sr^{-1}
(v)	$I[$ O I$]$ [erg cm^{-2} s^{-1} sr^{-1}]	1.4×10^{-3}	0.9×10^{-3}	0.8×10^{-3}	erg cm^{-2} s^{-1} sr^{-1}
(vi)	I_{tot} ions [erg cm^{-2} s^{-1} sr^{-1}]	5.5×10^{-3}	3.7×10^{-3}	2.5×10^{-3}	erg cm^{-2} s^{-1} sr^{-1}
(vii)	H_2 / Ions	2	4	6	

TABLE 1. Brightness of H_2 and Ion Lines

(ii) The brightness of the observed lines were corrected for extinction using the extinction derived from the H column density from X-ray absorption Rho & Petre 1998.

(iii) Next, using the ratio of the S(9)/S(3) line brightness, we calculated the excitation temperature of H_2.

(iv) Then, we calculated to total emission from all H_2 lines, assuming the same excitation temperature applies to all transitions. While this assumption will certainly fail for the higher-energy transitions, it is probably sufficient for the present purposes, because most of the energy comes out in rotational transitions similar to the ones that we observed.

(v) For comparison, we list the brightness of the [O I] 63 μm line, which is one of the brightest infrared lines from our remnants, and which is slightly brighter than the individual H_2 lines we observed.

(vi) The total brightness of infrared atomic fine structure lines is listed, including the bright [Si II] [Fe II], and [O III] lines.

(vii) Finally, we list the ratio of the total brightness of H_2 lines to the total brightness of fine structure lines. It is evident that the H_2 lines are brighter than the fine structure lines for the remnants we observed.

3.1. Conclusions

Molecular supernova remnants are bright sources of H_2 line emission. Based on existing theoretical models, the observed line brightnesses require both non-dissociative C-type shocks (to make the S(3) line from warm molecular gas) and faster, dissociative J-type shocks (to make the S(9) line from cooler molecules reformed behind the shock). Even faster shocks are required to explain the bright atomic fine structure lines, and of course faster-yet shocks are required to create the X-rays-emitting ions. Therefore there must be a mix of different shocks occurring in the same regions, due to the presence of a wide range of pre-shock densities in the parent molecular clouds. In order to disentangle the various types of shocks, we will need to zoom in with higher angular resolution, while maintaining enough spectral resolution to separate ion and molecular lines. For now, we can already see the different lines that trace different pre-shock densities, and we can see that the relatively dense gas which, after being shocked, is cooled by H_2 line emission, is one of the dominant phases of the molecular clouds.

REFERENCES

Arendt, R. G. 1996, *Astrophys J. Suppl.* **70**, 181–212.

Clegg, P. E., et al. 1996, *Astron. Astrophys.* **315**, L38–L42.

de Graauw, T., et al. 1996, *Astron. Astrophys.* **315**, L49–L54.

Draine, B. T., Roberge, W. G., & Dalgarno, A. 1983, ApJ, 264, 485

Frail, D. J. et al. 1996, *Astronom J.* **111**, 1651–1659.

Hollenbach, D. J., and McKee, C. F. 1989, *Astrophys. J.* **342**, 306–336

Reach, W. T., Rho, J. 1996, *Astron. Astrophys.* **315**, L277–L280.

Reach, W. T., Rho, J. 1996, *Astrophys. J.* **507**, L93–L97.

Rho, J.-H., Petre, R., 1998, *Astrophys. J.* **503**, L167–L170.

3D Integral Field H2 Spectroscopy in Outflows

By Jonathan A. TEDDS[1], Peter W. J. L. BRAND[1] AND Michael G. BURTON[2]

[1]Department of Physics & Astronomy, University of Leeds, UK

[2]Institute for Astronomy, University of Edinburgh, UK

[3]University of New South Wales, Australia

We describe the new capability provided by integral field spectroscopy for simultaneously mapping a wide range of shocked emission lines across outflows at high spatial resolution. We have used the MPE-3D near-IR integral field spectrometer on the AAT to carry out a detailed observational study of the physics of shocked H_2 and [Fe II] excitation *within* individual bow shocks. Simultaneous measurement of line ratio variations with position across and along bow shocks will strongly constrain shock models in a number of outflow sources. In Orion, where broad H_2 line widths had previously implied magnetically moderated C shocks, our higher resolution echelle observations of the H_2 velocity profiles in two of the bullets (Tedds et al. 1999) contradict *any* steady-state molecular bow shock models. This suggests that instabilities or supersonic turbulence may be important in this case. 3D measurements of the corresponding H_2 level populations will address this.

---

1. Introduction

The nature of molecular shocks, which play an important role in the processes of momentum and energy transfer within star forming molecular clouds (McKee 1989), is still uncertain (Draine & McKee 1993). In this paper we describe how new developments in integral field spectroscopy provide us with the opportunity to self-consistently distinguish between competing shock models. The Orion molecular cloud is the brightest known source of shocked H_2 emission and as such has been the primary test bed for theoretical models. We present preliminary results of 3D observations in the brightest, best-defined outflows associated with the Orion bullets.

2. Why Integral Field Spectroscopy?

To determine the excitation of shocked H_2 in outflows, we need to measure a range of near-IR emission lines at high spatial resolution over a reasonably large 2-D spatial field sufficient to cover a significant region at the curved heads of shocked outflows. There are two techniques normally employed to perform such a task: stepping the long slit of a spectrometer across the object in sequence or scanning a Fabry-Perot interferometer through the wavelength range while staring at the object (for velocity resolved studies). Both methods work well at spatial resolutions $\geq 1''$ but present significant problems when attempting subarcsecond scale measurements.

The fundamental difficulty is that both methods acquire only two of the required three dimensions of the datacube at any one time. Sequential stepping of a long slit spectrometer in either space or wavelength is inefficient and introduces significant errors due to variations in seeing, air mass and OH emission lines with time while using shorter integration times per frame introduces significant detector read-noise errors. To make more effective use of the field-of-view (FOV) provided by an instrument we wish to obtain a complete datacube in a single integration. If the FOV of such an instrument is larger than the object itself we may also relax the requirements for absolute positioning of a source and avoid any problems in reproducing telescope pointings.

The 3D instrument was developed at MPE (Krabbe et al. 1995a; Weitzel et al. 1996) to enable near-IR integral field spectroscopy on a 4m class telescope. The instrument includes an image slicer which transforms a 2-D image into a 1-D strip which exactly matches the entrance slit of a conventional cooled long slit grism spectrometer. Hence, the second dimension of the detector is used for spectral dispersion as normal. A cold closed-loop piezo-driven tilt mirror allows full spectral sampling. 3D can therefore *simultaneously* obtain 256 H or K band spectra at R=1100 or 2100 from a square 16×16 pixel field on the sky. Mounted on the 4m Anglo Australian Telescope (AAT) this provided a pixel scale of 0.4"/pixel and a 6.4" FOV. A Rapid Off-axis GUider Experiment (ROGUE) provides high sensitivity first order seeing corrections using a piezo controlled mirror in order to correct for image motion in real time.

We now go onto describe our latest observations of the Orion bullets and unique 3D observations to enable the measurement of variations in H_2 excitation *within* individual bow shocks.

3. Why Orion?

The discovery of Fe^+ bullets in the Orion outflow (Allen & Burton 1993) emphasised the importance of bow shock morphology and allowed the first high resolution measurements of bow-shock dynamics (Tedds et al. 1999) and excitation (Tedds et al., *in prep.*) in this bright source.

3.1. H_2 Dynamics

We have used the recently upgraded CGS4 spectrometer at UKIRT to observe [FeII] $a^4F_{5/2} - a^4D_{7/2}$ 1.644 μm velocity profiles of selected bullets in Orion and H_2 1-0 S(1) 2.122 μm velocity profiles for a series of positions along and across the corresponding bow-shaped wakes (Tedds et al. 1999). Integrated [FeII] velocity profiles of the brightest bullets are consistent with theoretical bow shock predictions. Background subtracted H_2 1-0 S(1) emission line profiles in the H_2 bullet wakes are dominated by a broad (intrinsic FWHM$\leq$27kms^{-1}) but singly-peaked profile. At no point on the wake is there any evidence of the double-peaked profile structure in the dominant emission profile centred near zero-velocity, as would be expected for bow shocks oriented close to the plane of sky. It is therefore very difficult to reconcile *any* steady-state molecular bow shock model with these observations in Orion and other mechanisms are being investigated (e.g. Bertoldi et al.; Pineau des Forets et al., *this conference*). To fit a single C shock absorber model to individual H_2 profiles implies a magnetic field strength and hence magnetic pressure much higher than observed estimates (Smith et al. 1991) and is not consistent with the bow-shaped wake morphology.

3.2. H_2 Excitation

Observations of a range of H_2 column densities in the K band using CGS4 at UKIRT (Tedds et al., *in prep.*) show a range of gas temperatures as expected for cooling, post-shock gas. Also, superimposed on the shock excited line ratios is a faint signature due to fluorescence. This is strongest off the wake (from the background nebular emission), but also present in the wake and may be attributable to UV-excitation from ionization at the bullet head. Uncertainties in the slit positioning and the relatively large 1.23" CGS4 pixel size do not allow us to accurately compare excitation at different positions

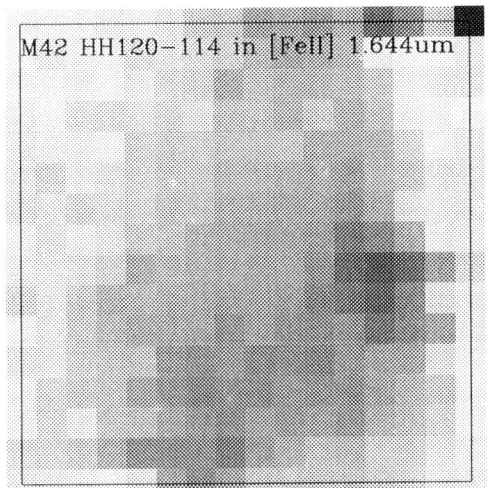 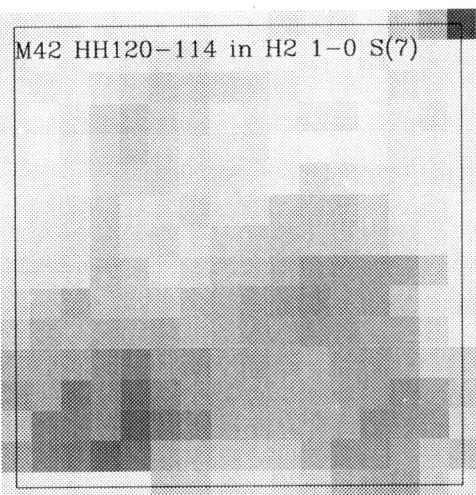

FIGURE 1. Two slices through a 3D H band datacube of the Orion bullet M42 HH120-114 showing [FeII] bullet and associated H$_2$ wake. *Left:* Cut at 1.644 µm shows highly excited [FeII] line emission associated with the tip of the bullet. *Right:* Cut at 1.748 µm shows shocked H$_2$ 1-0 S(7) [E$_{upper}$=12817K] line emission associated with the bow shock of H$_2$ created by passage of the bullet through the surrounding medium. Higher energy H$_2$ transitions were similarly measured in a corresponding K band 3D datacube at the same position.

within the H$_2$ wake. With 3D there is no uncertainty in the relative pointings of adjacent pixels and so we can look for variations of H$_2$ line ratios *within* this wake as are expected in bow C shock models previously used to model observations of shocked H$_2$ in Orion (Smith et al., 1991b). If instabilities are playing a prominent role we can measure any variations in the H$_2$ level populations in such a case.

4. 3D Observations

Observations of the Orion bullet M42 HH120-114 were made using 3D at the AAT. Fig 1 shows two cuts in different emission lines measured simultaneously through the 3D H band datacube. The tip of the bullet emits in lines of decreasing excitation, such as [OIII], Hα, [SII] as measured by HST (O'Dell et al., 1997) and then [FeII] as seen in these observations (Fig 1, left), followed by a wake of H$_2$ as seen in the H$_2$ 1-0 S(7) line (Fig 1, right). This progressively decreasing excitation is expected as the shock velocity decreases as the normal component of the bullet velocity along the wake. The ratio of the [FeII] lines we have measured at H from the head is sensitive to the gas density and also provides two pairs of lines from common upper state levels so that differential extinction inside the bullets may be estimated. This in turn will provide constraints on the amount of material in the bullet.

We also observed a 3D K band datacube in order to ratio more highly excited transitions of H_2. The pixel size samples the bullets well, and the spectral resolution and coverage allow key lines from the $v = 1$ to $v = 4$ energy levels to be observed simultaneously. The relative strengths of these lines is particularly sensitive to the mode of excitation (see Brand et al., 1988). By ratioing the bright $v = 1$ lines to the higher energy $v = 2 - 4$ lines we will construct H_2 column density ratio diagrams for *each* 3D pixel shown. We will then self-consistently test the proposed bow C shock models, in which significantly different line ratios occur depending on the shock velocity which varies in the bow. Further 3D observations in a well resolved bow shock in the HH212 outflow have been made for comparison.

5. Conclusions

The 3D integral field spectrometer provides an ideal tool with which to examine further the H_2 excitation within shocked outflows. Our 3D observations will unambiguously determine any variations of the H_2 excitation with position measured *within* individual bow shocks in the Orion bullets. If initial long slit observations (Tedds et al., in prep.) indicating near constancy of H_2 excitation with position are confirmed by 3D this will be inconsistent with bow C shock models previously fitted at OMC-1, in which significantly different line ratios occur depending on the shock velocity which varies in the bow. Furthermore, our observations of broad, singly-peaked H_2 velocity profiles within the most clearly resolved bullet wakes in Orion are extremely difficult to reconcile with existing steady-state shock models (Tedds et al. 1999). It may therefore be necessary to model the effects of instabilities and turbulence. Further 3D observations in other outflows such as HH212 will help constrain molecular shock models if the outflows in Orion prove to be dominated by instabilities.

We would like to acknowledge Lowell Tacconi-Garman for his expertise and enthusiastic help with the excellent 3D instrument. JAT gratefully acknowledges a University Research Fellowship from the University of Leeds and a Research Fellowship from the Royal Commission for the Exhibition of 1851 during this work.

REFERENCES

ALLEN, D.A. & BURTON, M.G. 1993 *Nature* **363**, 54.

BRAND, P.W.J.L. MOORHOUSE, A. BURTON, M.G. GEBALLE, T.R. BIRD, M. & WADE, R. 1988 *Ap. J. Lett.* **334**, L103.

DRAINE, B.T. & MCKEE, C.F. 1993 *Ann. Rev. Astron. Astrophys.* **31**, 373.

KRABBE, A. WEITZEL, L. KROKER, H. et al. 1995a *3D - A new generation imaging spectrometer*, in: *Proc. of SPIE conference on: Infrared imaging Systems, Orlando* **2457**, 172–183.

MCKEE, C.F. 1989 *Ap. J.* **345**, 782.

O'DELL, C.R HARTIGAN, P. LANE, W.M. WONG, S.K. BURTON, M.G. RAYMOND, J. & AXON, D.J. 1997 *Astron. J.* **114**, 730.

SMITH, M.D. BRAND, P.W.J.L. & MOORHOUSE, A. 1991b *MNRAS* **248**, 730.

TEDDS, J.A. BRAND, P.W.J.L. & BURTON, M.G. 1999 *MNRAS* **307**, 337.

WEITZEL, L. KRABBE, A. KROKER, H. THATTE, N. TACCONI-GARMAN, L.E. CAMERON, M. & GENZEL, R. 1996 *A&AS* **119**, 531.

Near-infrared Imaging and [OI] spectroscopy of IC 443 using 2MASS and ISO

By J. Rho, S. Van Dyk, T. Jarrett, R.M. Cutri, AND W.T. Reach

Infrared Processing and Analysis Center, California Institute of Technology, MS 100-22, Pasadena, CA, 91125, USA

We present near-infrared imaging of IC443, covering entire supernova remnant (50' diameter) from the Two Micron All Sky Survey (2MASS), which images are taken simultaneously in the J (1.25μm), H (1.65μm) and K$_s$ (2.17μm) bands. Emission from IC443 was detected in all 3 bands from most of the optically bright parts of the remnant, revealing a shell-like morphology. These are the first near-infrared images that covers entire remnant. The color and structure are very different between the northeastern and southern parts. Bright J and H band emission from the northeast rim can be explained mostly by [Fe II] and the rest by hydrogen lines of Pβ and Br10. We also report ISO LWS observation of [O I] (63μm) for 11 positions in the northeast. Strong lines were detected and the strongest line is in the northeastern shell, where 2MASS image showed filamentary structure in J and H. In contrast, the southern ridge is dominated by K$_s$ band light with knotty structure, and has weak J and H band emission. The shocked H$_2$ line emission is well known from the sinus ridge produced by an interaction with dense molecular clouds. The large field of view and color of the 2MASS images show that the H$_2$ emission extends to the east and the northeast. This H$_2$ emission suggests that the interaction with the molecular clouds extends to the front side in the northeast. The 2MASS color differences result because the emission of the northeastern rim are all ionized ionic lines while of the south are mostly shocked H$_2$. Comparison of the shock models imply that the dominant shock at the northeastern rim is a fast J-shock with a shock velocity of v_s = 80-100 km s^{-1} and a density of $10 << n_o \lesssim 1000$ cm^{-3}, while the dominant shock at the south is a slow C-shock with v_s = 30-40 km s^{-3} and $n_o = 10^4$ cm^{-3}.

1. Introduction

IC 443 is the only well-known case of supernova remnant-molecular cloud interaction: broad molecular lines, and the shocked H$_2$ emission have been detected along the southern sinus ridge (van Dishoeck et al. 1993; Burton et al. 1988). IRAS observations show bright emission from IC 443, which is interpreted as continuum of two components of thermal emission from heated small and large dust grains (Arendt 1989). However, recent ISO observations show no continuum due to heated small grains (Oliva et al. 1999; Cesarsky et al. 1999). While the ISO SWS observation detected only ionized ion lines such as [Ne II] and [Fe II] and [Si II], the ISOCAM observation from the south shows detection of only H$_2$: this raised a question if the primary contributor of 12 and 25μm IRAS emission is ionized gas or shocked H$_2$ emission.

2. Observations

The Two Micron All Sky Survey (2MASS) is being carried out by a pair of identical and dedicated 1.3 meter Cassegrain equatorial telescopes which are mapping the entire sky with 3.5" spatial resolution in the J (1.13-1.37μm), H (1.5-1.8μm), and K$_s$ (2-2.32μm) bands to a limit of 17.1, 16.4, and 15.6 mag, respectively. 2MASS has unique way of scanning the sky for the best efficiency: the telescope moves continuously in declination at $\sim$57" per second for 6 degree of tile, and the telescope's secondary mirror executes a sawtooth pattern of motion which freezes the image of sky on the focal plane. The details

FIGURE 1. Mosaiced 2MASS Atlas image of IC 443. J, H and K_s images show color contrast between northeastern rim (J and H emission) and southern ridge (K_s emission); northeastern rim is dominated by [Fe II] and Pβ (J) and [Fe II] (H), while southern sinus ridge emission is mostly H_2 lines in all J, H and K_s band.

on 2MASS project are described in Skrutskie et al. (1997) and 2MASS Explanatory Supplement (Cutri et al. 1999). The observations toward IC443 took place on Nov. 23, 1997, and twenty-four survey atlas images covered IC443 (in Fig. 1). The background variation between the images was less than 1%.

ISO LWS observation of [O I] (63μm) was performed for 11 positions, which cross the northeastern rim with an interval of 160 arcsec. The LWS beam size is 80″. The observation took place 1998, Feb 27 and approximately 1 minute per spectrum was taken; the total observation including calibrations took 20 minutes.

3. Results

Near IR-emission from IC443 was detected in all three 2MASS bands from most of the optically bright parts of the remnant, revealing a shell-like morphology as shown in Fig. 1. These are the first near-infrared images that cover entire remnant. The color and structure are in sharp contrast between the northeastern and southern parts. While the northeastern rim is dominated by J and H band emission, all of the southern ridge and the east is dominated by K_s band emission. The surface brightness in Fig. 1 ranges from (0.4 - 3.2), (0.13 - 3.6), (0.56 - 18) with units of 10^{-4} erg s^{-1} cm^{-2}sr^{-1} for J, H and K_s band, respectively. The K_s band image after the stars are subtracted using DAOPHOT package is shown in Fig. 2a, which shows the brightest emission among the 3 band images, in particular, at the southern sinus ridge.

The northeastern shell shows sheet-like filamentary structure, similar to that of optical

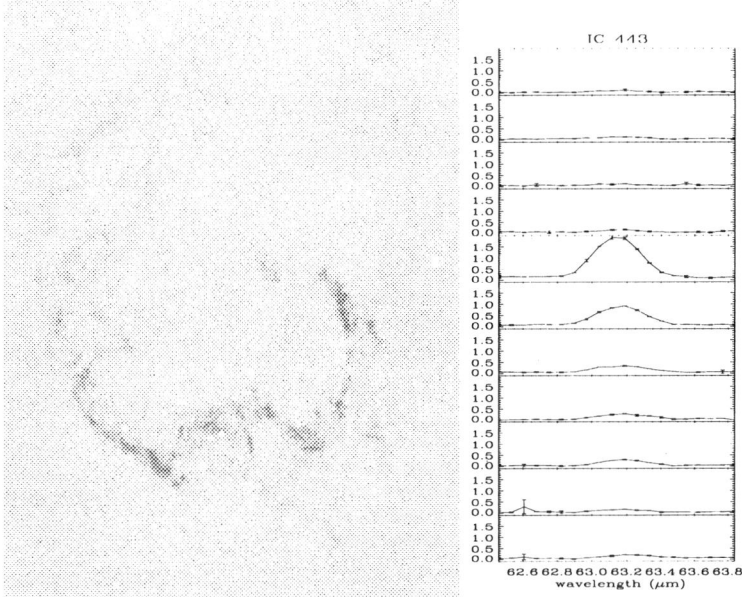

FIGURE 2. (a)K_s band image after subtracting stars. (b)ISO LWS [OI] Spectra of 11 positions. The unit of y-axis is 1×10^{-10} erg s^{-1}cm$^{-2}\mu$m^{-1}. The brightest line has a flux density of 4.8×10^{-4} erg cm^{-2}s^{-1}sr^{-1}. Spectra from top to bottom are from interior to outside the remnant. The peak position coincides with northeastern filaments shown in 2MASS image.

emission; the J: H: K_s ratio is 1:1:0.2. The possible emission sources that contribute to the near IR emission include hydrogen recombination, molecular hydrogen, forbidden ionic lines like [Fe II]. We assume there is no continuum within the 2MASS bands because ISO observations show no continuum for 5-14μm, and the estimated non-thermal continuum by interpolation from the radio flux is too small to contribute to 2MASS bands. First we examined H recombination lines within the 2MASS bands: J, H, K_s bands cover Pβ (1.28μ), Br10 (1.74μ) and Brγ (2.17μ), respectively. The ratio of J: H : K_s line is 1:0.17:0.2 with 3% changes depending on the temperature (2000 - 20000K) for case B. The [Fe II] line 1.64μm has been shown to be strong in a position on the northeast rim (Graham et al. 1987). We have calculated line intensities of [Fe II] for 13 levels including more than 45 lines, by solving the radiative transfer equations as a matrix using the atomic data from Nussbaumer & Storey (1988). With this model, the line ratio of I(17.9μm)/I(26μm) of 0.5 and I(1.64μm)/I(17.9μm) of 3 (Oliva et al. 1999) suggest a density of 600 cm^{-3} and temperature of 8,500 K. The ratio of [Fe II] intensities for J/H is 1.4 with 6 lines in J, 4 lines in H, and no lines in K_s; the ratio is largely based on the primary line ratio of [FeII](1.25μm)/[FeII](1.64μm)=1.35. The 2MASS colors suggest very weak K_s emission at the northeastern rim and, where there has been previously no evidence of H$_2$ emission; spectroscopy toward the northeastern rim (a few times more sensitive than 2MASS) did not detect the H$_2$ line (Graham et al. 1987). Therefore, K_s band emission for the northeastern rim can be explained by Brγ line alone, if the ratio of H$_2$/Brγ is constant within the rim. The J band emission is composed of [Fe II] (88%) and P β (12%), and H band is [Fe II] (>99%) and Br 10 (<1%).

In contrast to the northeastern rim, the south ridge is dominated by K_s band light with knotty structure (Fig. 2a). The J and H band emission is weak in the south, while K_s emission is bright. The shocked H$_2$ lines are well known from the sinus ridge (Burton et al. 1988), produced by an interaction with dense molecular clouds, e.g. H$_2$ 1-0 S(1)

(2.12μm) in K$_s$ band. We have estimated H$_2$ intensities within the 2MASS bands by deriving temperatures from the column density and energy of upper level using the data from Cesarsky et al. (1999) and Richter et al. (1995); three temperatures of 650, 1300, and 2580 K are required. From the H$_2$ line intensities of 9 lines in J, 6 lines in H and 5 lines in K$_s$, the J:H:K$_s$ ratios are predicted to be 0.08:0.1:1, which is approximately observed ratios from the images in the southern part. Therefore, most of 2MASS emission in the south can be explained by H$_2$ lines. The near-infrared line intensities of H$_2$ are largely consistent with the C-shock model of v$_s$=30-40km s^{-1} and n$_o$=10^4 cm^{-3}.

Strong [O I] (63μm) lines were detected using ISO LWS toward 11 positions in the northeast as shown in Fig. 2b, which peaks at the northeastern rim where the 2MASS image showed filamentary structure in J and H. The peak brightness of the [O I] line is 4.8$\times$10^{-4} erg s^{-1} cm^{-2} sr^{-1}.

4. Discussion

Using a line brightness list combined from [Fe II] lines, [OI] and other detected infrared and optical lines (Oliva et al. 1999; Fesen & Kirshner 1980) from the northeastern rim, we have compared various shock models. The ratio of [Ne III]/[Ne II], and the intensities of [Fe II](25.98μm) and [Si II](34.88μm) vary depending on the shock velocity (Hartigan et al. 1991). The ratio of [Ne III]/[Ne II]$\sim$0.55 implies two solutions of v$_s$ of 100 km s^{-1}, and 250 km s^{-1} with n$_o$ of 100-1000 cm^{-3}. However, [Fe II] and [Si II] intensities of 1.14, and 4.45 (10^{-4} erg s^{-1} cm^{-2} sr^{-1}) are not consistent with that for v$_s$ = 250 km s^{-1}. A fast J shock model (Hollenbach & McKee 1989; Hartigan 1992) with n$_o\sim$10^3 cm^{-3} and v$_s\sim$ 80 km s^{-1} is consistent with most of detected lines. Comparison of the line intensities, [OI], [Fe II] and [NeII], with the shock models as a function of the density, implies the density ranges 300-1000 cm^{-3}, and is not consistent with a low density (10cm^{-3}) J-shock model. In summary, the 2MASS color differences result because the emission in the northern rim is only ionic lines while in the south it is mostly shocked H$_2$, which are consistent with previous ISO observations. 2MASS color differences are also due to differences in the dominant shock type, shock velocities and densities: J-shock, v$_s$ of 80-100 km s^{-3}, and 10$<<$n$_o$ $\lesssim$ 1000 cm^{-3} can reproduce the detected lines and their intensities in the northeastern rim, while C-shock, v$_s$ of 30-40 km s^{-3}, and n$_o$=10^4 cm^{-3} in the southern sinus ridge.

The large field of view and color of the 2MASS image shows that the H$_2$ emission extends to the east and the front side of the northeastern part. This H$_2$ emission suggests that the interaction with the molecular clouds extends to the front side in the northeast where the remnant is abutting a molecular cloud shown in CO maps (Cornett et al. 1976), as well as in the southern ridge.

REFERENCES

Burton, M.G., Geballe, T.R, Brand, P.W., & Webster, A.S., 1988, MNRAS , 231, 617
Cesarsky, D, Cox, P., Pineau des Forets, G.P. et al., 1999, A&A, 348, 945
Cornett, R.H., Chin, G. and Knapp, G.R., 1977, ApJ, 54, 889
Cutri, R. M. et al., http://www.ipac.caltech.edu/2mass/releases/first/doc/explsup.html
Fesen, R.A, & Kirshner, R.P., 1980, ApJ, 242, 1023
Graham, J.R., Wright, G.S., & Longmore, ApJ, 1987, ApJ, 313, 847
Hartigan, P., Curiel, S. Raymond, J., 1987, APJ, 316, 323
Hollenbach, D. & McKee, C. F., ApJ, 1989, 342, 306
Oliva, E., Lutz, D., Drapatz, S., & Moorwood, A. F.M., 1999, A&A, 341, 75
Richter, M.J, Graham, J.R., & Wright, G. S., ApJ, 1995, 454, 277
Skrutskie, et al., "The impact of large scale near-IR sky surveys", ed.by F. Garzon et al., p25
van Dishoeck, E. F., Jansen, D.J, & Phillips, T.G., 1993, A&A, 279, 541

ISOCAM spectro-imaging of the supernova remnant IC 443

By Pierre COX[1], D. CESARSKY[2] AND G. PINEAU DES FORÊTS[1,3]

[1]Institut d'Astrophysique Spatiale, Université de Paris XI, F-91405 CEDEX Orsay, France

[2]Max-Planck-Institut für extraterrestrische Physik, Garching, Germany

[3]DAEC, Observatoire de Paris, F-92195 Meudon Principal CEDEX, France

We describe spectro-imaging observations of the bright western ridge of the supernova remnant IC 443 obtained with the ISOCAM circular variable filter (CVF) on board the *Infrared Space Observatory (ISO)*. The CVF data show that the 5 to 14 μm spectrum is dominated by the pure rotational lines of molecular hydrogen (v = 0–0, S(2) to S(8) transitions). We compare the data to a new time-dependent shock model.

1. Introduction

The supernova remnant IC 443 is a prime example of the interaction of a supernova blast wave with an ambient molecular cloud. On optical plates, IC 443 appears as an incomplete shell of filaments (Fig. 1) with a total extent of about 20 arcmin, i.e. ~ 9 pc for an adopted distance of 1,500 pc. The shock generated by the supernova explosion, that occurred $(4-13) \times 10^3$ years ago, encountered nearby molecular gas which is mainly found along a NW-SE direction across the face of the optical shell. IC 443 has been the subject of numerous studies from X-rays, visible, infrared to radio wavelengths (e.g., Mufson et al. 1986 and references therein). Studies of the interaction between the shock and the ambient molecular gas were done by observing molecular hydrogen in the rotational–vibrational transitions (Burton et al. 1988, 1990 - see Fig. 1 - and Richter et al. 1995a), in the pure rotational S(2) transition (Richter et al. 1995b).

Recently, Cesarsky et al. (1999) reported mapping results of the pure rotational lines of H_2 using the ISOCAM CVF over the western ridge of IC 443, a position corresponding to clump G in the nomenclature of Huang et al. (1986). This ridge corresponds to a location where the interaction between the blast wave of the supernova and ambient molecular gas is amongst the strongest. These observations reveal the details of the structure and the physical conditions of the shocked molecular gas in IC 443 with a pixel field of view of 6″ and at unprecedented sensitivity (mJy).

2. Observations and results

ISOCAM was pointed towards a position corresponding to the center of the molecular clump G which is also a peak in the H_2 emission (Fig. 1). The observations and the data reduction are described in Cesarsky et al. (1999). Figure 2 shows the total emission of the H_2 lines between 5 and 13.5 μm, i.e. the sum of the S(2) to S(7) lines, together with the integrated line intensity of the S(2) and S(7) transitions (top panels). The 5 to 14 μm spectra towards three peak positions are shown in the middle panels. They are dominated by the series of the pure ($v = 0 - 0$) rotational lines of molecular hydrogen from the S(2) to the S(8) transitions. In particular, there is no indication of any atomic fine structure line. Note that a weak contribution from dust bands, unrelated to the

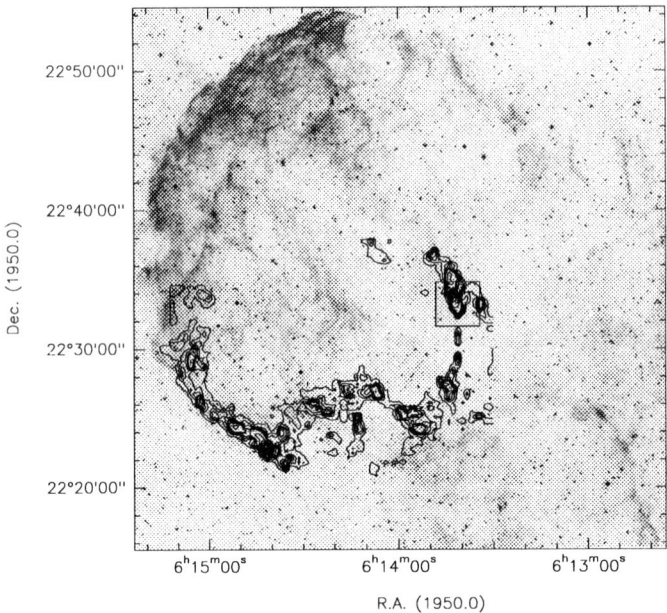

FIGURE 1. The footprint of ISOCAM (box) depicted against the DSS image of IC 443 together with the H_2 1–0 S(1) emission (contours) from Burton et al. (1990).

supernova remnant, has been subtracted as explained in Cesarsky et al (1999). The H_2 emission is found along a ridge of about $30'' \times 80''$ (0.25 pc × 0.65 pc) running SW to NE, a structure which is comparable to that seen in CO or HCO^+ (van Dishoeck et al. 1993, Tauber et al. 1994). The higher spatial resolution of the ISOCAM data clearly reveal a series of knots sitting on a plateau. The H_2 knots are very bright with peak values of a few 10^{-3} erg s^{-1} cm^{-2} sr^{-1}. The eastern side of the molecular ridge, facing the origin of the supernova explosion, appears sharper than the opposite side where weak emission is found extending westwards.

3. Discussion

Altogether there are about 130 pixels that show H_2 emission with intensities above the $10\,\sigma$ level, i.e. $> 2 \times 10^{-4}$ erg s^{-1} cm^{-2} sr^{-1} for all the six rotational transitions S(2) to S(7). This fact allows the construction of as many H_2 excitation diagrams which plot the logarithm of the column density, corrected for statistical weight, in the upper level of each H_2 transition vs. the energy of that level, E_u. When calculating the column densities, we corrected the observed line fluxes for extinction, using a screen model and the extinction curve from Draine & Lee (1984). The middle and bottom panels in Fig. 2 present the CVF spectra and the corresponding excitation diagrams of the three emission peaks labeled A, B, C. Peak A corresponds to a position studied by Moorhouse et al. (1991) and Richter et al. (1995a - their Position 3). The statistical weights used in Fig. 2 include a factor of 3 for ortho-H_2 and 1 for para-H_2.

The excitation diagrams for Peaks A, B, and C show that a single excitation temperature does not reproduce the H_2 lines observed in IC 443 and that emission from gas with a range of temperatures is required. The results of a simple LTE two-component H_2 model are shown in Fig. 2: a 'warm' H_2 component with an excitation temperature of ~ 500 K and typical column densities in between 10^{20} and 10^{21} cm^{-2}, and a 'hot' H_2

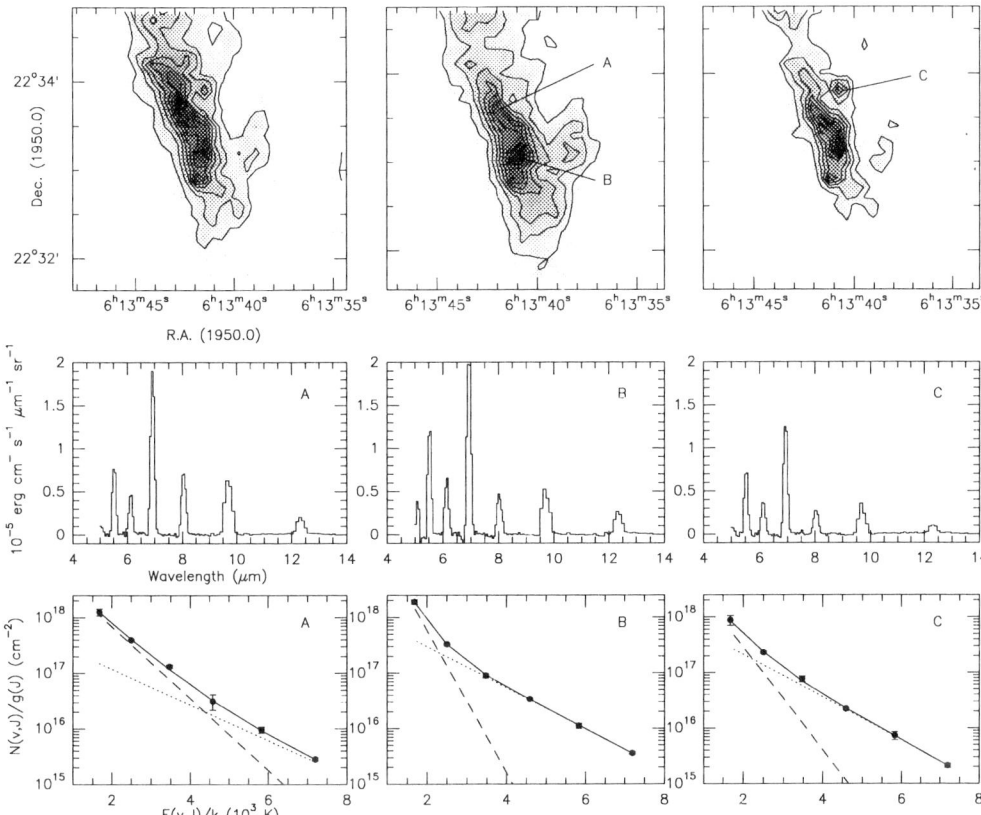

FIGURE 2. The distribution of the emission of the S(2) to S(7) H$_2$ lines towards Clump G in IC 443 (upper left panel). The next top panels show the emission in the S(2) and S(7) H$_2$ lines. Contours are drawn at the 10%, 20%, etc. level. The corresponding peak strengths are 8.1×10^{-3}, 7.9×10^{-4} and 2.2×10^{-3} erg s^{-1} cm^{-2} sr^{-1} from left to right. The middle panels show the ISOCAM CVF spectra towards the three emission peaks A, B, and C. The lower panels present the excitation diagrams after correcting for an extinction of $A_{2.12\,\mu m} = 0.6$ mag. Each excitation diagram has been fitted with a two-component model involving 'warm' (dashed line) and 'hot' (dotted line) H$_2$. The full lines represent the sum of both H$_2$ components - see text for details.

component with $T_{ex} \sim 1200$ K and N_{H_2} a few 10^{19} cm^{-2}. The parameters of the 'warm' component are determined almost entirely by the intensities of the S(2) and S(3) lines and the 'hot' component dominates the H$_2$ transitions S(4) to S(7). Although the evidence for a 'warm' component is very strong, the ISOCAM data only poorly constrain its properties because of the lack of measurement of the S(1) and S(0) H$_2$ transitions. Furthermore, the uncertainty in the extinction correction (especially for the S(3) line whose position coincides with the peak of the silicate 9.7 μm band) introduces an additional uncertainty in the temperature determination. Using different extinction laws and adopting values for $A_{2.12\,\mu m}$ between 0.5 and 1 mag., we derive typical uncertainties of ± 100 and ± 250 K for the warm and the hot components, respectively.

ISO spectroscopy of regions where shocks dominate the excitation has revealed that the shocked gas has a range of temperatures from a few 100 K to several 1000 K: Cepheus A (Wright et al. 1996), Orion (Rosenthal et al. 1999) and bipolar outflows (Cabrit et al. 1999). The H$_2$ lines cannot be explained by a single shock model and combinations

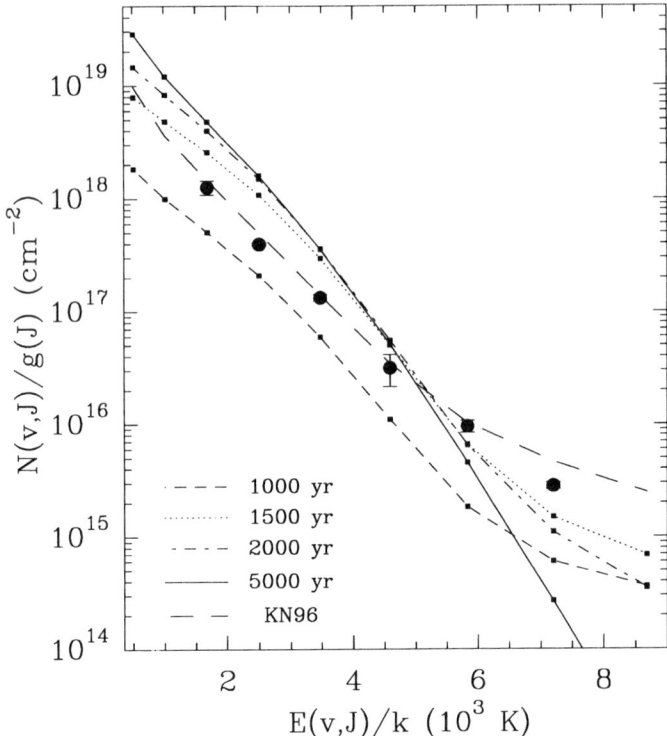

FIGURE 3. The excitation diagram for Peak A (filled circles) compared to the predictions of the time-dependent shock model of Chièze et al. (1998) with a pre-shock density of 10^4 cm^{-3}, a shock velocity of 30 km s^{-1} and four evolutionary times. The long-dashed curve labelled (KN96) presents the predictions of a model with two C-shocks from Kaufman & Neufeld (1996) - see text.

of J- and/or C-type shocks with different velocities and pre-shock densities have been invoked to account for the observed line fluxes. Similar conclusions have been reached for IC 443. Based on an analysis of the [O I] 63 µm fine structure line and near-infrared H_2 lines, Burton et al. (1990) concluded that the infrared line emission of IC 443 can only be modeled as a slow (10-20 km s^{-1}), partially dissociating J shock where the oxygen chemistry is suppressed, i.e. the cooling is dominated by [O I] emission and not by H_2O cooling - see also Richter et al. (1995a, 1995b).

Following the interpretation of Wright et al. (1996) for the shocked H_2 gas in Cepheus A, the H_2 lines in IC 443 can be fit by a combination of two C-shocks from the models of Kaufman & Neufeld (1996). A good match to the data towards Peak A is obtained combining a first shock with a pre-shock density of 10^4 cm^{-3}, a velocity of 20 kms^{-1} and a covering factor Φ of 0.85 with a second shock of 10^6 cm^{-3}, 35 kms^{-1} and Φ of 0.008 (see Fig. 3). Such a steady-state model requires at least two C-shocks with a set of 3 free parameters and relatively high pre-shock densities for the high velocity component.

As pointed out by Chièze et al. (1998), the intensities of the ro-vibrational H_2 lines are sensitive to the temporal evolution of a shock wave. In many astrophysical situations, shock waves are unlikely to have reached steady-state which occurs at approximately 10^4 yr. At times scales of a few 10^3 yr, the shocked gas may show both C- and J-type characteristics: within the C-shock, a J-type shock is established heating a small fraction of the gas to high temperatures (Flower & Pineau des Forêts 1999). In the case

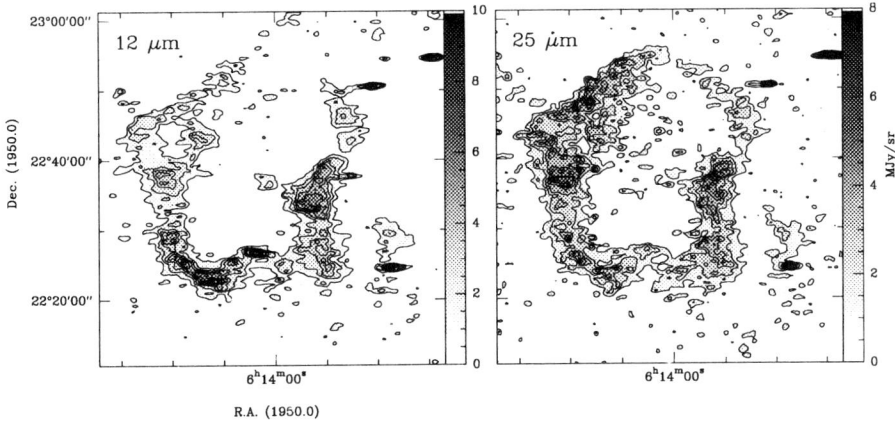

FIGURE 4. The distribution of the *IRAS* 12 and 25 μm emissions in IC 443. The contour levels go from 1 to 8 by steps of 1 MJy/sr.

of IC 443, the typical size of the molecular clump G is $\leq 20''$, i.e. $\leq 2 \times 10^{17}$ cm, but some of the molecular clumps have typical sizes of about a few arcsec (Richter et al. 1995a). For a shock velocity of $20-30$ kms^{-1}, the crossing time of the shock wave is thus $\leq 2,500-4,000$ yr indicating that the shock wave in clump G is not in steady-state.

Figure 3 shows the predictions of the time-dependent shock model for an observation along the direction of the shock propagation. The model results are given for four evolutionary times with the following parameters: pre-shock gas density $n_H = 10^4$ cm^{-3}, shock velocity $v_s = 30$ kms^{-1}, magnetic field strength B = 100 μG and ortho-to-para ratio of 3. The filling factor in this model is equal to 1. A smaller filling factor could be compensated by a larger line of sight path across the shocked H$_2$ gas for a non face-on shock. The post-shock densities are 10^5-10^6 cm^{-3} comparable with the values derived by van Dishoeck et al. (1993). The agreement between the model predictions and the observations is best for earlier epochs when the C-shock intermediate times, i.e. $\leq 2,000$ yr. At earlier epochs when the J-shock dominates, the low excitation H$_2$ lines are too weak. And after 5,000 yr, when the C-shock steady-state is reached, the higher excitation H$_2$ lines (above 10^4 K) are much too weak. In between, the intensities of the H$_2$ rotational lines are predicted within factors of 2 and the coexistence of the 'hot' and 'warm' H$_2$ components is well explained within a single model. Models with other parameters (e.g., $n_H = 3 \times 10^3$ cm^{-3}, $v_s = 35$ kms^{-1}) provide less good fits. The best fits are obtained for early evolutionary times ($\sim 1,000$-$2,000$ years) and densities of $\sim 10^4$ cm^{-3} with shock velocities $v_s = 30-40$ kms^{-1}. Higher shock velocities will predict too large intensities for the high-excitation H$_2$ lines.

The agreement between these time-dependent model predictions and the data of IC 443 is very encouraging in view of some of the simplistic underlying assumptions. In particular, the assumed geometry (plane parallel) is oversimplified and does not describe the molecular filament which is seen edge-on and consists of numerous small clumps. Clearly a more thorough study should be done to explore the entire parameter space of the model and compare the predictions with additional data available on IC 443.

In the model, the S(2) to S(7) lines account for about 70% of the luminosity in all the H$_2$ lines. At peak A, the measured S(2) to S(7) flux is $\sim 4.6 \times 10^{-12}$ erg s^{-1} cm^{-2} (~ 0.4 L$_\odot$) and, according to the model, the H$_2$ lines alone would thus carry ~ 0.6 L$_\odot$. Towards clump G, the S(3) and S(2) H$_2$ lines account for almost the entire IRAS 12 μm-band emission. The mean value of these lines in that band is 5.4 MJy sr^{-1} comparable to the IRAS peak value, i.e. 6 MJy sr^{-1} (Figure 4). Similarly, the IRAS 25 μm-band (Fig. 4) could also be due to H$_2$ line emission. The model predictions for the S(0) and S(1) line fluxes (at an evolutionary time of 2000 years) are 1.2×10^{-5} and 7.7×10^{-4} erg s^{-1} cm^{-2} sr^{-1}, respectively. Taking into account the 20% transmission at 17.03 μm of the IRAS 25 μm band, the strong S(1) line would thus contribute $\sim$ 4 MJy sr^{-1} at 25 μm, which is comparable to the measured 25 μm IRAS flux at Peak A ($\sim$ 4.5 MJy sr^{-1} (Fig. 4). These results strongly suggest that the excitation of the gas in IC 443 is entirely collisional.

Oliva et al. (1999) found towards the optical filaments of IC 443, which trace the low density atomic gas, that most of the 12 and 25 μm IRAS fluxes is accounted for by ionized line emission (mainly [Ne II] and [Fe II]). Our results show that this conclusion cannot be generalised towards the molecular hydrogen ring (Fig. 1) where the dense molecular gas essentially cools via the H$_2$ lines in the near- and mid-infrared (see also contribution by J. Rho, this volume) and via the [O I] emission line in the far-infrared.

Michael Burton is kindly thanked for providing his H$_2$ map of IC 443. We would like to acknowledge J. Rho and W. Reach for useful discussions.

REFERENCES

BURTON, M.G., GEBALLE, T.R., BRAND, P.W.J.L. & WEBSTER, A.S. 1988 *MNRAS* **231**, 617-634

BURTON, M.G., HOLLENBACH, D.J., HAAS, M.R. & ERICKSON E.F. 1990 *ApJ* **355**, 197-209

CABRIT, S., BONTEMPS, S., LAGAGE, P.O. ET AL. 1999 in *The Universe as seen by ISO* (ed. P. Cox & M.F. Kessler). ESA SP-427, pp. 449-452

CESARSKY, D., COX, P., PINEAU DES FORÊTS, G., VAN DISHOECK, E.F., BOULANGER, F., & WRIGHT, C.M. 1999 *A&A* **348**, 945-949

CHIÈZE, J.-P., PINEAU DES FORÊTS, G. & FLOWER D.R. 1998 *MNRAS* **295**, 672-682

DRAINE, B.T. & LEE, H.M. 1984 *ApJ* **285**, 89-108

FLOWER, D.R. & PINEAU DES FORÊTS G. 1999 *MNRAS* **338**, 271-280

HUANG, Y.-L., DICKMAN, R.L. & SNELL R.L. 1986 *ApJ* **302**, L63-L66

KAUFMAN, M.J. & NEUFELD, D.A. 1996 *ApJ* **456**, 611-630

MOORHOUSE, A., BRAND, P.W.J.L., GEBALLE, T.R. & BURTON M.G. 1991 *MNRAS* **253**, 662-668

MUFSON, S.L., MCCULLOUGH, M.L., DICKEL, J.R., PETRE, P., WHITE, R., & CHEVALIER, R. 1986 *AJ* **92**, 1349-1357

OLIVA, E., LUTZ, D., DRAPATZ, S. & MOORWOOD A.F.M. 1999 *A&A* **341**, L75-L78

RICHTER, M.J., GRAHAM, J.R., WRIGHT, G.S., KELLY, D.M. & LACY J.H. 1995a *ApJ* **449**, L83-L86

RICHTER, M.J., GRAHAM, J.R., & WRIGHT G.S. 1995b *ApJ* **454**, 277-292

ROSENTHAL D., BERTOLDI F., DRAPATZ S. & TIMMERMANN R. 1999in *The Universe as seen by ISO* (ed. P. Cox & M.F. Kessler). ESA SP-427, pp. 561-564

TAUBER, J.A., SNELL, R.L., DICKMAN, R.L. & ZIURYS, L.M. 1994 *ApJ* **421**, 570-580

VAN DISHOECK, E.F., JANSEN, D.J. & PHILLIPS, T.G. 1993 *A&A* **279**, 541-566

WRIGHT, C.M., DRAPATZ, S., TIMMERMANN, R., ET AL. 1996 *A&A* **315**, L301-L304

Spatial Structure of a Photo-Dissociation Region in Ophiuchus

By F. BOULANGER [1], E. HABART [1], A. ABERGEL[1], E. FALGARONE[2], G. PINEAU DES FORETS[1,3] AND L. VERSTRAETE[1]

[1]Institut d'Astrophysique Spatiale, Université Paris XI, 91405, Orsay Cedex, France

[2]Ecole Normale Supérieure, 24 rue Lhomond, 75005 Paris, France

[3] Observatoire de Meudon

We present spectroscopic and imaging observations of dust and gas emission from the western edge of the ρ Ophiuchus molecular cloud facing the B2 III/IV star HD 147889. The emissions from dust heated by the external UV radiation, from collisionally excited and fluorescent H_2 are resolved and observed to coincide spatially. The spectroscopic data allows to estimate the gas temperature to 350 ± 30 K in the H_2 emitting layer. In the framework of a steady state model of the photo-dissociation region, a high formation rate: $2\,10^{-16} cm^3 s^{-1}$ at 350 K, seems to be required to account for this temperature. For smaller formation rates the H_2 emitting layer moves into the cloud where the gas is colder due to radiation attenuation.

1. Introduction

ISO observations of the dust emission and H_2 rotational lines are bringing a new perspective on the structure and physical conditions in regions of H_2 photodissociation (PDRs) at the surface of molecular clouds illuminated by hot stars. Spectroscopic observations of bright PDRs such as NGC 2023 have allowed to build detailed excitation diagrams of H_2 with numerous lines to test physical models (Draine this conference). In this paper, we present observations of a fainter PDR on the western edge of the ρ Ophiuchus molecular cloud heated by the B2III/IV star HD 147889. This is a nearby PDR (d = 135 ± 15 pc from the star parallax) with an edge-on geometry where the observations allow to spatially resolve the layer of UV light penetration and of H_2 photo dissociation. The observations presented in Section 2 are used in Section 3, in relation to a physical model of PDRs, to discuss the formation of H_2 in warm gas.

2. The Ophiuchus Photo-Dissociation Region

In the mid-IR image made with the ISO camera (Abergel et al. 1996), the western edge of the nearby star forming cloud ρ Ophiuchus is delineated by a long filament located at the edge of the dense molecular as traced by its $^{13}CO(1-0)$ emission (Loren 1989, Figure 1). Spectral observations carried out with the Circular Variable Filter (CVF) of the ISO camera show that the mid infrared emission from the cloud is dominated by the dust features considered to be characteristic of aromatic hydrocarbons (Boulanger et al. 1998). The interstellar particles at the origin of this emission are hereandafter referred to as PAHs. This a generic term which can encompass large molecules and small dust grains with up to a few 1000 atoms. The CVF observations also provided a map of the emission in the v=0-0 S(3) line of H_2 at 9.66μm. We have observed several H_2 emission lines with the ISO Short Wavelength Spectrometer (SWS) at the positions marked on Figure 1 and obtained from the ground an image of the near-IR 1-0 S(1) line of H_2 over a small section of the filament. In Figure 2, the 1-0 S(1) H_2 emission is shown

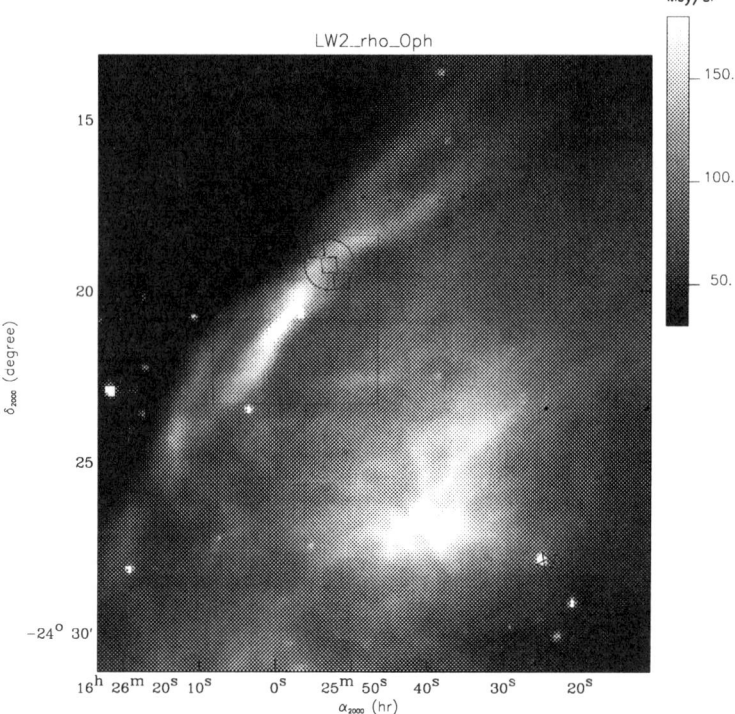

FIGURE 1. Image of the western Ophiuchus PDR in the $5-8.5\mu m$ ISOCAM filter (Abergel et al. 1996). The star marks the exciting star HD 147889 at the origin of the bright diffuse emission. The filament marks the edge of this emission in the direction of the dense molecular cloud. The three SWS positions (squares), the LWS position (big circle) and that of the 1-0 S(1) ESO observations (big square) are marked. The symbol sizes represent the field of view of each of these observations.

to delineate a surprisingly smooth and well resolved filament spatially coincident with the dust emission. This is unlike observations of the brighter PDRs NGC 7023 and NGC 2023 which show contrasted structures down to scales as small as 75 AU (0.5″ at the distance of Ophiuchus) (Lemaire et al. 1996, Rouan et al. 1997).

For the physical conditions of the Ophiucus PDR, the photo-electric effect on dust grains is the main heating source for the gas. It is an important part of models of gas emission in PDRs which can be tested with ISO observations. In Table 1, the power radiated by the main gas coolants is compared with the emission from PAHs in the wavelength range 5-16μm. These measurements have been averaged over the 80″ beam of the Long Wavelength Spectrometer (LWS) shown in Figure 1. Model estimates of the photo-electric yield as a function of grain size suggest that the photo-electric heating is dominated by PAHs and small dust grains (Bakes and Tielens 1994). For the diffuse interstellar medium, the PAH emission accounts for about 25% of the bolometric dust emission. The ratio between the gas cooling and the PAH emission, 2%, compares well with the estimates of the photo-electric efficiency on dust grains required to explain the heating of the diffuse interstellar medium (Wolfire et al. 1995).

In the ISOCAM image (Figure 1), the filament marks the edge of the diffuse emission to the North West in the direction of the dense gas in the Ophiuchus molecular cloud. We believe that it represents the illuminated surface of this dense cloud seen edge-on. A

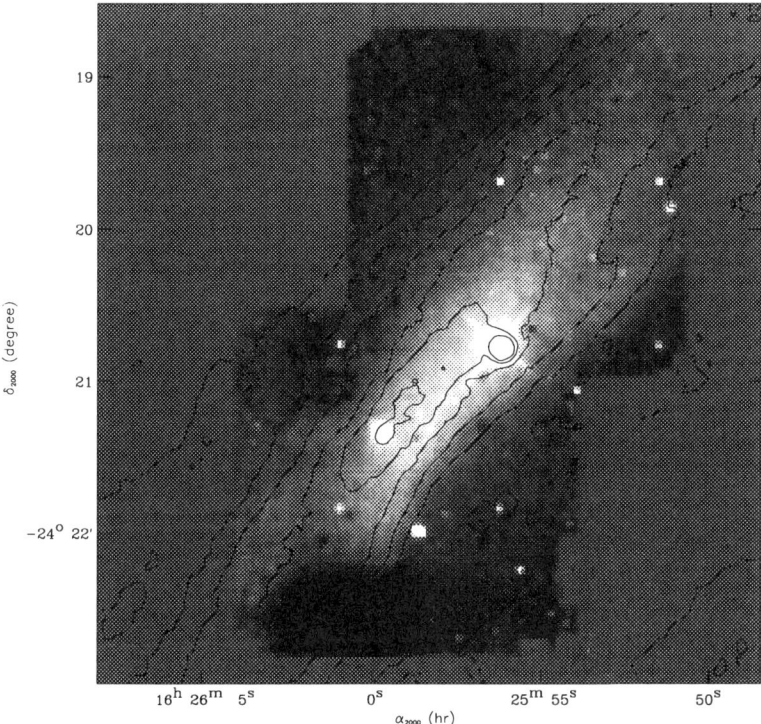

FIGURE 2. Contours of PAH emission on grey scale of 1-0 S(1) H_2 emission obtained at ESO. The dark hole at the bottom left of the image is due to extended emission in the reference fields used to subtract the sky emission.

TABLE 1. Power radiated by Gas Coolants and Small Dust Grains

Species	Power
	$Wm^{-2}sr^{-1}$
CII	$2.4\,10^{-7}$
OI	$>1.6\,10^{-7}$
H_2	$9\,10^{-7}$
SiII	$6.4\,10^{-7}$
Total Gas Cooling	$1.4\,10^{-6}$
PAHs (5-16μm)	$7.8\,10^{-5}$

profile of the PAH emission across the filament is presented in Figure 3; a comparison with the dashed line shows that beyond the emission peak the PAH brightness decreases exponentially into the cloud. The most straightforward explanation of this exponential decrease is the attenuation of the radiation field. Assuming an homogeneous medium and a dust mean UV extinction per H, $<k_{UV}> = 1.5\,10^{-21}$ cm^2 H^{-1}, we get a density $n_H = 8\,10^3$ cm^{-3} at distances > 0.07 pc in Figure 3. More generally, we have derived a gas density profile across the interface from the PAH emission profile using the following approximate relation: $I_{PAH}(z) \propto n_H \times e^{-\int_0^z <k_{UV}>\,n_H(u)\,du}$, which assumes constant abundance, absorption and emission properties of PAHs and constant cloud depth across the interface.

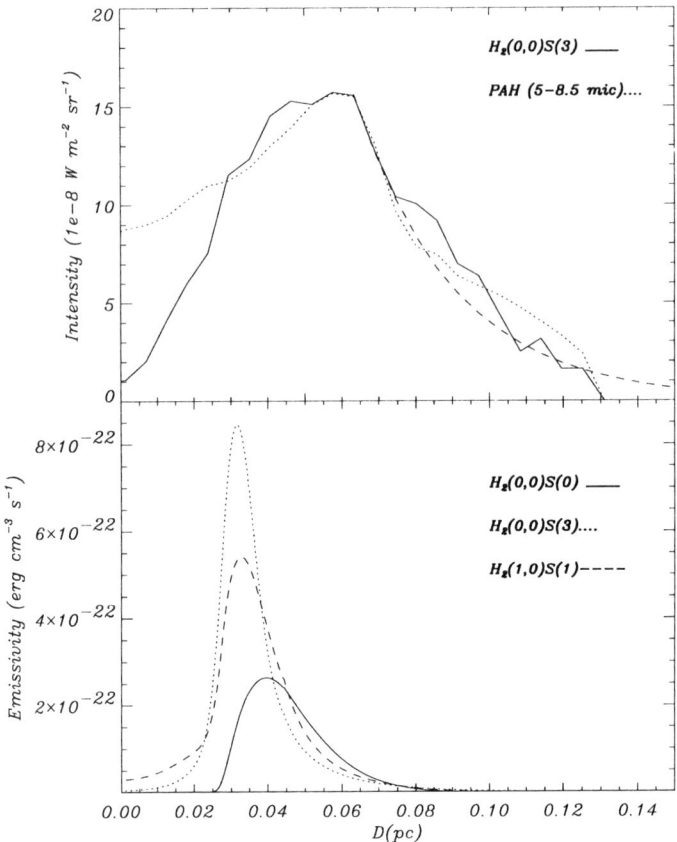

FIGURE 3. Top: Brightness cut of the v=0-0 S(3) H_2 line (solid line) compared with the PAH emission in a 5–8.5μm filter (dotted line). The cut goes through the SWS pointings marked in Figure 1 and goes into the cloud from left to right. The excess PAH emission relative to the S(3) line emission on the outer edge corresponds to the extended halo of emission seen all around HD 147889 (Figure 1). The dashed line shows the exponential decrease of the UV radiation for an homogeneous medium with $n_H = 8\,10^3\,cm^{-3}$. Bottom: Model H_2 emission profiles as a function of distance from the PDR outer edge. The H_2 formation rate in this model is that which accounts for the temperature of the emitting gas (see text and Figure 4).

3. H_2 Formation in PDRs

Because it is nearby and seen edge-on, the Ophiuchus PDR appears well suited to study the spatial structure of PDRs. A physical model is necessary to interpret the H_2 excitation which results both from collisions and fluorescence. The model calculations presented in Figures 3 and 4 are based on an updated version of the model of Le Bourlot et al. (1993). In this model, the abundance of H_2 is that given by the steady-state solution to the system of equations defining the chemical and physical state of the gas across the PDR. Ignoring advection of gas by turbulent motions, the steady state is reached in a time scale of $\sim (2\,R\,n_H)^{-1}$ where R is the H_2 formation rate and n_H the gas density (Goldshmidt and Sternberg (1995)). The far-UV radiation field intensity, G=300 in units of the mean Solar Neighborhood radiation ($1.6 \times 10^{-3}\,erg\,cm^{-2}\,s^{-1}$), has been estimated from the spectral type of HD 147889 and the projected distance between the star and the interface. The stellar light is assumed to arrive at the origin of the position axis in Figure 3 with no extinction; at positive z, we have used the density profile derived from the PAH emission profile (Section 2). For these physical parameters, the gas heating is mainly provided by the photoelectric effect on small dust particles and the temperature profile across the interface is little dependent on the H_2 formation rate.

Figures 3 shows that the H_2 emission starts to peak at d = 0.03 pc as in the model but that the width of the H_2 emitting layer is significantly wider in the observations than in the model. This could be accounted for by projection effects in a non perfectly edge-on geometry and/or clumping. Model and data are also inter-compared in the excitation diagram of Figure 4. The model results show that within the emission layer, the excitation of the first few H_2 rotational lines is dominated by collisions and can be used to determine the gas kinetic temperature. Outside this region the contribution of UV pumping followed by fluorescent cascade becomes important even for the lowest energy levels. Qualitatively, the contribution of collisional excitation relative to UV pumping decreases inwards because the gas becomes too cold and outwards because the gas density drops. In the excitation diagram of figure 4, all column densities up to the J=7 level line up on the 330 K line provided that we use an ortho to para ratio of 1; a similar excitation temperature is found at the two other SWS positions. For the J=6 and 7 levels (T_up = 3474 and 4586 K), the gas density is below the critical density but UV pumping happens to compensate the drop in collisional excitation.

The model results presented in Figure 4 show the effect of the H_2 formation rate on the gas temperature over the line emitting region. A high formation rate, $2\,10^{-16} cm^3 s^{-1}$ at 330 K, appears necessary to account for the H_2 temperature of the emitting gas. For the lower formation rate considered in the Figure, the warm gas is fully photo-dissociated, the H I/H_2 transition is moved inwards where the gas is colder due to the radiation attenuation.

The high H_2 formation rate proposed here to account for the H_2 emission temperature does not fit with formation from physisorbed H atoms on grain surfaces which is effective only at low temperatures (Pirronello, this volume). It suggests that H_2 formation, at least in PDRs, comes from chemically attached H atoms. One possibility which still needs to be experimentally and/or theoretically validated is the reaction of free H atoms with H atoms attached on the periphery/surface of PAHs.

The ortho to para ratio suggested by the excitation diagram is much lower than the equilibrium value computed in the model to be ~ 3 at 330 K. This difference could be a signature of advection of molecular hydrogen from colder layers of gas. For the physical conditions in the H_2 emitting layer, the dominant conversion process between ortho and para H_2 is the proton exchange with H atoms with a time scale $1.9\,10^4$yr at 330 K (Schofield 1967). Note that this is also about the life time of H_2 molecule before photodissociation within the approximations of the steady-state model. Based on the 0.02 pc thickness of the H_2 emitting layer in the model, we find that the advection speed has to be of the order of 1 km s^{-1} or larger for the ortho to para ratio not to be at the local equilibrium value. This is a reasonable value which could be accounted for by turbulent motions and/or a progression of the dissociation front into the cloud. It is presently not easy for us to quantify the effect of this interpretation of the ortho to para ratio on the estimate of the H_2 formation rate based on the steady state model.

Acknowledgments: Based on observations with ISO, an ESA project with instruments funded by ESA Member States with the participation of ISAS and NASA. We thank Pierre Cox for the LWS observation.

REFERENCES

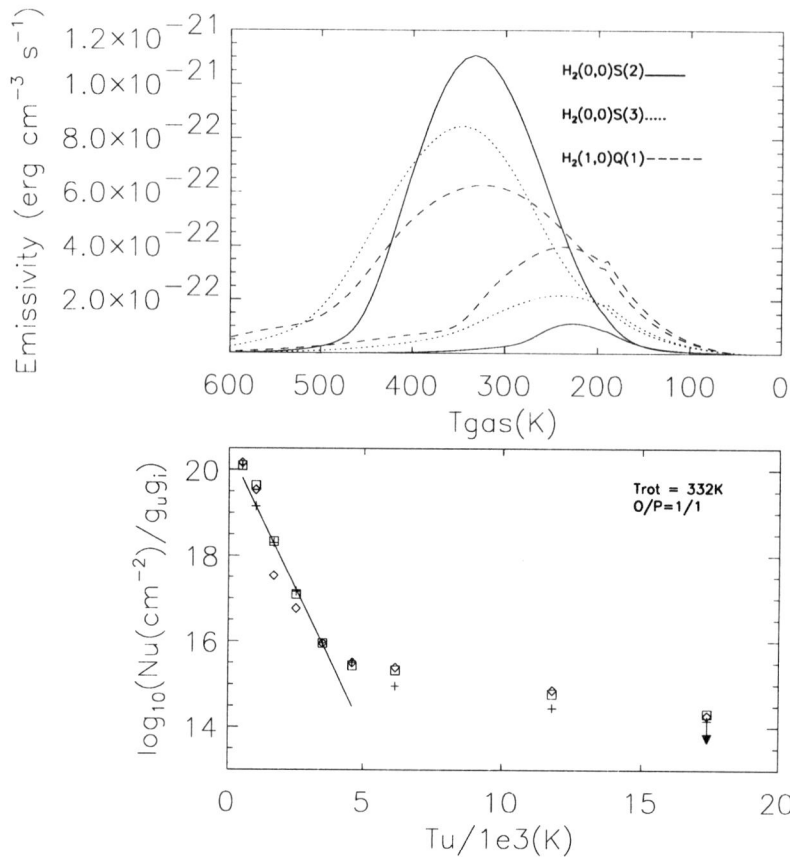

FIGURE 4. Upper panel: Line emission profiles as a function of gas temperature for two H2 formation rates $2.\,10^{-17}\,T^{0.5}/(1+T/400K)\,\mathrm{cm}^3\mathrm{s}^{-1}$ (set of curves to the left) and $4.3\,10^{-18}\,T^{0.5}/(1+T/400K)\,\mathrm{cm}^3\,\mathrm{s}^{-1}$ (set of curves to the right) as a function of the gas temperature. The lower rate gives the reference value inferred from Copernicus data: $3\,10^{-17}\,cm^3\,s^{-1}$ at 70 K. In the lower panel, the excitation diagram of H_2 at the peak of the H_2 emission for the observations (plus signs) and the two models of the upper panel (diamonds and squares for the low/high H_2 formation rate) are compared. N_u is the column density of the transition upper level, g_u is the degeneracy of the upper level, g_I the nuclear spin degeneracy and T_u is the upper level energy in Kelvin. For the observations we have used an ortho to para ratio of 1.

ABERGEL, A. ET AL. 1996 *A&A* **315**, 329.

BAKES, E.L.O. AND TIELENS, A.G.G.M. 1994 *Ap. J.* **427**, 822.

BOULANGER F., BOISSEL P., CESARSKY D. AND RYTER C. 1998 *A&A* **339**, 194.

GOLDSHMIDT O. AND STERNBERG, A. 1994 *Ap. J.* **439**, 256.

LE BOURLOT, J. ET AL. 1993 *A&A* **267**, 233.

LEMAIRE, J.L., GERIN, M., FIELD, D., GERIN, M., LEACH, S., PINEAU DES FORÊTS, G., ROSTAS, F. AND ROUAN, D. 1996 *A & A* **308**, 895.

LOREN, R.B. 1989 *Ap. J.* **338**, 902.

SCHOFIELD, K. 1967 *Pl. & Spa. Sc.* **15**, 643.

ROUAN, D., FIELD, D., LEMAIRE, J.L., LAI, O., PINEAU DES FORÊTS, G., FALGARONE, E. AND DELTORN, J.M. 1997 *MNRAS* **284**, 395.

WOLFIRE, M.G., HOLLENBACH, D., MC KEE, C.F., TIELENS, A.G.G.M. AND BAKES, E.L.O. 1995 *Ap. J.* **443**, 152.

Tracing H$_2$ Via Infrared Dust Extinction

By João Alves[1], C. Lada[2], and E. Lada[3]

[1]European Southern Observatory, Garching, Germany

[2]Harvard-Smithsonian Center for Astrophysics, Cambridge MA, USA

[3]University of Florida, Gainsville FL, USA

Most of the H$_2$ in our Galaxy resides in the cold interiors of molecular clouds. The most reliable way to trace the H$_2$ content of a molecular cloud is, in principle, to measure the distribution of dust through it. In this contribution we present a new observational approach that uses infrared dust extinction of starlight to construct high resolution maps of the distribution of dust (H$_2$) inside molecular clouds over unprecedented ranges of cloud depth: $1 < A_V < 40$ magnitudes. We also present a comparison of our results with conventional molecular-line column density tracer C^{18}O and conclude that for cloud depths of $A_V > 10$ magnitudes this species is a very poor tracer of H$_2$.

1. Introduction

Molecular clouds are the reservoirs of H$_2$ in the Galaxy. They contain about half of the mass of the Interstellar Medium and hence an important fraction of the mass of the Galaxy. By far the most important characteristic of molecular clouds is that they are the nurseries out of which stars like our Sun were born. This creation process not only determines the origins of stars and planetary systems in our Galaxy but also regulates the structure and evolution of galaxies on the large scale. To understand star and planet formation is to understand how cold H$_2$ clouds evolve.

Unfortunately, about 99% of the mass of a molecular cloud is virtually inaccessible to direct observation, hindering the required detailed knowledge of the physical and chemical structure of these molecular clouds. This unfortunate situation arises from the fact that the primary mass component of molecular clouds is in the form of H$_2$ and two factors render this molecule generally unobservable in this clouds: First, being a homonuclear molecule, H$_2$ lacks a permanent dipole moment and has extremely weak rotational transitions. Second, being the lightest interstellar molecule, its lowest energy rotational transitions are at mid–infrared wavelengths which are both inaccessible to observation from earth and also too energetic to be collisionally excited at the cool temperatures (T ~ 10 K) that characterize these clouds. For this reason, the traditional method used to derive the basic physical properties of these objects (e.g., sizes, masses, temperatures) is the spectroscopic observation of rare but detectable trace molecules (CO, CS, NH$_3$) which are several orders of magnitude ($\sim$ 4–9) less abundant than the primary mass component, molecular hydrogen (H$_2$). However, the interpretation of such molecular–line observations is not always straightforward. Several poorly constrained effects (e.g., deviations from local thermodynamic equilibrium, opacity variations, chemical evolution, small scale cloud structure, depletion) compromise the derivation of the distribution of mass and structure of a molecular cloud. We need then a less complicate and more robust tracer of H$_2$ not only to access cloud structure but also to be able to calibrate molecular abundances inside these clouds and investigate their chemical structure.

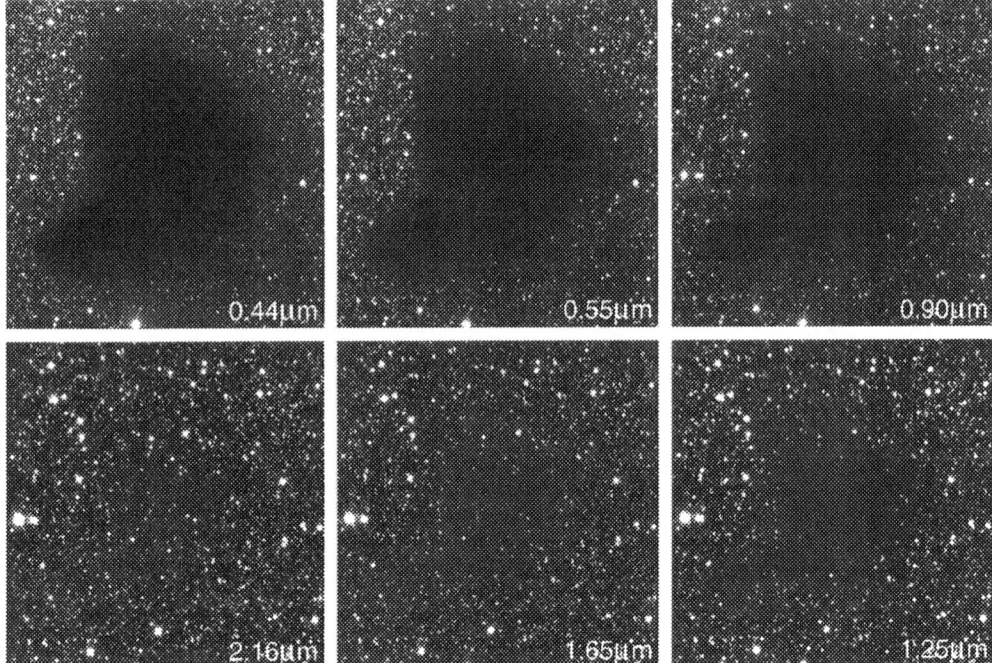

FIGURE 1. Deep BVIJHK imaging of dark molecular cloud B68 done with ESO's VLT and NTT telescopes. These data allowed the construction of the first 10″ resolution map of dust extinction and the most finely sampled density profile ever made for a dark cloud (from Alves et al. 2000).

2. The Method and Results So Far

The most straightforward and reliable way to measure molecular cloud density is to measure dust extinction of background starlight. We have developed a new powerful technique for measuring and mapping the distribution of dust through a molecular cloud using data obtained in large-scale, multi-wavelength, infrared imaging surveys. This method combines direct measurements of near-infrared color excess and certain techniques of star counting to directly measure extinctions and derive and map the dust column density distribution through a cloud (Lada et al. 1994). *It is the most straightforward and unambiguous way of determining the density structure in dark molecular clouds.* Moreover, the measurements can be made at significantly higher angular resolutions and substantially greater optical depths than previously thought possible. We have conclusively demonstrated the efficacy of this technique with our study of the dark cloud complex L977 (Alves et al. 1998), IC 5146 (Lada et al. 1999), and B68 (Alves et al. 2000) where we detected nearly 6000 infrared sources background to these clouds and produced detailed maps of the extinction across the cloud to optical depths an order of magnitude higher than previously possible (Av ~ 40 magnitudes) (see Figure 1 and 2 for molecular cloud B68).

We have used our extinction observations to measure the masses, density structure, extinction laws and distances to these objects. We found the radial density profiles of these clouds to be well behaved and smoothly falling with a power-law index of $\alpha = -2$, significantly shallower than predicted by early theoretical calculations of Ostriker (1964) ($\alpha = -4$). Moreover because we are using pencil beam measurements of dust column density along the line of sight to background stars we were able to demonstrate

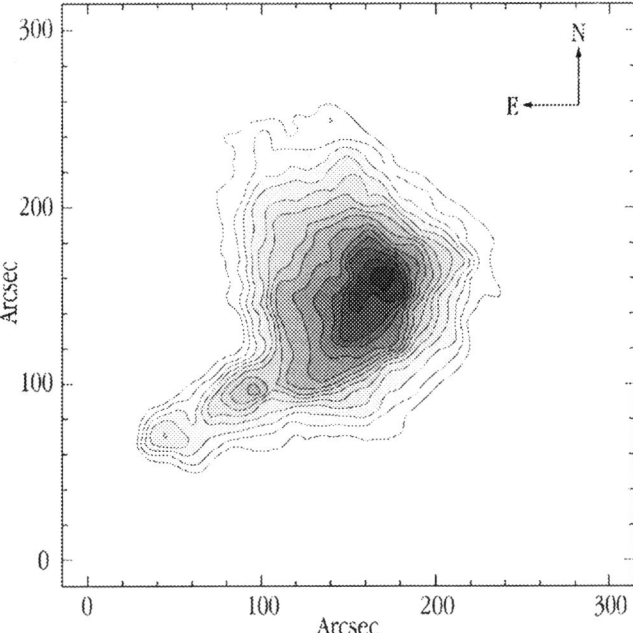

FIGURE 2. Extinction map of B68. The contours start at $A_V = 4$ mag and increase in steps of 2 mag. This is the first $10''$ extinction map ever made of a cold molecular cloud. The peak extinction measured through the very center of the cloud is 33 magnitudes of extinction (from Alves et al. 2000).

that the small-scale structure of the clouds is surprisingly smooth with random density fluctuations ($\delta A_V / A_V$) present at very small levels ($<< 20\%$). When convolved to the appropriate spatial resolution our maps showed structure in the dust distribution which was strikingly well correlated with millimeter-wave CO and CS emission maps of the cloud, although showing crucial differences at high optical depths where these other tracers of column density become unreliable (see Figure 3). This enabled us to directly derive CO, CS, and N_2H^+ abundances over an extinction range of 1–30 magnitudes, a range nearly an order of magnitude greater than achieved previously with optical star counting techniques (indeed, our CS and N_2H^+ abundance determination is the first direct abundance determination for these molecule to be made in a dark molecular cloud). In a recent experiment we were able to make the first direct measurement of molecular depletion in a cold cloud core (Kramer et al. 1999). Moreover, and using millimeter continuum emission measurements, we obtained the most accurate and direct measurement ever obtained of the ratio of dust absorption coefficients at millimeter and near-infrared wavelengths (Kramer et al. 1998). In a more recent study on dark cloud B68 with ESO's NTT we were able to construct the first $10''$ resolution map of dust extinction in a dark cloud. These observations produced the highest resolution, most finely sampled density profile of a cloud ever constructed and showed the cloud to be extremely smooth with the random component of small-scale spatial density fluctuations $\leq 3\%$! These observations also lead us to the unexpected discovery of diffuse NIR light arising from this cold molecular cloud.

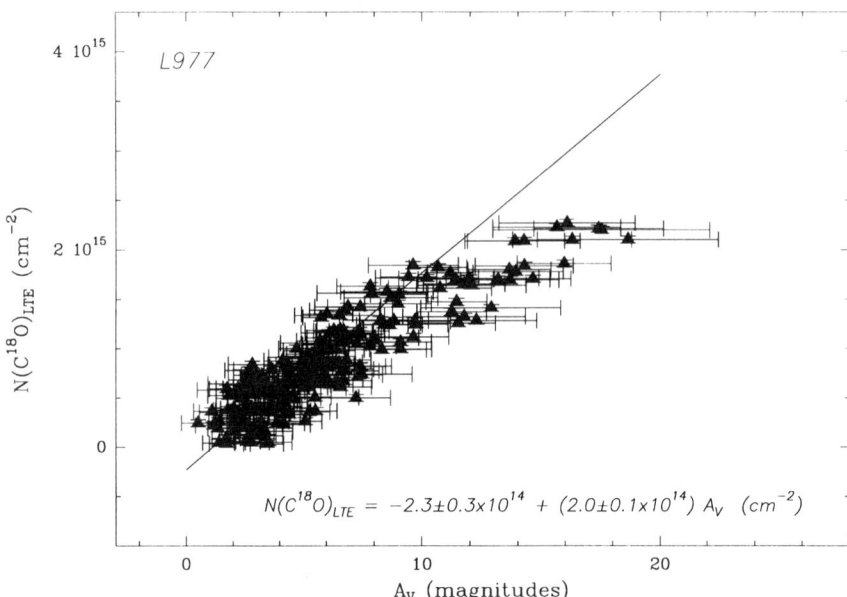

FIGURE 3. Relation between $N(C^{18}O)_{LTE}$ and visual extinction A_V for molecular cloud L977. There is a clear deviation from the linear relation at extinctions ≥ 10 magnitudes above which $C^{18}O$ becomes a very poor tracer of H_2 (from Alves et al. 1999).

3. The Future

Significant progress in extinction mapping studies will result when the 2MASS all sky near-infrared survey imaging survey is completed and released. This survey will be sufficiently sensitive to produce moderate depth extinction maps (i.e., $5 \leq A_V \leq 25$ mag) of many nearby dark clouds in those directions of the Galaxy where field stars suffer little extraneous extinction. The new generation of large aperture optical telescopes, such as the VLT, will provide the additional capability to perform deeper surveys of the higher extinction regions ($25 \leq A_V \leq 60$ mag) in these clouds. Finally, space-based infrared telescopes, such as NGST, should enable the regions of deepest extinction ($A_V > 60$ mag) to be probed. Together such observations promise to render a very complete description of the structure of nearby star forming clouds.

JA is grateful to the organizing committee of the "H_2 in Space" International Conference for financial support. JA is especially thankful to Françoise Combes for fruitful discussions and very kind assistance during the conference.

REFERENCES

Alves, J., Lada, C.J., Lada, E.A., Kenyon, S., & Phelps, R. 1998, ApJ, 506, 292
Alves, J., Lada, C.J., & Lada, E.A. 1999, ApJ, 515, 265
Alves, J., Lada, C.J., & Lada, E.A. 2000, ApJ submitted
Lada, C.J., Lada, E.A., Clemens, D.P., & Bally, J. 1994, ApJ, 429, 694
Lada, C.J., Alves, J., Lada, E.A. 1999, ApJ, 512, 250
Kramer, C. et al. 1997, A&A, 329, 33
Kramer, C. et al. 1999, A&A, 342, 257
Ostriker, J. 1964, ApJ, 140, 1056

The small scale structure of H$_2$ clouds

By P. BOISSÉ[1], S. THORAVAL[1], J.C. CUILLANDRE[2],
G. DUVERT[3], AND L. PAGANI[4]

[1]Radioastronomie/ENS, DEMIRM/CNRS, 24 rue Lhomond, F-75231 Paris, France

[2]CFHT Corporation, P.O. Box 1597, Kamuela, HI 96743, USA

[3]Laboratoire d'Astrophysique, Observatoire, B.P. 53X, F-38041 Grenoble Cedex, France

[4]DEMIRM/Observatoire de Paris, 61 Avenue de l'observatoire, F-75014 Paris, France

We present recent results from a program devoted to the study of small scale structure in translucent molecular clouds, using dust as a tracer. Several methods have been employed: i) statistical analysis of stellar fields, ii) studies of background galaxies; iii) searches for time variations of the extinction and reddening. For each method, we summarize the principles, the type of constraints provided (scales, sensitivity) and present the results obtained so far. We conclude by some prospects concerning direct studies of the distribution of H$_2$ itself.

1. Motivations

Originally, we wished to constrain in a direct way the level at which the penetration of the stellar UV and visible radiation is enhanced by structure effects (cf Boissé 1990). Studies of spatial variations of dust extinction and reddening offer a powerful way to address this question. Indeed, in the presence of density fluctuations, the analysis of extinction in stellar fields directly provides a measure of the effective opacity (Thoraval et al. 1997). Further, in the visible, the required observations are easy to perform and provide an excellent spatial resolution.

Another motivation comes from the detection of very small scale structure both in atomic (Frail et al. 1994) and molecular gas (Moore & Marscher 1995). In the assumption of a uniform dust to gas ratio, one should observe corresponding variations of the amount of dust, resulting in local variations of the extinction.

Our motivation for investigating the structure of *diffuse* or *translucent* clouds is twofold: i) if interstellar material is made of building blocks with more or less universal characteristics (column density in particular), the latter should be more apparent in low column density clouds where the average number of fragments per line of sight is small; ii) these clouds dominate the mass of the ISM and thus play an important role in its evolution. Of course, it is also important to investigate the structure of dense clouds since the latter are the site of stellar formation (see Alves et al., this volume).

2. Extincted stellar fields

Unextincted stars have a relatively well defined average color (e.g. δ(H - K) $\simeq$ 0.1 and δ(V - I) $\simeq$ 0.6), which can be used to map the reddening induced by a foreground cloud, if the surface number density of background stars is high enough. This method has been successfully applied in the near IR (Lada et al. 1999, Alves et al., this volume)) and in the visible (Thoraval et al. 1997).

These two domains are quite complementary; in the NIR, due to a low extinction, it is possible to probe the structure of dense clouds which are completely opaque to UV/visible radiation whereas the visible range is well adapted to probe diffuse or translucent clouds (for which NIR color excesses become inaccurate). In both domains, the dynamical range (largest measurable value/1σ uncertainty) turns out to be about the same (20 - 40).

Since the stellar intrinsic NIR (e.g. H - K) colors are nearly universal, NIR color excesses can be derived for individual lines of sight. In the visible, the spread in intrinsic colors is larger and to get a good relative accuracy, it is necessary to consider the average (or median) color excess computed over a few (5 - 10) stars.

One feature of the method which is to be outlined here, is that it avoids any "transverse averaging" effect. Measuring star colors is equivalent to performing observations with a very small beam ($\simeq$ 10 μarcsec for a sun-like star at 1 kpc): averaging occurs only *along* the line of sight and if dusty fragments inducing significant reddening were present, they should show up.

We have studied a field with $A_V \simeq 2$ (Thoraval et al. 1997), selected for its apparent uniformity in star number density in order to detect more clearly (if present) local stochastic small scale fluctuations. Small amplitude color variations at scales of 0.05 pc or larger are seen but when individual star colors are corrected for these smooth variations, the color distribution for the whole extincted field is indistinguishable from the intrinsic one, indicating that fluctuations, $\delta A_V / A_V$, at scales less than 0.05 pc (or 10^4 AU) are smaller than 10 %. In higher density regions, Lada et al. (1999) obtain qualitatively similar results; the scatter noticed in their 1994 data appear to be mostly due to gradients occuring around the densest parts of the filamentary cloud and not to the presence of unresolved fragments.

In order to fill the gap between the opacity ranges probed by the Lada et al. NIR data and our own visible observations, we have recently performed deep VRI imaging at the CFHT, benefiting from the large field of the UH8k camera. In the V band, the limiting magnitude reached is $m_V \simeq 25$, which allows us to detect stars suffering $A_V > 10$. For easier comparison, we selected a 30 x 30 arcmin^2 field comprising the region imaged in the NIR by Lada et al. (1999). The densest parts of the cloud still appear starless which is consistent with a "compact" and smooth structure. The large density of background stars will be used to probe in detail the edges of the opaque regions. Our images also reveal impressive variations of the background light over the field, especially at the boundary of opaque regions. These variations are likely due to a combination of occultation and backscattering of the diffuse Galactic light, which we here detect with a high S/N. Modeling of these effects could provide valuable information on the outer structure of the dense regions.

3. Galaxies seen through molecular clouds

Background galaxies allow us to probe clouds in a different manner. The transverse extent of the "equivalent beam" is, this time, defined by the seeing, which implies a lower angular resolution (10^{-3} pc or 200 AU for 1 arcsec seeing and a cloud at 200 pc). On the other hand, using galaxies, one can map the foreground cloud in a continuous way (and thus compare to data obtained for tracers other than dust grains, e.g. CO emission). When seen through translucent clouds (A_V = 2 - 3), galaxies are bright enough to reveal extinction fluctuations with an amplitude as low as a few %; these should appear as "granularity" at the seeing scale or larger, and should be more pronounced at blue wavelengths. The symmetry properties of the background galaxy can also be used to investigate variations at the scale of the galaxy itself: since the cloud structure and background galaxy morphology are uncorrelated, a brightness profile symmetrical about the center implies i) that the galaxy is intrinsically symmetrical and ii) that no extinction variation is present.

We have studied in B, V, R and I a few obscured galaxies, mostly located at intermediate or high b where confusion is not a problem (Thoraval et al. 1999). A representative example is CGCG 525-46 seen through the Taurus complex: the brightness distribution is smooth ($\delta A_V/A_V < 3\%$) and symmetrical (($\delta A_V/A_V < 10\%$), while ^{12}CO emission from the foreground cloud displays a significant gradient ($\delta I(CO)/I(CO) = 30$ %) over the same extent. Towards Maffei 1 (located at b = - 0.55 deg.), we detect a localized extinction peak, probably associated with a cloud at d = 3.3 kpc. The rest of the image is again essentially smooth, although significant fluctuations ($\delta A_V/A_V \simeq 1.6$ %) are detected.

4. Searches for variable reddening/extinction

The method adopted by Thoraval et al. (1996) is similar to that proposed by Moore & Marscher (1995) but background stars are used instead of extragalactic sources. It is based on the motion of the sun ($V_{LSR} \simeq 4$ AU/yr) and of the earth which induces a drift of the line of sight through the cloud. Since we do not know the transverse velocity of the cloud and of the background stars nor their distances, it is not possible to compute the exact value of the scale probed when comparing observations performed at two given epochs. For stars located far away behind the cloud and time intervals of the order of 1 yr, this scale will typically be 1 - 10 AU.

We have performed differential photometry for about ten bright stars in a field around 3C111 (where $A_V \simeq 2$) and found no evidence for time variations: $\delta A_V/A_V < 1.5\%$ (Thoraval et al. 1996; Boissé & Thoraval 1997). Assuming no transverse motion of the cloud, the largest scale probed is 8.5 AU.

The extinction induced by the clouds we studied is partly due to atomic gas surrounding the molecular phase; for $A_V \simeq 1 - 2$, the relative contribution of this diffuse gas should be around 50 %. Since atomic gas also displays fluctuations at the level $\delta N(HI)/N(HI) \simeq 10 - 20$ % (Frail et al. 1994; Heiles 1997) at scales 5 - 100 AU, we can conclude than dust grains are distributed more smoothly than the gas in the atomic phase too.

Our observations constrain the variations of A_V, a quantity integrated over the line of sight. One might imagine that if radio studies led to positive detections, this is because the high spectral resolution make it possible to distinguish the various velocity components, of which only one may vary. This is however not true: Moore & Marscher (1995) detect variations of the "total equivalent width" exceeding 10% while, for similar scales, we get upper limits of about 2%. Furthermore, Moore & Marscher (1995) detect variations against all sources selected while we observe no fluctuation towards about ten stars. We thus conclude that the relative amount of H_2CO display variations from one line of sight to another separated by a few AU which are at least a factor a five larger than those of the relative amount of dust grains responsible for the visible extinction.

One explanation for the apparently distinct behavior of the gas and dust grains might be that, due to scattering, the size of the region probed in the optical is (although smaller than the PSF) larger than expected from the solid angle subtended by stellar discs. However, from the analysis of reflection nebulae by Witt & Cottrell (1980), it can be seen that an unlikely geometrical configuration is required to get a nebular flux larger than the stellar flux. In our case, most of the stars used as probes are unassociated with material from the cloud and the "effective beam" is therefore truly defined by stellar disc sizes.

5. Discussion and prospects

Results inferred from the three methods discussed above all point towards the absence (or weakness) of small scale structure in the distribution of the dust (in any event, it is clear that A_V fluctuations would be too low to significantly affect the transfer of UV light inside clouds). For scales ranging from 1 to 10^4 AU (with some gaps, especially in the interval 10 - 100 AU), we get upper limits between 1 and 10 % for the amplitude of relative variations. Of course, given the small size of the regions studied, this is not in contradiction with the known presence of dust concentrations e.g. where stars are forming, nor with the existence of marked structure at larger scales, as revealed by far infrared maps provided by IRAS or ISO. The meaning of our results is rather that *the kind of ubiquitous small scale structure seen in atomic and molecular gas is not present in the distribution of dust grains*. An immediate consequence is that the dust to gas ratio, although relatively well defined at large scales, is not uniform at small scales. This could have important consequences on the heating of the gas and then on the time evolution of these gaseous fragments.

In Thoraval et al. (1999), we briefly considered the dynamics of dust grains relative to the gas and conclude that random motions are very efficient at decoupling dust grains from the gas (especially the large ones). Further, due again to the limited coupling between gas and grains, it is not easy to imagine a physical process that would generate simultaneously a large overdensity in the gas and in dust particles; in this regard, it could be useful to investigate the behavior of dust grains in a specific model describing structure formation, such as the one recently proposed by Hennebelle & Pérault (1999) for the atomic phase.

One important question that remains open concerns the distribution of H_2 itself in molecular regions. Indeed, the distribution of tracers like H_2CO, CO and its isotopomers, etc may not reflect that of H_2, due to different self-shielding effects, different formation/destruction mechanisms. In order to determine in a reliable way whether or not a significant fraction of the mass is enclosed in tiny fragments that occupy a small volumic fraction, it is essential to probe H_2 itself. FUSE does offer that opportunity: for instance, repeated spectral observations of large velocity O stars seen through molecular material should provide valuable constraints on the variations of $N(H_2)$ between lines of sight separated by 1 - 100 AU. Such observations are planned in the future.

We would like to thank A. Witt for helpful comments concerning the effect of scattering on the transverse extent of the region probed by a star.

REFERENCES

BOISSÉ, P. 1990 *Astron. Astrophys.* **228**, 483.
BOISSÉ, P. & THORAVAL, S. 1997 *La lettre de l'OHP* **17**, 1.
FRAIL, D., WEISBERG, J., CORDES, J., MATHERS, C. 1994 *Astrophys. J.* **436**, 144.
HEILES C. 1994 *Astrophys. J.* **481**, 193.
HENNEBELLE, P. & PÉRAULT, M. 1999 *Astron. Astrophys.* **351**, 309.
LADA, C., ALVES, J., LADA, E. 1999 *Astrophys. J.* **512**, 250.
MOORE, E. & MARSCHER, A. 1995 *Astrophys. J.* **452**, 671.
THORAVAL, S., BOISSÉ, P., STARK, R. 1996 *Astron. Astrophys.* **312**, 973.
THORAVAL, S., BOISSÉ, P., DUVERT, G. 1997 *Astron. Astrophys.* **319**, 94.
THORAVAL, S., BOISSÉ, P., DUVERT, G. 1999 *Astron. Astrophys.* **351**, 1051.
WITT, A. N. & COTTRELL, M. J. 1997 *Astrophys. J.* **235**, 899.

Hot chemistry in the cold diffuse medium: spectral signature in the H$_2$ rotational lines

By E. FALGARONE[1], L. VERSTRAETE,[2] P. HILY-BLANT,[3], AND G. PINEAU DES FORÊTS,[2]

[1] Ecole Normale Supérieure, Paris, France
[2] Institut d'Astrophysique Spatiale, Orsay, France
[3] IRAM, Granada, Spain

Most of the diffuse interstellar medium is cold, but it must harbor pockets of hot gas to explain the large observed abundances of molecules like CH$^+$, OH and HCO$^+$. Because they dissipate locally large amounts of kinetic energy, MHD shocks and coherent vortices in turbulence can drive endothermic chemical reactions or reactions with large activation barriers. We predict the spectroscopic signatures in the H$_2$ rotational lines of MHD shocks and vortices and compare them to those observed with the ISO-SWS along a line of sight through the Galaxy which samples 20 magnitudes of mostly diffuse gas.

1. The trigger of hot chemistry in the cold diffuse medium

The large observed abundances of CH, CH$^+$, HCO$^+$, and OH in the (mainly cold) diffuse medium (T ≈ 50 K) imply that activation barriers and endothermicities of several 10^3 K are overcome. Pockets of hot gas must therefore exist. Large ion-neutral drift speeds, of several km s^{-1}, can equally contribute to triggering certain ion-neutral reactions in cold gas, such as the endothermic reaction C$^+$ + H$_2$ which forms CH$^+$ ($\Delta E/k$=4640 K). Two phenomena, operating at very different scales, are able to reproduce the observed abundances. These are MHD shocks (Flower & Pineau des Forêts, 1998 and references therein) and intense vortices, thought to be responsible for a large fraction of the viscous dissipation of supersonic turbulence (Joulain et al. 1998).

In MHD shocks, the neutrals are heated by the ion-neutral drift which occurs in the magnetic precursor. The peak kinetic temperature of the neutrals depends, among other parameters, on the shock velocity: it is $T_n \approx 10^3$ K for a 10 km s^{-1} shock propagating in a gas at $n = 20$ cm^{-3}. The hot layers have a thickness of about 0.1 pc or 20 000 AU and the time for a fluid particle to cross the hot layers is of the order of 10^4 yr.

In dissipative vortices with a magnetic field, the main heating mechanism is viscous dissipation in the layers of strong shear at the outer boundaries of the vortices. The streaming of the neutrals relative to the ions contributes no more than 10% of the total heating rate. The size of such vortices is ill-defined because no laboratory experiment is able yet to establish the dissipative scales in compressible MHD turbulence. For the sake of simplicity, and to allow an analytic description of these vortices with only two independent parameters, Joulain et al. (1998) adopted $r \approx 20$ AU, a radius close to the Kolmogorov scale (*i.e.* that of the viscous dissipation in the turbulent cascade). This radius corresponds to a balance between the stretching action of the large scales, which increases the vorticity, and the diffusion of the vorticity. In laboratory experiments on incompressible turbulence (Cadot et al. 1995; Belin et al. 1996), coherent vortices have been found to have radii which range between the Kolmogorov and the Taylor microscale. We therefore assume that dissipative vortices in interstellar turbulence have radii ranging from ≈ 20 AU up to ten times larger. For a rotational velocity of 3.5 km s^{-1} (imposed by the rms velocity dispersion of the ambient turbulence, Jimenez 1997), the temperature of the neutrals in the outer layers, which are heated by viscous dissipation, is $T_n \approx 10^3$ K and their thickness is comparable to the vortex radius. The timescale for a fluid particle to cross the hot layers is much shorter than for MHD shocks, a few 100 yr only in the case of the thinnest vortices.

The size and the crossing time are therefore the main differences between MHD shocks and dissipative vortices. Non-equilibrium effects are thus expected to be more pronounced in the vortices. The column densities of molecules formed specifically in the hot layers (or layers with large ion-neutral drift) are much larger in a single MHD shock than in a single vortex because of their widely different sizes.

The column densities of CH and CH$^+$ observed in absorption in the direction of nearby stars tend to increase linearly with the visual extinction to the star (A_v) and therefore with the total gas column density sampled by the line of sight (N_H) since $N_H = 1.9 \times 10^{21} A_v$ cm^{-2}. We have merged the data of Crane et al. (1995), which sample low column densities, with those of Gredel (1997), which sample a larger range of A_v (up to 4.5 mag), and obtain the following scaling relation:

$$N(\text{CH}^+) \approx 1.5 \times 10^{13} \left(\frac{A_v}{1\,\text{mag}}\right) \text{cm}^{-2}$$

This relation enables an estimate of the number of shocks and vortices, per magnitude of diffuse gas, which are required to produce the observed column densities of CH$^+$ and CH: a few MHD shocks at 10 km s^{-1} or about a thousand vortices. Note that 1000 vortices per magnitude correspond to only $\approx 1\%$ of the total column density of the gas, and 6 shocks of 10 km s^{-1} to only 3%.

Figure 1 illustrates the good agreement between the column densities of OH and HCO$^+$ observed by Lucas & Liszt (1996) in absorption against extragalactic sources (*i.e.* on lines of sight which sample the diffuse galactic medium) and model predictions. In Fig. 1a, the predictions for MHD shocks of various velocities are shown. In Fig. 1b, those for 1000 and 3000 vortices on the line of sight, and various degrees of shielding from the ambient UV field are shown.

Thus, on the sole basis of the molecular abundances they produce, a few MHD shocks or a much larger number of small vortices cannot be distinguished. Whether the chemically active structures are shocks or vortices has yet to be established. The outcome has a bearing on the lifetimes of molecular clouds and their star formation efficiency: shocks dissipate the supersonic turbulent energy of a cloud on shorter time scales and larger distance scales than do vortices (Porter et al. 1994). If small-scale vortices are the main sites of the dissipation of interstellar turbulence, then the cascade has to develop without much dissipation on intermediate scales, between 100 pc (the outer scale) and the size of the solar system, or over more than five orders of magnitude in scale.

2. ISO-SWS observations of the diffuse galactic medium

To test the existence and determine the characteristics of the small fraction of hot gas in the cold interstellar medium, we have searched for H$_2$ rotational lines on a long line of sight across the Galactic plane, avoiding star-forming regions and giant molecular clouds.

The selected target was a line of sight in the direction of the molecular ring, characterized by a minimum of 5 GHz continuum emission (to ensure the absence of large star-forming regions associated with shocks on the line of sight), and a minimum in CO line emission. The CO($J = 1 - 0$) line integrated area ($W(\text{CO}) = 60$ K km s^{-1}, Sanders et al. 1986) provides the column density of H$_2$ along this line of sight, $N(\text{H}_2) = 1.2 \times 10^{22}$ cm^{-2}. The 100 μm emission $I(100\mu m) = 2080$ MJy sr^{-1} (Beichman et al. 1988) provides the

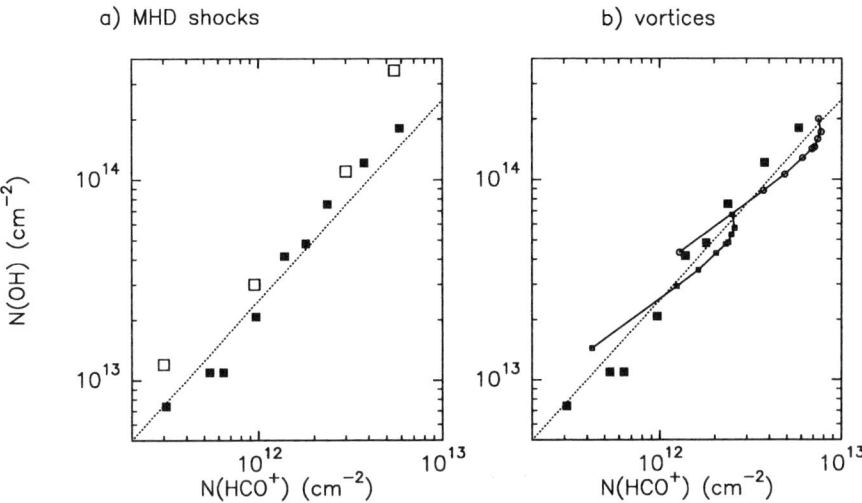

FIGURE 1. The correlation between the observed column densities of OH and HCO^+ (filled squares are data from Lucas & Liszt,1996). The error bars are smaller than the size of the symbols. The dotted line corresponds $N(OH)/N(HCO^+) = 25$. (a) Column densities predicted by the MHD shock model (open squares) for shock velocities of 9, 10, 12 and 15 km s^{-1}, a pre-shock density of 20 cm^{-3} and a shielding from the UV field of 0.1 mag. The predictions for several shocks of same velocity would align along the dotted line. (b) Column densities predicted by the vortex model for 10^3 (lower curve) and 3×10^3 standard vortices (upper curve) on the line of sight. Each point along these curves corresponds to a different value of the UV shielding, at the location of the vortex, from 0.1 to 1 mag.

column density of atomic and ionized gas $N(H) + N(H^+) = 1.3 \times 10^{22}$ cm^{-2}, using the IR luminosity per H nucleus $4\pi\nu I_\nu(100\mu m) = 6 \times 10^{-23}$ erg s^{-1} H^{-1} estimated above the molecular ring by Pérault (1987). The total column density sampled by this line of sight is about 20 mag.

The 5 GHz-brightness temperature, $T_{cont}(5GHz) = 0.5$ K, (Altenhoff et al. 1970) if due to free-free emission, may be converted into a flux of ionizing photons $N^\star_{Lyc} = 1.5 \times 10^{45}$ photon s^{-1}; this corresponds to a single main sequence star of type between B1 and B2, and to a local radiation field $\chi \approx 10$ times more intense than in the solar neighborhood, if the B star is 1 pc from the edge of the molecular cloud observed in CO.

Figure 2 shows all the lines observed with the ISO-SWS with a spectral resolution $R \approx 2000$. Our detection of the S(0) line is marginal and we quote an upper limit to its line flux as well. In any case, the S(0) line emission traces predominantly the cold diffuse gas, which is not our concern here. The S(4) line was not detected.

Line fluxes and errors have been estimated from Gaussian fits to the line profiles shown in Figure 2. We used the aperture sizes given in de Graauw et al. (1996) to convert to integrated brightnesses. The detected line intensities are $3.0(\pm 0.6) \times 10^{-5}$, $1.8(\pm 0.5) \times 10^{-5}$ and $5.6(\pm 2.0) \times 10^{-6}$ erg s^{-1} cm^{-2} sr^{-1} for the S(1), S(2) and S(3) lines respectively (the errors quoted include the absolute calibration error of the SWS from Schaiedt et al. 1996). Upper limits of 1.1×10^{-4} and 2.4×10^{-6} erg s^{-1} cm^{-2} sr^{-1} are found for the S(0) and S(4) lines respectively.

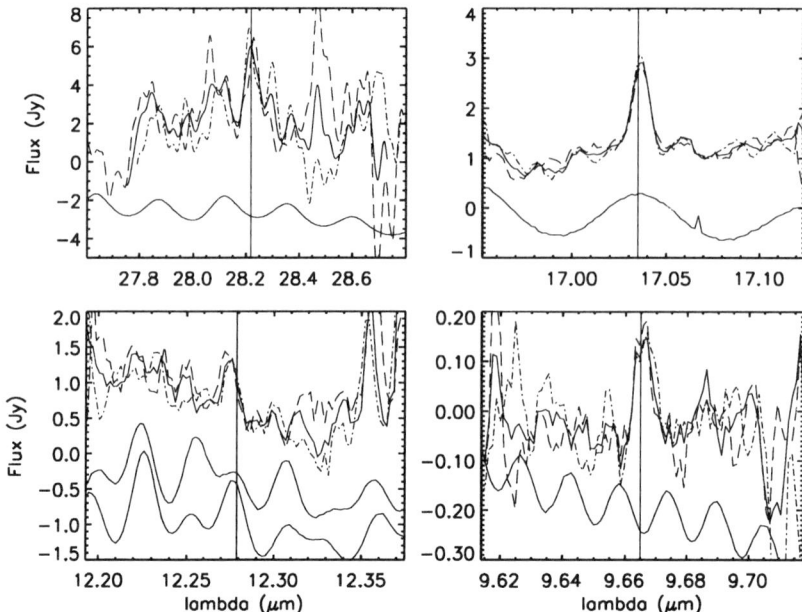

FIGURE 2. SWS spectra of the H_2 rotational lines. The full scan (solid) is superposed on the UP (wavelength increasing with time, dot-dashed line) and DOWN (wavelength decreasing with time, dashed) scans. At the bottom of each plot, we display the Relative Spectral Response Function (RSRF) of the instrument for a given detector.

3. Comparison with model predictions

There is a contribution to the observed H_2 line emission from *(i)* the 20 mag of cold diffuse gas and *(ii)* an expected interface (PDR) between the H II region of the B star and the molecular cloud, in the likely possibility that they are related. To compute the contribution of the 20 mag of gas, we have assumed that all the diffuse gas along the line of sight is PDR material, *i.e.* the line of sight crosses a large number of edges of cold neutral clouds ($n = 30$ cm^{-3}) illuminated by an ambient UV field with $\chi = 1$ in the Solar neighborhood up to $\chi \approx 10$ at a galactocentric radius of 5 kpc. The adopted depth of each individual PDR is $A_v = 0.3$ mag, and the total number of interfaces along the line of sight is 20/0.3=67. We find that the H_2 line emission of the ambient gas depends only weakly on the above assumptions. Increasing the ambient UV field reduces the H_2 fraction but increases the gas temperature of the diffuse gas, so that the H_2 line intensities have only a small dependence on χ in this framework. The 20 mag of cold gas contribute only to the S(0) line at a level slightly smaller than the upper limit derived from the observations. All the other lines are about one order of magnitude weaker than observed. The line emission of the possible PDR ($n = 100$ and 10^4 cm^{-3} and $\chi = 10$) is two orders of magnitude weaker than the observed intensities, even in the S(0) line. The contributions of the cold diffuse gas and the PDR to the 5 lower rotational transitions of H_2 are shown in the Figure 3 of Verstraete et al. (1999). Thus, the large observed intensities of the H_2 S(1), S(2) and S(3) lines relative to the undetected S(0) line cannot be explained by the emission of the 20 mag of diffuse gas nor by that of a possible PDR excited by a B star.

In Figure 3 we compare the observed H_2 line intensities to those computed for MHD shocks and vortices. The number of shocks and vortices assumed to be present on the line of sight are those given in Section 1 (*i.e.* number per magnitude) multiplied by

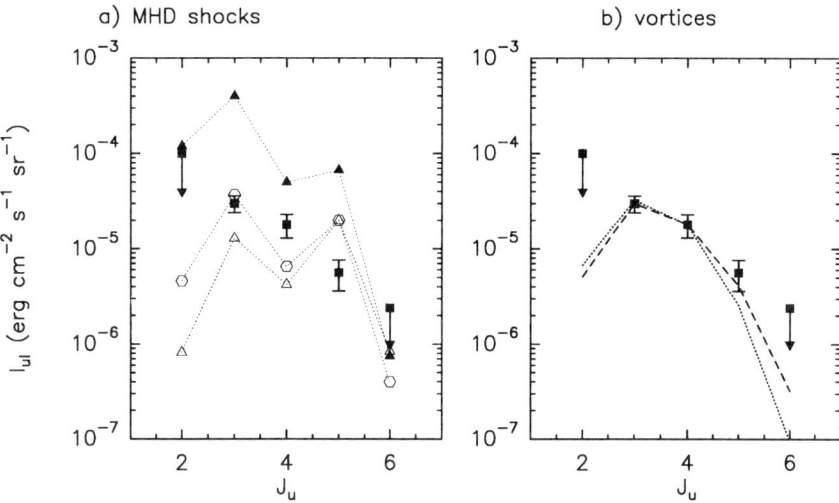

FIGURE 3. Comparison of the observed line intensities (solid squares) with model predictions. (a) Predictions of the H_2 line intensities in the MHD shock models. The emission is that of MHD shocks scattered in 20 mag of diffuse gas and corresponds to 20×170 shocks of 8 km s^{-1} (solid triangles), 20×6 shocks of 10 km s^{-1} (empty hexagons) and 20×0.7 shocks of 16 km s^{-1} (empty triangles). (b) Predictions of the H_2 line intensities of the vortex model. The emission corresponds to an ensemble of 20×1000 isolated vortices scattered in 20 mag of diffuse gas. The dotted line connects the predictions when statistical equilibrium is assumed for the H_2 rotational levels. The dashed line corresponds to the case when the level populations are out of statistical equilibrium.

20. As a whole, each set would be able to reproduce the anticipated value of $N(CH^+)$ derived from the scaling between $N(CH^+)$ and A_v. This figure shows that the relative intensities of the ortho and para lines produced by the MHD shocks do not quite resemble the observed values. This is because the ortho/para ratio of H_2 reaches 3 in the MHD shocks. The timescales are long enough in the MHD shocks for this ratio to reach its thermal value of 3 at 10^3 K (see Wilgenbus et al., this conference). At the opposite, the relative intensities of the ortho and para lines predicted in the case of the vortices is closer to that observed because the H_2 ortho/para ratio does not have time in the short-lived vortices to reach the thermal value at 10^3 K and stays close to the initial adopted value of 1 (the thermal value at ≈ 80 K).

In the case of the vortices, we have released the assumption of statistical equilibrium for the rotational levels of H_2, an assumption which is only marginally valid for the short timescales involved in the vortices. It is interesting to note that the release of the equilibrium assumption has a non-negligible consequence since it increases the intensity of the S(3) line by almost a factor 2, and the S(4) line even more (dashed line in Fig. 3).

4. Conclusion

The H_2 rotational emission of the diffuse medium detected along a line of sight across the Galaxy, avoiding star forming regions, does not arise in the cold diffuse gas nor in the

PDR possibly associated with the low brightness free-free emission and the small fraction of dense gas. The observed H_2 emission might in turn be produced by a large number of localized hot regions (MHD shocks or dissipative vortices) which are the most plausible sites of formation of CH^+, CH, OH, HCO^+, C_2H ..., observed in the diffuse medium.

Further observations are necessary in order to distinguish the shocks from the vortices:
- high spectral resolution observations of molecular emission
- further tests of departures from equilibrium, which are more pronounced in the short-lived vortices than in the MHD shocks.

Last, we note that the intensity of the S(1) line is comparable to that detected by Valentijn et al. (1999) in the outer disk of NGC 891. It may then be possible that the bulk of the S(1) emission detected in the outer disk of this galaxy be due to a large number of MHD shocks or dissipative vortices scattered in the turbulent diffuse medium (see van der Werf, this conference).

REFERENCES

Altenhoff, W.J., Downes, D. et al. 1970: A&A Supp. 1, 319

Beichman, C.A., Neugebauer, G., Habing, H.J. et al. 1988 IRAS explanatory supplement

Belin, F., Maurer, J., Tabeling, P., Willaime, H., 1996: J. Physique II 6, 573

Cadot, O., Douady, S., Couder, Y., 1995: Phys. Fluids 7, 630

Crane, P., Lambert, D.L., Sheffer, Y., 1995: ApJS 99, 107

Flower, D.R., Pineau des Forêts, G., 1998: MNRAS 297, 1182

de Graauw T. et al., 1996: A&A 315, L49

Gredel R., 1997: A&A 320, 929

Jimenez J. 1997: in *Dynamics and statistics of concentrated vortices in turbulent flows*, EuroMech Coll. 384

Joulain, K., Falgarone, E., Pineau des Forêts, G., Flower, D., 1998: A&A 340, 241

Lucas, R., Liszt, H.S., 1996: A&A 307, 237

Pérault, M., 1987: PhD dissertation, Université Paris VII

Porter, D.H., Pouquet, A., Woodward, P.R., 1994: Phys. Fluids 6, 2133

Sanders D.B., Clemens D.P. Scoville N.Z., Solomon P.M., 1986: ApJS 60, 1

Schaiedt, S.G. et al., 1996: A&A 315, L55

Valentijn, E.A., van der Werf, P.P, 1999: ApJ 522 L29

Verstraete, L., Falgarone, E., Pineau des Forêts, G., Flower, D., Puget, J.-L. 1999: in *The Universe as seen by ISO*, eds. P. Cox & M. Kessler

H₂ Observations of the OMC-1 Outflow with the ISO-SWS

By Dirk Rosenthal[1], Frank Bertoldi[2]
AND Siegfried Drapatz[1]

[1]Max-Planck-Institut für Extraterrestrische Physik, D-85740 Garching, Germany

[2]Max-Planck-Institut für Radioastronomie, D-53121 Bonn, Germany

Using the Short-Wavelength-Spectrometer on the Infrared Satellite Observatory (ISO), we obtained near- and mid-infrared spectra toward the brightest H₂ emission peak of the Orion OMC-1 outflow. A wealth of emission and absorption features were detected, dominated by 60 H₂ ro-vibrational and pure rotational lines reaching from H₂ 0-0 S(1) to 0-0 S(25).

The total H₂ luminosity in the ISO-SWS aperture is (17 ± 5) $L_\odot$, and extrapolated to the entire outflow, (120 ± 60) $L_\odot$. The H₂ level column density distribution shows no signs of fluorescent excitation or a deviation from an ortho-to-para ratio of three. It shows an excitation temperature which increases from about 600 K for the lowest rotational and vibrational levels to about 3200 K at level energies $E(v,J)/k > 14\,000$ K.

1. Introduction

The Orion molecular cloud, OMC-1, located behind the Orion M42 Nebula at a distance of ∼450 pc (Genzel & Stutzki 1989), is the best-studied massive star forming region. This cloud embeds a spectacular outflow arising from some embedded young stellar object, which can possibly be identified as the radio source "I" 0.49 arcsec south of the infrared source IRc2-A (Menten & Reid 1995; Dougados et al. 1993). The outflow shocks the surrounding molecular gas, thereby giving rise to the strongest H₂ infrared line emission appearing in the sky. Peak 1 (Beckwith et al. 1978) is the brighter of the two H₂ emission lobes of the outflow. Although the outflow has been studied extensively for nearly two decades, the nature of the emission mechanism remains unclear.

Molecular hydrogen, through it infrared rotational and rotation-vibrational transitions, is an important coolant in shocks and photodissociation regions, and thereby a particularly well suited tracer of the fluorescently- and/or shock-excited gas. The Short Wavelength Spectrometer (SWS, de Graauw et al. 1996) aboard the Infrared Space Observatory (ISO, Kessler et al. 1996) offered the first opportunity to observe pure rotational and rotation-vibrational H₂ lines from 2.4 µm to 28 µm with one instrument, un-hindered by the Earth's atmosphere.

2. Observations

We observed OMC-1 in the SWS 01 (2.4 − 45 µm grating scan) and SWS 07 (Fabry-Pérot) modes of the short wavelength spectrometer (de Graauw et al. 1996) on board ISO on October 3, 1997, and in the SWS 02 ($\approx 0.01\lambda$ range grating scan) mode on September 20, 1997 and February 15, 1998 (Rosenthal et al. 1999, 2000). The observations were centered on $\alpha_{2000.} = 5^h 35^m 13\overset{s}{.}67$, $\delta_{2000.} = -5° 22` 8.5"$.

2.1. Results and discussion

The 2.4 − 45 µm full scan spectrum of Fig. 1 is dominated by a large number of rotational and ro-vibrational H₂ lines. The pure rotational lines arise from levels with energies ranging from $E_u/k =1015$ K for the 0-0 S(1) line to $E_u/k = 42\,515$ K for the 0-0 S(25)

line. They represent gas with excitation temperatures ranging from 600 K for the low energy levels to over 3000 K for level energies $E/k > 14\,000$ K.

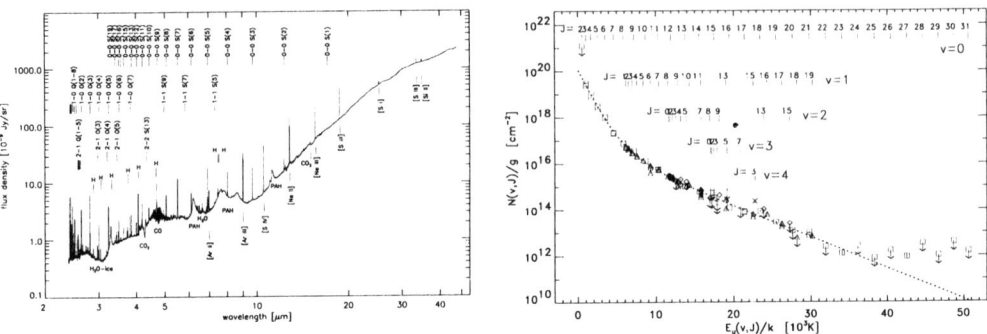

FIGURE 1. Left:2.5–45 μm spectrum of Orion Peak 1 obtained in the SWS 01 grating scan observing mode. A number of detected lines, bands and features are labeled. The continuum levels of the individual bands, which differ due to aperture changes, were arbitrarily shifted to match continuously. Right:Extinction-corrected, observed H_2 level column densities, divided by their degeneracy, plotted against the upper level energy $E(v, J)$. Vibrational levels are distinguished by different symbols: squares, triangles, diamonds, +, and × represent $v = 0$, 1, 2, 3, and 4, respectively. The dotted line represents the fit Eq. 2.2. Error bars represent 1σ flux uncertainties from the flux determination against the continuum, but do not include $\sim 30\%$ calibration uncertainties.

2.1.1. Excitation of H_2

All molecular hydrogen lines, due to the small radiative transition probabilities, remain optically thin. Therefore the corresponding upper level column density can be computed from the observed line flux,

$$N(v, J) = \frac{4\pi\lambda}{hc} \frac{I_{\rm obs}(v, J \to v', J')}{A(v, J \to v', J')} 10^{0.4 A_\lambda}, \qquad (2.1)$$

where $I_{\rm obs}(v, J \to v', J')$ and $A(v, J \to v', J')$ are the observed line flux and the Einstein-A radiative transition probability of the transition from level (v, J) to (v', J'), respectively. The Einstein coefficients are adopted from Turner et al. (1977) and Wolniewicz et al. (1998). The transition energies we computed from level energies kindly provided by E. Roueff (1992, private communication). A_λ accounts for dust extinction at the transition wavelength (Bertoldi et al. 1999).

A convenient way to visualize the level column densities is divide them by the level degeneracy g_J, and plot this against the upper level energy $E_u(v, J)/k$; the degeneracy $g_J \equiv g_s(2J+1)$, where $g_s = 3$ for ortho (odd J) H_2 and $g_s = 1$ for para (even J) H_2. For the lines we observed toward Peak 1 we found that in such a "Boltzmann diagram" the level columns shows a smooth distribution, where the level columns line up irrespective of their quantum numbers. There is no sign of fluorescent excitation or of a deviation from the ortho-to-para H_2 ratio of three.

Fluorescently excited H_2, as seen in photodissociation regions (Timmermann et al. 1996) would produce a level distribution in which the "rotational temperature" derived from levels at given v is lower than the the "vibrational temperature" derived from levels of the same J (e.g. Draine & Bertoldi 1996). Fluorescent excitation therefore shows a characteristic jigsaw distribution of the $v > 1$ levels, unlike the smooth line-up we observed here, where N/g appears not to depend on the state quantum number. Furthermore, fluorescently excited gas usually shows ortho-to-para ratios in vibrationally excited levels smaller than the total ortho-to-para ratio of the gas along the line of sight. This is due to the enhanced self-shielding, and therefore reduced excitation rate, of the more abundant ortho-H_2.

To describe the apparent range of excitation temperatures, $T_{ex} \sim 600$ K near the lowest levels and $T_{ex} \sim 3200$ K for $E(v,J)/k \geq 14\,000$ K, we decomposed the distribution of column densities to a sum of five Boltzmann distributions of different excitation temperatures:

$$N(v,J)/g_J = \sum_{i=1}^{5} C_i \, e^{-E(v,J)/kT_{ex,i}}, \qquad (2.2)$$

where we chose $T_{ex,i} = (628, 800, 1200, 1800, 3226)$ K, and the C_i were determined by a least-squares-fit to the observed level columns. In Fig. 1 right the dotted line shows the five-component fit. By summing the column densities over all levels following the interpolated level column distribution we compute for the total warm H_2 column density $N_{H_2,tot} = (1.9 \pm 0.5) \times 10^{21}$ cm^{-2}. Adopting a distance of 450 pc (Genzel & Stutzki 1989), this column corresponds to a warm H_2 mass of (0.06 ± 0.015) $M_\odot$ in the ISO-SWS aperture.

By summing from $J = 0$, we extrapolated the observed H_2 v = 0 level populations, $J \geq 3$, to the unobserved $J = 0$ to 2 levels. Note that thereby we estimate the total *warm* H_2 column density, but we do not account for the *total* H_2 column along the line of sight, which includes an additional $\approx 10^{22}$ cm^{-2} cold gas from the molecular cloud which embeds the outflow. Most of this cold H_2 resides in the ground states $J = 0$ and $J = 1$, and does not contribute to the emission observed from the shock-excited gas in the outflow.

With Eq. 2.2 we can also compute the column densities corresponding to the five excitation temperature components. The components' columns, $N_{H_2,i}$, are graphically displayed in Fig. 2 which shows that only a small fraction of the total H_2 column density is at high excitation temperature. On the other hand, the $J = 27$ level observed through the 0-0 S(25) line appears overpopulated by a factor of seven over what would be expected from the least-squares fit of Fig. 1 right. Bertoldi et al. (this proceedings) addresses the question how the highest-energy levels are excited. Considered are time-dependent C-shock models, pumping due to H_2 formation, and non-thermal collisions between molecules and ions in a magnetic shock. There the data are also compared to the prediction of current shock models.

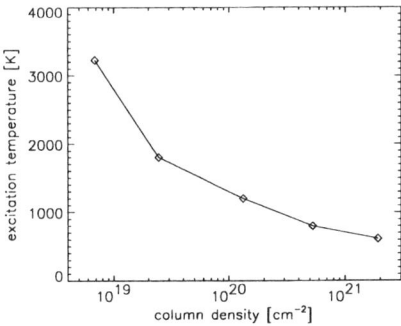

FIGURE 2. Decomposition in excitation temperature of molecular hydrogen toward Peak 1. The total column of warm H_2 is divided into five components. Only a small fraction of the total column is at high excitation temperature.

With the interpolated excitation distribution Eq. 2.2 the column densities of all H_2

energy levels can be estimated, even those from which no lines were observed. Then the total H_2 fluorescent emission from the electronic ground state extrapolates to (0.28 ± 0.08) erg s^{-1}cm^{-2}sr^{-1}. Over the ISO-SWS aperture this amounts to (17 ± 5) $L_\odot$. Compared with the total observed H_2 line emission (after extinction correction) of (0.16 ± 0.05) erg s^{-1}cm^{-2}sr^{-1}, we find that our line spectra account for more than half of the total H_2 emission.

Our observations target the brightest field in the Orion outflow. The outflow covers an area of about 2 arcmin x 2 arcmin. The average H_2 brightness over this area we estimate from the 1-0 S(1) map of Garden (1986) to approximately 20% of that in our observed field, so that the total H_2 luminosity of the OMC-1 outflow is estimated to be (120 ± 60) $L_\odot$. This value is consistent with the 94 $L_\odot$ estimated by Burton & Puxley (1990).

REFERENCES

Beckwith, S., Persson, S., Neugebauer, G., & Becklin, E. 1978, ApJ, 223, 464

Bertoldi, F., Timmermann, R., Rosenthal, D., Drapatz, S., & Wright, C. .M. 1999a, A&A, 346, 267

Burton, M. G. & Puxley, P. 1990, in The Interstellar Medium in External Galaxies: Summaries of Contributed Papers, ed. D. Hollenbach & H. Thronson, NASA CP, 3084,238

Cernicharo, J., González-Alfonso, E., Sempere, M. J., Leeks, S., van Dishoeck, E., Wright, C., Lim, T., Cox, P., & Pérez-Martínez, S. 1999, in *The Universe as seen by ISO*, ESA proc., P. Cox & M. Kessler (eds), 565

de Graauw,Th., Haser, L., Beintema, D. et al. 1996, A&A, 315, L49

Dougados, C., Léna, P., Ridgway, S., Christou, J., & Probst, R. 1993 ApJ, 406, 112

Draine, B. 1980a, ApJ, 241, 1081

Draine, B., Roberge, W., & Dalgarno, A. 1980b, ApJ, 241, 1081

Draine, B., Roberge, W., & Dalgarno, A. 1983, ApJ, 264, 485

Draine, B. & Bertoldi, F. 1996, ApJ, 468, 269

Garden, R. 1986, PhD Diss., Univ. Edinburgh

Genzel, R. & Stutzki, J. 1989, ARA&A, 27,41

Harwit, M., Neufeld, D. A., Melnick, G. J., & Kaufman, M. J. 1998, ApJ, 497, L105

Kessler, M., Steinz, J., Anderegg, M. et al. 1996, A&A, 315, L27

Menten, K. M. & Reid, M. J. 1995, ApJ, 445, L157

Moorhouse, A., Brand, P., Geballe, T., & Burton, M. 1990, MNRAS, 242, 88

Rosenthal, D., Bertoldi, F., Drapatz, S., & Timmermann, R. 1999, in *The Universe as seen by ISO*, ESA proc., P. Cox & M Kessler (eds), 561

Rosenthal, D., Bertoldi, & Drapatz, S. 2000, A&A, submitted

Timmermann, R., Bertoldi, F., Wright, C., Drapatz, S., Draine, B., Haser, L., & Sternberg, A. 1996, A&A, 315, L281

Turner, J., Kirby-Docken, K., & Dalgarno A. 1977, ApJS, 35, 281

Wolniewicz, L., Simbotim, I., & Dalgarno, A. 1998, ApJS, 115, 293

4. Extragalactic and Cosmology

David Wilgenbus and Emilie Habart at the organisers' stand.

The Role of H₂ Molecules in Cosmological Structure Formation

By T. Abel[1] AND Z. Haiman[2]†

[1]Harvard Smithsonian Center for Astrophysics, 60 Garden Street, Cambridge, MA 02138, USA

[2]Princeton University Observatory, Princeton, NJ 08544, USA

We review the relevance of H₂ molecules for structure formation in cosmology. Molecules are important at high–redshifts, when the first collapsed structures appear with typical temperatures of a few hundred Kelvin. In these chemically pristine clouds, radiative cooling is dominated by H₂ molecules. As a result, H₂ "astro–chemistry" is likely to determine the epoch when the first astrophysical objects appear. We summarize results of recent three–dimensional simulations. A discussion of the effects of feedback, and implications for the reionization of the universe is also given.

1. Introduction

In current "best–fit" cosmological models, cold dark matter (CDM) dominates the dynamics of structure formation, and processes the initial density fluctuation power spectrum $P(K) \propto k^n$ with $n = 1$ to predict $n = 1$ on large scales and $n \approx -3$ on small scales (Peebles 1982). The r.m.s. density fluctuation σ_M then varies inversely with the mass–scale ($\sigma_M \propto M^{-2/3}$ for $M \gg 10^{12} M_\odot$, while the dependence is only logarithmic for $M \ll 10^{12} M_\odot$). The more overdense a region, the earlier it collapses, implying that the present structure was built from the bottom up, with smaller objects appearing first, and subsequently merging and/or clustering together to assemble the larger objects (Peebles 1980). The predicted formation epochs of "objects" (i.e. collapsed dark matter halos) with various masses in the so–called standard CDM cosmology (Bardeen et al. 1986) are shown in Figure 1. Galaxies, which have masses around $10^{11-12} M_\odot$, are expected to have formed when the universe had approximately 10% of its present age (redshift $z \sim 3$), just around the limit of the deepest present–day observations (i.e. the Hubble Deep Field, HDF, Williams et al. 1996; or Lyα emission line detections, Weymann et al. 1998). Clusters of galaxies with masses around $10^{14-15} M_\odot$ are predicted to have formed as recently as 80% of the current age, with the more massive clusters still assembling at the present time.

Objects with the masses of globular clusters, $10^{5-6} M_\odot$, are predicted to have condensed as early as ~1% of the current age, or $z \sim 20$. It is natural to identify these condensations as the sites where the first "astrophysical" objects (stars, or quasars) might be born. Although the CDM model in Figure 1 predicts still smaller condensations at even earlier times, the cosmological Jeans mass in the smooth gas after recombination is $\sim 10^4 M_\odot$ (Peebles 1965), implying that gas pressure inhibits the collapse of gas below this scale (see Haiman, Thoul & Loeb 1996 on collapse on somewhat smaller scales due to gas/DM shell–crossing, hereafter HTL96).

What happens in a newly collapsed halo? Formally, in the absence of non–gravitational forces, a perfectly spherical top–hat perturbation simply collapses to a point (Peebles 1980). According to more accurate treatments describing self–similar solutions of spherical, but inhomogeneous secondary infall for a mixture of cold dark matter and baryons (Gunn & Gott 1972, Fillmore & Goldreich 1984, Bertschinger 1985), the evolution is

† Hubble Fellow

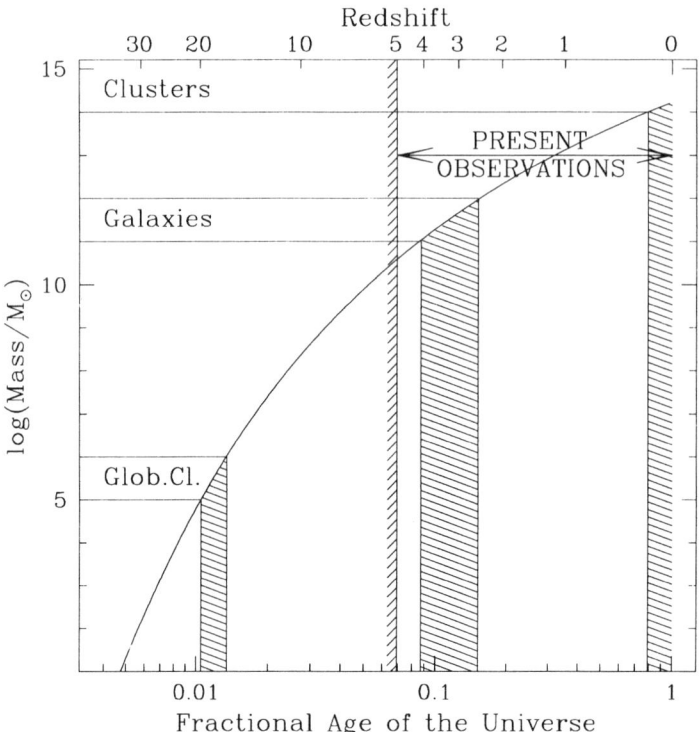

FIGURE 1. The formation epoch of objects with various masses in a standard CDM model ($\Omega = n = \sigma_{8h^{-1}} = 1$), shown as a fraction of the current age of the universe. The upper labels indicate redshifts. The first objects form at redshift $z \sim 20$, at a "lookback" fraction of $\sim$99%.

as follows. In the initial stages, while the bulk of the cloud is still turning around or expanding, the central, densest region of the cloud collapses and forms a dense virialized clump. The dark matter component virializes through violent relaxation (Lynden–Bell 1967), while the kinetic energy of the gas is converted into thermal energy through a hydrodynamic shock that raises the gas temperature to its virial value. As subsequent gas shells fall towards the center, they encounter an outward-propagating shock, and are brought to a sudden a halt. Continuous accretion onto the center then establishes a stationary, virialized object with a $\rho \propto r^{-2.25}$ density profile.

Early discussions of the formation of galaxies and clusters have argued that the subsequent behavior of the gas in such a virialized object is determined by its ability to cool radiatively on a dynamical time (Rees & Ostriker 1977; Dekel & Silk 1986). The same ideas apply on the smaller scales expected for the first collapsed clouds (Silk 1977; Kashlinsky & Rees 1983). Objects that are unable to cool and radiate away their thermal energy maintain their pressure support and identity, until they become part of a larger object via accretion or mergers. On the other hand, objects that can radiate efficiently will cool and continue collapsing.

The cooling time is determined by the virial temperature, $T_{\rm vir} \sim 10^4{\rm K} \, (M/10^8{\rm M}_\odot)^{2/3}$ $[(1+z)/11]$. For the largest halos, above $T_{\rm vir} \gtrsim 10^6$K, the most important cooling mechanism is Bremsstrahlung; for galaxy–sized halos (10^4K $\lesssim T_{\rm vir} \lesssim 10^6$K), cooling is possible via collisional excitation of neutral H and He. As first pointed out by Saslaw & Zipoy (1967) and Peebles and Dicke (1968), the virial temperatures of the first clouds are below 10^4K. In the primordial gas, the only molecule that could have a sufficient abundance is H_2, allowing cooling between 10^2K $\lesssim T \lesssim 10^4$K (see Dalgarno & Lepp 1987 for a review of astrochemistry in the early universe, and Stancil, Lepp and Dalgarno 1996 on the possible importance of other molecules, such as HD and LiH). Below $T \lesssim 10^2$K, the collapsed clouds are unable to cool, and remain pressure–supported for longer than a Hubble time.

Although H_2 molecules are unimportant in the formation of structures on galactic scales, they likely play a key role in the formation of the first, smaller structures. In particular, the abundance of H_2 controls the minimum sizes and formation times of the very first systems (HTL96, Tegmark et al. 1997).

The main issues regarding H_2 molecules in structure formation, addressed in this review, are:

• What is the H_2 abundance in the first collapsed objects?

• What is the parameter space (in mass, redshift and metallicity) when H_2 cooling dominates over all other cooling mechanisms?

• What is the parameter space when the H_2 cooling time is shorter than the dynamical time, so that H_2 can effect the dynamics of a system?

• What feedback mechanisms effect the H_2 abundance once the first stars or quasars light up?

2. H_2 Chemistry and Cooling

Because the cosmological background density of baryons $\sim (\Omega_B h^2/0.01)(1+z)^3 \, 10^{-7}$cm^{-3} is very small, chemical reactions in the smooth background gas occur on long timescales. As a consequence, for dynamical situations of structure formation chemical equilibrium is rarely an appropriate assumption. The dominant H_2 formation process in the gas phase,

$$\text{H} + \text{e}^- \rightarrow \text{H}^- + h\nu, \tag{2.1}$$

$$\text{H}^- + \text{H} \rightarrow \text{H}_2 + \text{e}^-, \tag{2.2}$$

relies on the abundance of free electrons to act as catalysts. At temperatures low enough to inhibit collisional dissociation by collisions with neutral hydrogen such electrons can only exist due to non-equilibrium effects. Electrons are also produced by photoionization of neutral hydrogen from an external UV radiation field. It is required to solve the time dependent chemical reaction network including the dominant chemical reactions. A very fast numerical method to solve this set of stiff ordinary differential equations has been developed by Anninos et al. (1997). The number of possible chemical reactions involving H_2 is large, even in the simple case of metal free primordial gas. The tedious work of selecting the dominant reactions and their reaction rates has been done by many authors, some recent examples being HTL96, Abel et al. (1997) and Galli & Palla (1998, see also Galli and Palla in this volume).

Cooling (defined here as the radiative loss of internal energy of the gas) is either due to reaction enthalpy released by a photon, or due to the radiative decay of collisionally excited atomic or molecular levels. For typical densities and proto-galactic scales it is accurate to assume the gas to be optically thin and only consider excitations from atomic and molecular ground states. The latter fact is due to the low densities resulting in collisional excitation time scales much longer than the corresponding radiative decay times (sometimes referred to as the coronal limit). For H_2 molecules this assumption breaks down for neutral hydrogen number densities in excess of $\sim 10^{2-3}$cm^{-3}. In comparison, for hydrogen atoms the coronal limit is reached only at electron densities of $\sim 10^{17}$cm^{-3} (Abel et al. 1997, and references therein). The calculations of the appropriate cooling function for molecular hydrogen seem to be converging. See Flower (2000, this volume) for a discussion and further references.

3. H_2 and the First Structures

Studies that incorporate H_2 chemistry into cosmological models and address issues such as non–equilibrium chemistry, dynamics, or radiative transfer, have appeared only in the past few years. However, pioneering works on the effect of H_2 molecules during the formation of ultra–high redshift structures go back to the 1960's. Saslaw & Zipoy (1967) first mentioned the importance of H_2 in cosmology. Peebles & Dicke (1968) speculated that globular clusters formed via H_2 cooling constitute the first building blocks of subsequent larger structures. Several papers soon constructed complete gas–phase reaction networks, and identified the two possible ways of gas–phase formation of H_2 via the H_2^+ or H^- channels. These were applied to derive the H_2 abundance in the smooth gas in the post–recombination universe (Lepp & Shull 1984; Shapiro, Giroux & Babul 1994), and under densities and temperatures expected in collapsing high–redshift objects (Hirasawa 1969; Matsuda et al 1969; Ruzmaikina 1973). Palla et al. (1983) combined the molecular chemistry with simplified dynamics, assuming a uniform sphere in free–fall, finding that three–body reactions significantly increase the H_2 abundance at the later (dense) stages of the collapse. The significance of non–equilibrium H_2 chemistry was realized by Shapiro & Kang (1987), who studied H_2 formation in a shock–heated gas, and found that the high electron fraction in the post–shock region leads to a significantly enhanced H_2 abundance. Based on a self–consistent treatment of radiative transfer of the diffuse radiation field (Kang and Shapiro 1992), this H_2 enhancement, regulated by photodissociation inside proto–galaxies, was suggested to lead to the formation of globular clusters (Kang et al. 1990).

The basic picture that emerged from this papers is as follows. The H_2 fraction after recombination in the smooth 'intergalactic' gas is small ($x_{H2} = n_{H2}/n_H \sim 10^{-6}$). At high redshifts ($z \gtrsim 100$), H_2 formation is inhibited even in overdense regions because the required intermediaries H_2^+ and H^- are dissociated by cosmic microwave background (CMB) photons. However, at lower redshifts, when the CMB energy density drops, a sufficiently large H_2 abundance builds up inside collapsed clouds ($x_{H2} \sim 10^{-3}$) at redshifts $z \lesssim 100$ to cause cooling on a timescales shorter than the dynamical time. This last conclusion was found to hold when the rotation of a collapsing sphere was also included (Hutchins 1976). Using a different approach, Silk (1983) explicitly demonstrated that a thermal instability exists for a collapsing gas–cloud forming H_2 molecules, leading to fragmentation. In summary, these early papers identified the most important reactions for H_2 chemistry, and established the key role of H_2 molecules in cooling the first, relatively metal–free clouds, and thus in the formation of population III stars.

The minimum mass of an object that can collapse and cool as a function of redshift has been studied by Tegmark et al. (1997) assuming constant density objects, and by HTL96 (see also Bodenheimer and Villere 1986), using spherically symmetric one–dimensional Lagrangian hydrodynamical models (see also Bodenheimer and Villere 1986). Sufficient H_2 formation and cooling requires the gas to reach temperatures in excess of a few hundred Kelvin; or masses of few $\times 10^5$ $M_\odot [(1+z)/11]^{-3/2}$. Initially linear density perturbations are followed as they turn around and grow in mass. When the DM dynamics are included, the center of the collapsing object contracts adiabatically into the growing DM potential well. As the object grows in mass, a weak accretion shock is formed. One might imagine a case where the adiabatic central core forms molecules early and allows rapid collapse before the accretion shock is formed; however this is not seen in the simulations. These temperatures are much larger than the temperature of the intergalactic medium ($T_{IGM} \sim 0.014(1+z)^2$) at the redshifts of interest. In other words, CDM models predicts the existence of numerous virialized objects with temperatures $\lesssim 500$ K that cannot cool. The object that contained the first star in the universe grows by the infall of both intergalactic DM and of gas, heated in an accretion shock. The residual fraction of free electrons catalyze the formation of H_2 molecules in its central region.

4. Three-dimensional Numerical Simulations

Cosmological hydrodynamical simulations of hierarchical models of structure formation have proven very successful in explaining cosmic structure on sub-galactic scales (Cen et al. 1994, Zhang et al. 1995, Hernquist et al. 1996, Davé et al. 1999), galactic scales (e.g. Kauffmann et al. 1999, Katz et al. 1999) and galaxy clusters (e.g. Frenk et al. 1999, Bryan and Norman 1998). With a realistic primordial chemistry model (e.g. Abel et al. 1997) and efficient numerical methods (e.g. Anninos et al. 1997) it is possible to also simulate the formation of the first cosmological objects. The first two–dimensional simulations of structure formation studied the collapse and fragmentation of cosmological sheets (Anninos and Norman 1996). In these simulations, the collapsing gas is heated in the accretion shock, and subsequently found to cool isobarically. The non–equilibrium abundance of electrons behind the strong accretion shock with $T \gg 10^4$K reaches its maximum value, and fast H_2 formation up to an abundance of a few 10^{-3} (number fraction) is observed, in agreement with the study of H_2 formation in shock–heated gas by Shapiro & Kang (1987). A similar calculation has been carried out by Abel et al. (1998b) for the study of the two dimensional collapse of long cosmic string induced sheets. In this cosmic string scenario, the dominant mass component was assumed to be hot dark matter (HDM, e.g. massive neutrinos). One then envisions long, fast moving cosmic strings to induce velocity perturbations, causing sheets to collapse, cool and fragment (Rees 1986). The feedback from the newly formed stars might have then induced further structure formation as in Ikeuchi and Ostriker (1986). It turns out, however, that H_2 formation is inefficient at high redshift where the CMBR dissociates the intermediaries H^- and H_2^+. Furthermore, the shocks caused by the string–induced velocity perturbations are too weak to enhance H_2 formation. Consequently Abel et al. (1998b) concluded that the cosmic string plus HDM model cannot develop luminous objects before the HDM component becomes gravitationally unstable.

Numerical calculations of more "mainstream" structure formation scenarios have been presented by Abel (1995), Gnedin and Ostriker (1997, GO97 hereafter), Abel et al. (1998a, AANZ98), and Abel, Bryan, and Norman (2000, ABN00). The simulations presented in GO97 focused on simulating the thermal history of the intergalactic medium and included star formation and feedback mechanisms. Their simulations were designed to accurately compute the number of the first objects able to cool and collapse and hence had to sacrifice numerical resolution within the collapsed objects. Results such as the typical fragment masses, typical temperatures, etc. are also found in multi–dimensional simulations that start from more idealized initial conditions (e.g. Bromm, Coppi, and Larson 1999)

In general cosmological hydrodynamical simulations treat the dynamics of the assumed collisionless cold dark matter using N–body techniques. This is typically coupled to a hydrodynamic grid code, solving the fluid equations for a gas of primordial composition. The simulations are initialized at a high redshift ($\gtrsim 100$) where density and velocity perturbations are small and in the linear regime. The simulation volume typically needs to be chosen rather small ($\lesssim 1$ comoving Mpc) to make sure that the simulation is at least capable of resolving the baryonic Jeans Mass prior to reionization,

$$M_{J,IGM} \approx 1.0 \times 10^4 M_\odot \left(\frac{1+z}{10}\right)^{3/2} \left(\frac{\Omega_B h^2}{0.02}\right)^{-1/2}. \quad (4.3)$$

The use of periodic boundary conditions for such small box sizes is only accurate at relatively high redshifts where they do model a representative piece of the model universe. Even in the lowest resolution simulations it becomes clear that an intricate network of sheets, and dense knots at the intersection of filaments is found. To the eye this structure is very similar to the simulation results on much larger scales.

Sufficient H_2 molecules, enabling the gas to cool, form only in the dense spherical knots at the intersection of filaments (AANZ98). These knots consist of virializing dark matter halos that accrete gas mostly from the nearby filaments but also from the neighboring voids. An accretion shock transforms the kinetic energy of the incoming gas into internal thermal energy. This shock tends to be spherical towards the directions of the voids but is often disturbed and more complex in morphology at the interface to the filaments. For the first objects that show any molecular hydrogen cooling the accretion shock is too weak to raise the ionization level of the gas over its residual primordial fraction of $n_{H+}/n_H \approx 2.4 \times 10^{-4} \Omega_0^{1/2} 0.05/(h\Omega_B)$ Peebles (1993). However, the associated raise in temperature of the post-shock gas allows molecule formation to proceed at time scales smaller than the Hubble time.

For the very first (i.e. least massive) objects with virial temperatures $\lesssim 1000$ K molecule formation is relatively slow, and the cooling time remains longer than the free fall time. As the objects merge and accrete, the higher virial temperature allows the chemistry and cooling to operate on faster than dynamical time scales. This further merging induces a rather complex velocity and density field in the gas, as well as the dark matter. Typical cosmological hydrodynamic methods can not follow the further evolution of the fragmentation of the gas clouds due to lack of numerical resolution and it was not possible to asses the nature of the first luminous objects by direct simulation. This drawback was recently overcome by the simulations presented in Abel, Bryan and Norman (2000, ABN00) by exploiting adaptive mesh refinement techniques (Berger and Collela, 1989, Bryan & Norman 1997, 1999). This numerical scheme allows to follow the gas dynamics to smaller and smaller scales by introducing new finer grids as they are needed. Since this is done at a scale much below the local Jeans length one is confident to capture the essential scale of the fragmentation due to gravitational instability.

These simulations clearly show how a region at the center of the virialized halo, containing approximately 200 $M_\odot$ in baryons, collapses rapidly. This "core" is formed via the classical Bonnor–Ebert instability of isothermal spheres. It contracts faster than the dynamical time scales in its parent halo. Hence these simulations indicate that the first luminous object(s) (perhaps a massive star) will form before most of the gas in the halo can fragment (ABN00). This core might still fragment further when it turns fully molecular via the three body formation process (Palla et al. 1983, Silk 1983)†. If the core forms stars at 100% efficiency an the ratio of produced UV photons per solar mass is the same as in present day star clusters than about 6×10^{63} UV photons would be liberated during the average life time of massive star ($\sim 5 \times 10^7$ yr). This is a few million times

---

† Abel, Bryan, and Norman (unpublished) have carried out a simulation which includes the three–body H_2 formation covering a dynamic range of 3×10^7. A preliminary analysis suggests that the core does not fragment further when it turns full molecular. This suggests that most likely a massive star will form in the collapsing core.

more than the $\sim 10^{57}$ hydrogen molecules within the virial radius, further suppressing H$_2$ cooling and fragmentation. Hence, the first star(s) may halt star formation until the massive star(s) die(s). Only a small fraction of primordial gas might be able to condense into PopIII stars of pristine primordial composition.

5. Feedback Issues

The first stars formed via H$_2$ cooling are expected to produce UV radiation, and explode as supernovae (if they are more massive than $\sim 8 M_\odot$), producing significant prompt feedback on the H$_2$ abundance in their own parent cloud. In addition, any soft UV radiation produced below 13.6eV and/or X–rays above $\gtrsim$ 1keV from the first sources can propagate across the smooth H background gas, possibly influencing the chemistry of distant regions. Soft UV radiation is expected either from either a star or an accreting black hole, with a black hole possibly contributing X–rays, as well. (Although recent studies find that metal–free stars have unusually hard spectra, these do not extend to $\gtrsim$ 1keV. See, e.g. Tumlinson & Shull 1999). The most important (and uncertain) quantity for assessing a stellar feedback is the IMF of the first stars. Several authors have argued that the IMF might be (see, e.g. Larson 1999 and references therein) biased towards massive stars. The lack of zero–metallicity stars, the so–called G–dwarf problem is resolved if the first generation of stars were short–lived; while the relatively inefficient cooling of metal–free gas could impose a minimum mass. A similar conclusion was reached by a recent 3D simulation (Bromm, Coppi, and Larson 1999).

The key question is whether the H$_2$ abundance in a collapsed region is effected shortly after the first few sources turn on (either in the same collapsed region, or elsewhere in the universe), i.e. before the H$_2$ abundance becomes irrelevant either because objects with $T_{\rm vir} \lesssim 10^4$K are already collapsing, or because metal enrichment has reached sufficiently high levels that H$_2$–cooling no longer dominates ($\sim 1\%$ solar, Böhringer & Hensler 1989). Ferrara (1998) considered the internal feedback from supernova explosions, and found that the non–equilibrium chemistry in the shocked gas can increase the H$_2$ abundance. However, Omukai & Nishi (1999) argued that a single OB star can photo-dissociate the H$_2$ molecules inside the whole $M \sim 10^6 M_\odot$ cloud. Even if molecules re-form after SN explosions, they found a net negative feedback.

External feedback from an early soft UV background were considered by Haiman, Rees & Loeb (1997). It was found that H$_2$ molecules are fragile, and easily photo–dissociated even inside large collapsed clumps via the two–body Solomon process (cf. Field et al. 1966)† – even when the background flux is several orders of magnitude smaller than the level $\sim 10^{-21}$ erg cm^{-2} s^{-1} Hz^{-1} sr^{-1} inferred from the proximity effect at $z \sim 3$ (Bajtlik et al. 1988), and needed for cosmological reionization at $z > 5$. These results were confirmed by a more detailed, self-consistent calculation of the build-up of the background and its effect on the contributing sources (Haiman, Abel & Rees 1999). The implication is a pause in the cosmic star–formation history: the buildup of the UVB and the epoch of reionization are delayed until larger halos ($T_{\rm vir} \gtrsim 10^4$K) collapse. (This is somewhat similar to the pause caused later on at the 'H-reionization' epoch, when the Jeans mass is abruptly raised from $\sim 10^4$ M$_\odot$ to $\sim 10^{8-9}$ M$_\odot$.) An early background extending to the X-ray regime would change this conclusion, because it catalyzes the formation of H$_2$ molecules in dense regions (Haiman, Rees & Loeb 1996, Haiman, Abel & Rees 2000). If quasars with hard spectra ($\nu F_\nu \approx$ const) contributed significantly to

† Note that these objects might also explain the large number of Lyman Limit Systems observed in high redshift quasar spectra (Abel & Mo 1998).

the early cosmic background radiation then the feedback might even be positive, and reionization can be caused early on by the small halos.

6. Conclusions

In popular CDM models, the first stars or quasars likely appeared inside condensed clumps with virial temperatures $T{\rm vir} \lesssim 10^4$K at redshifts $z \sim 20$. Because of the low virial temperatures, H_2 cooling (or lack thereof) played a dominant role in the gas dynamics inside these condensations. State–of–the art numerical simulations identify the sites for the first star–formation with the intersection of dense filaments (Abel et al. 1998, 2000). Although not directly visible in H_2 emission, these star–formation sites could be detected out to $z \sim 15$ with the Next Generation Space Telescope, provided they have a star formation efficiency of $\gtrsim$ one percent (Haiman & Loeb 1998), or they form quasar black holes with an efficiency of $\sim 10^{-3}$ and shine at the Eddington luminosity (Haiman & Loeb 1998). In the latter case, early mini–quasars from H_2 cooling would reionize intergalactic hydrogen by $z \sim 10$. The average star–formation efficiency in collapsed halos before $z \sim 3$ can be estimated by matching the average metal enrichment of the Ly α forest. If the latter is $\sim 10^{-3}$ solar, this implies that $\sim 2\%$ of the gas mass in collapsed regions are processed through stars (assuming a ratio in the number of high–mass to intermediate–mass stars as in a Scalo IMF, see Haiman & Loeb 1997). This value is not far from the efficiency of $\sim 1\%$ suggested by 3D simulations that resolve sub–parsec scales (ABN00); however it is unlikely that the early IMF was similar to the one observed in the local universe (e.g. Larson 1999). Arguably, further progress towards answering this question will have to come from a combination of more accurate simulations of star–formation in a metal–free plasma; including realistic radiative transfer.

We thank M. Rees for helpful comments. This work was supported by NASA through a Hubble Fellowship. TA acknowledges partial support by NSF grant AST-9803137 and NASA grant NAG5-3923.

REFERENCES

ABEL, T. 1995, *Thesis*, University of Regensburg, Germany
ABEL, T., ANNINOS, P., ZHANG, Y., NORMAN, M. L. 1997, *NewA* **2**, 181
ABEL, T., ANNINOS, P., NORMAN, M. L., ZHANG, Y. 1998a, *ApJ* **508**, 518 (AANZ98)
ABEL, T., STEBBINS, A., ANNINOS, P., NORMAN, M.L. 1998, *ApJ* **508**, 530
ABEL, T., MO, H. J. 1998, *ApJ* **494**, L151
ABEL, T., NORMAN, M. L., MADAU, P. 1999, *ApJ* **523**, 66
ABEL, T., BRYAN, G. L., NORMAN, M. L. 2000, *ApJ* in press (ABN00)
ANNINOS, P., NORMAN, M.L. 1996, *ApJ* **460**, 556
ANNINOS, P., ZHANG, Y., ABEL, T., NORMAN, M. L. 1997, *NewA* **2**, 209
BAJTLIK, S., DUNCAN, R. C., & OSTRIKER, J. P. 1988 *ApJ* **327**, 570
BARDEEN, J. M., BOND, J. R., KAISER, N., & SZALAY, A. S. 1986 *ApJ* **304**, 15
BERTSCHINGER, E. 1985 *ApJS* **58**, 39
BÖHRINGER, H., & HENSLER, G. 1989 *A&A* **215**, 147
BODENHEIMER, P. AND VILLERE, K. R. 1986 *Final Technical Report, UC Santa Cruz*
BROMM, V., COPPI, P. S., & LARSON, R. B. 1999 *ApJL* **527**, 5
BRYAN, G. L., NORMAN, M. L. 1997, in *Computational Astrophysics*, eds. D.A. Clarke and M. Fall, ASP Conference #123

Bryan, G. L., Norman, M. L. 1998, *ApJ* **495**, 80
Bryan, G. L., Norman, M. L. 1999, in *Workshop on Structured Adaptive Mesh Refinement Grid Methods*, IMA Volumes in Mathematics No. 117, ed. N. Chrisochoides, p. 165
Cen, R., Miralda-Escudé, J., Ostriker, J. P., & Rauch, M. 1994, *ApJ* **437**, L9
Dalgarno, A., & Lepp, S. 1987, in *Astrochemistry*, eds. Vardya, M.S. & Tarafdar, S.P. (Dordrecht:Reidel), 109
Davé, R., Hernquist, L., Katz, N., Weinberg, D. H. 1999, *ApJ* **511**, 521
Dekel, A., & Silk, J. 1986 *ApJ* **303**, 39
Field, G. B., Somerville, W. B., & Dressler, K. 1966 *ARA&A* **4**, 207
Fillmore, J. A., & Goldreich, P. 1984 *ApJ* **281**, 1
Frenk, C. S., White, S. D. M., Bode, P. et al. 1999, *ApJ* **525**, 554
Gnedin, N. Y., & Ostriker, J. P. 1997 *ApJ* **486**, 581
Gunn, J. E., & Gott, J. R. III. 1972 *ApJ* **176**, 1
Haiman, Z., Abel, T., & Rees, M. J. 2000 *ApJ*, in press, astro-ph/9903336
Haiman, Z., & Loeb, A. 1997 *ApJ* **483**, 21
Haiman, Z., & Loeb, A. 1998 *ApJ* **503**, 505
Haiman, Z., Rees, M. J., & Loeb, A. 1996 *ApJ* **467**, 522
Haiman, Z., Rees, M. J., & Loeb, A. 1997 *ApJ* **476**, 458
Haiman, Z., Thoul, A., & Loeb, A. 1996 *ApJ* **464**, 523 (HTL96)
Hernquist, L., Katz, N., Weinberg, D. H., & Miralda-Escudé, J. 1996, *ApJ*, **457**, L51
Hirasawa, T. 1969 *Prog. Theor. Phys.* **v. 42**, no. 3, p. 523
Hutchins, J. B. 1976 *ApJ* **205**, 103
Ikeuchi, S., Ostriker, J. P. 986, *ApJ* **301**, 522
Kashlinsky, A., & Rees, M. J. 1983 *MNRAS* **205**, 955
Kang, H., & Shapiro, P. R. 1992 *ApJ* **386**, 432
Kang, H., Shapiro, P. R., Fall, S. M., & Rees, M. J. 1990 *ApJ* **363**, 488
Kauffmann, G., Colberg, J. M., Diaferio, A., White, S. D. M. 1999 *MNRAS* **303**, 188
Katz, N., Hernquist, L., Weinberg, D. H. 1999, *ApJ* **523**, 463
Kolb, E. W., & Turner, M.S. 1990 *The Early Universe*, Addison-Wesley, Redwood City, CA
Larson, R. B. 1999 in *Star Formation From the Small to the Large Scale*, eds. F. Favata, A.A. Kaas, & A. Wilson, (ESA Special Publications Series, Noordwijk, Holland)
Lepp, S., & Shull, J. M. 1983 *ApJ* **270**, 578
Lepp, S., & Shull, J. M. 1984 *ApJ* **280**, 465
Lynden-Bell, D. 1967 *MNRAS* **136**, 101
Matsuda, T., Sato, H., & Takeda, H. 1969 *Prog. Theor. Phys.* **v. 42**, no. 2, p. 219
Omukai, K., & Nishi, R. 1999 *ApJ* **518**, 64
Palla, F., Salpeter, E. E., & Stahler, S.W. 1983 *ApJ* **271**, 632
Peebles, P. J. E. 1965 *ApJ*, **142**, 1317
Peebles, P. J. E. 1980 *Large-Scale Structure of the Universe* **Princeton Univ. Press, NJ**
Peebles, P. J. E. 1993 *Principles of Physical Cosmology* **Princeton Univ. Press**
Peebles, P. J. E. 1982, *ApJ* **263**, L1
Peebles, P.J.E., & Dicke, R.H. 1968 *ApJ* **154**, 891
Rees, M. J., & Ostriker, J. P. 1977 *MNRAS* **179**, 541
Rees, M. J. 1986, *MNRAS* **222**, 27
Ruzmaikina, T. V. 1973 *Sov. Astron.* **v. 16**, no. 6, p. 991
Saslaw, W. C., & Zipoy, D. 1967 *Nature* **216**, 976

SHAPIRO, P. R., GIROUX, M. L., & BABUL, A. 1994 *ApJ* **427**, 25
SHAPIRO, P. R., & KANG, H. 1987 *ApJ* **318**, 32
SILK, J. 1983 *MNRAS* **205**, 705
STANCIL, P.C., LEPP, S., DALGARNO, A. 1996, *ApJ* **458**, 401
TEGMARK, M., SILK, J., REES, M. J., BLANCHARD, A., ABEL, T., & PALLA, F. 1997 *ApJ* **474**, 1
TUMLINSON, J., & SHULL, J. M. 1999 *ApJL*, in press, astro-ph/9911339
WEYMANN, R. J., ET AL. 1998, *ApJ* **505**, 95
WILLIAMS, R. E., ET AL. 1996, *AJ* **112**, 1335
ZHANG, Y., ANNINOS, P., & NORMAN, M. L. 1995, *ApJ*, **453**, L57

The rôle of H$_2$ molecules in primordial star formation

By Francesco PALLA AND Daniele GALLI

Osservatorio Astrofisico di Arcetri, Largo E. Fermi, 5, 50125 Firenze, Italy

H$_2$ and HD molecules provide the cooling needed for the fragmentation and collapse of the first structures in the universe. In this review, we describe the main chemical and physical processes occurring in the primordial gas after the recombination epoch. We also highlight the areas where improvements in the determination of reaction rates and excitation coefficients are necessary to reduce the remaining uncertainties in the predictions of the numerical models. The interaction of primordial molecules with the CBR and the role of H$_2$ and HD cooling in the early universe are discussed. Finally, we comment on the results of recent simulations of the fragmentation and collapse of primordial clouds, with an emphasis on the typical mass scale of the first objects.

1. Introduction

The formation of the first stellar objects in the universe is a fascinating, yet little understood process. Although we have now observational data on bright quasars and galaxies out to redshifts of about 5 (corresponding to 10^9 yr after the Big Bang) and on the density fluctuations at redshifts about 1000 (or an age of $\sim 10^6$ yr), there is no direct evidence as to when and how the first structures formed. This unique epoch and the nature of the primordial objects define what has been called the end of the "Dark Ages" (Rees 1999).

According to Big Bang cosmology, there must have been an epoch in the history of the universe during which the original gas mixture was altered by the manufacture of heavy elements inside stars. The mass fraction of metals synthesized before decoupling of matter and radiation was insignificant ($Z < 10^{-10}$) compared to the lowest limits currently determined on the surface of stars in the galactic halo ($Z \gtrsim 10^{-4}$; e.g. Beers 1999). Clearly, one or more generation of stars must have enriched the pristine gas mixture. Massive stars are the obvious candidates for efficient and fast nucleosynthesis. The same objects may have been responsible for reionizing and reheating the intergalactic medium (e.g. Haiman & Loeb 1997; Ferrara 1998). From the absence of the Gunn-Peterson absorption in the spectra of high redshifted quasars, we know that the universe was reionized before a redshift of $\sim$5. Were massive stars the preferred products of the star formation process, or did a full spectrum of stellar masses down to subsolar values (and beyond perhaps) develop instead? And if so, is there any chance of discovering these long-lived, low-mass stars? The first generation of stars represent a special case in which the gas mixture is extremely simple, magnetic fields were likely dynamically unimportant, and the effect of the environment could be neglected. These characteristics make the problem quite different from the present-day case and lead to simplification of all the relevant physics (e.g. Palla 1988; Larson 1998). Many of the topics presented here, and more, can be found in the proceedings of a recent conference entirely dedicated to *The First Stars* (Weiss et al. 2000).

2. The cosmological background

Models of structure formation are based on the growth of small primordial density fluctuations by gravitational instabilities. In both cold dark matter (CDM) and baryonic dark matter (BDM) models of structure formation, the first objects predicted to go nonlinear are the smallest ones. In CDM models, it is expected that overdense regions with masses 10^5–10^7 M$_\odot$ would first collapse in the redshift range $10 \lesssim z \lesssim 100$ (Cen et al. 1993). The crucial question is whether cooling will enable the baryonic clouds to dissipate their kinetic energy and collapse more than the dark matter, to eventually become self-gravitating and form interesting objects such as a galaxy, a very massive object, a black hole, or a normal star (e.g., Couchman & Rees 1986; de Arújo & Opher 1991). For low-mass objects, the smaller they are, the less efficiently they dissipate energy and cool. Thus, a detailed treatment of gas dynamical processes will predict the characteristic mass scale M_c such that objects with $M > M_c$ can cool rapidly, whereas smaller lumps will merely remain pressure-supported and not form anything luminous. In other words, M_c *is the mass scale of the first luminous objects*. In order to compute realistic values of M_c, both the abundances of the main molecular species and their cooling properties must be accurately known. It is the purpose of this contribution to review the main chemical and physical properties of the primordial gas in the post-recombination epoch.

3. The first molecules

Saslaw & Zipoy (1967) and Peebles & Dicke (1968) were the first to realize the importance of gas phase reactions for the formation of the simplest molecule, H$_2$. They showed that trace amounts of molecular hydrogen, of order 10^{-6}–10^{-5}, could indeed form via the intermediary species H$_2^+$ and H$^-$ once the radiation field no longer contained a high density of photons with energies above the threshold of dissociation (2.64 and 0.75 eV, respectively).

The presence of even a trace abundance of H$_2$ is of direct relevance for the cooling properties of the primordial gas which, in its absence, would be an extremely poor radiator: cooling by Ly-α photons is in fact ineffective at temperatures less than ~ 8000 K, well above the matter and radiation temperature in the post-recombination era. Since the evolution of primordial density fluctuations is controlled by the ability of the gas to cool down to low temperatures, it is very important to obtain a firm picture of the chemistry of the dust-free gas mixture, not limited to the formation of H$_2$, but also to other molecules of potential interest (Lepp & Shull 1984, Puy et al. 1993, Dalgarno et al. 1998).

The chemical composition of the primordial gas consists of e$^-$, H, H$^+$, H$^-$, D, D$^+$, He, He$^+$, He^{++}, Li, Li$^+$, Li$^-$, H$_2$, H$_2^+$, HD, HD$^+$, HeH$^+$, LiH, LiH$^+$, H$_3^+$, and H$_2$D$^+$. The fractional abundances of these species are calculated as function of redshift, starting at $z = 10^4$ where He, H, D, and Li are fully ionized. The chemical evolution of the gas in the expanding universe resulting from the solution of the full network of rate equations is shown in Figure 1 (Galli & Palla 1998). The standard model is characterized by a Hubble constant $h = 0.67$, a closure parameter $\Omega_0 = 1$, and a baryon-to-photon ratio $\eta_{10} = 4.5$ (Galli et al. 1995). The initial fractional abundance of H, D, He and Li are taken from the standard big bang nucleosynthesis model of Smith et al. (1993). The residual ionization fraction at $z = 1$ for the standard model is [e/H] $= 3.02 \times 10^{-4}$.

As for the physics of recombination, several improvements in the detailed processes have been introduced since the seminal work of Peebles (1968) and Zel'dovich et al. (1968), but little has changed. One important correction has been found by Sasaki & Takahara (1993) who used the prescriptions for the rate of radiative recombination given by Grachev & Dubrovich (1991). The new calculations yield an abundance of electrons and H$^+$ at freeze-out which is a factor 2–3 lower than previous studies (e.g., Peebles 1993). The smaller residual ionization has a negative effect on the chemistry of the primordial gas in which electrons and protons act as catalysts in the formation of the first molecules. For an improved recombination calculation, see the recent work of Boschán & Biltzinger (1998) and Seager et al. (1999) who find a delayed He I recombination and an even lower residual ionization fraction (by ∼10%).

The evolution of the abundance of H_2 follows the well known behaviour (e.g., Lepp & Shull 1984, Black 1991) where the initial steep rise is determined by the H_2^+ channel followed by a contribution from H^- ions at $z \simeq 100$. The freeze-out value of H_2 is $[H_2/H] = 1.1 \times 10^{-6}$, a value somewhat lower than found in similar studies, as a consequence of the net reduction of the ionization fraction.

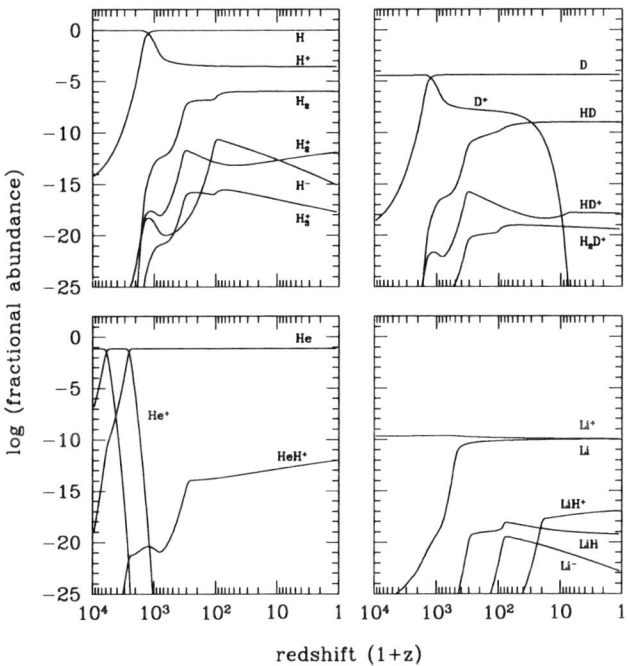

FIGURE 1. The evolution of the chemical species considered in the standard model as a function of redshift. The four panels show the results for H (upper left), D(upper right), He (lower left), and Li (lower right). (From Galli & Palla 1998)

The formation of H_2^+ is controlled by radiative association and photodissociation at $z \geq 100$, whereas at lower redshifts there is a contribution from HeH^+ that explains the gentle rise up to $[H_2^+/H_2] \simeq 10^{-6}$. An additional channel for H_2^+ formation is the associative ionization reaction $H + H(n = 2) \to H_2^+ + e^-$ (Rawlings et al. 1993; see also Latter & Black 1991 for H_2 formation).

The evolution of H_2^+ and, to a large extent, the abundance of H_2 depend critically on the adopted photodissociation rate. This is clearly shown in Fig. 2 where the results obtained with various choices of the rate are compared with the results of the standard model (shown by the solid line). Photodissociation of H_2^+ from $v = 0$ (dashed line) would result in an enhancement of a factor ∼ 200 of the final H_2 abundance, whereas photodissociation from $v = 9$ (long-dashed line), would delay the redshift of formation of H_2^+ from $z \simeq 10^3$ to $z \simeq 300$ and that of H_2 from $z \simeq 600$ to $z \simeq 250$. Note that in either case the asymptotic abundances of these species are left unchanged. However, there are good reasons to believe that the uncertainty in the redshift evolution of H$_2$ is not as large as that shown in Fig. 2, because photodissociation from $v=0$ is extremely unlikely to occur at $z \simeq 10^3$.

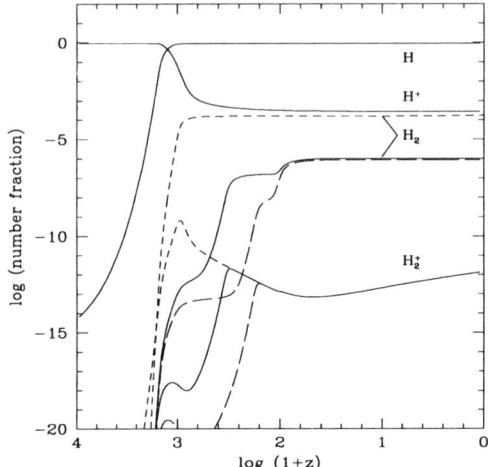

FIGURE 2. Effects of varying the photodissociation rate of H$_2^+$ on the evolution of H$_2$ and H$_2^+$. The solid curves are for the standard models, whereas the results obtained with photodissociation from the $v=0$ and $v=9$ vibrational levels of H$_2^+$ are shown by the *dashed* and *long dashed* lines, respectively. (From Galli & Palla 1998)

As for deuterium–bearing species, the only molecule formed in significant amount is HD, whose evolution with redshift follows closely that of H$_2$ (Stancil & Dalgarno 1997b; Stancil et al. 1998). However, its abundance at freeze-out is [HD/H$_2$] $\simeq 10^{-3}$, with an enhancement factor of $\sim$30 from the initial [D/H] abundance.

The main molecular species containing helium is HeH$^+$, formed by the radiative association of He and H$^+$. As emphasized by Lepp & Shull (1984), this reaction is slower than the usual formation mode via association of He$^+$ and H, but since the abundance of He$^+$ is quite small, radiative association takes over. The HeH$^+$ ions are removed by CBR photons and at $z < 250$ also by collisions with H atoms to reform H$_2^+$. This explains the abrupt change in the slope of the curve of HeH$^+$ shown in Fig. 1. The final abundance is small, [HeH$^+$/H] $\sim 6 \times 10^{-13}$ at $z = 1$.

Finally, the chemistry of lithium is complicated, but the molecular abundances are indeed very small. The more abundant complex is LiH$^+$ whose formation is controlled by the radiative association of Li$^+$ and H (Dalgarno & Lepp 1987; Stancil & Dalgarno 1997a). As shown in Fig. 1, lithium remains more ionized at low redshifts and this explains why LiH is less abundant than LiH$^+$. The final abundance of LiH$^+$ is $\simeq 10^{-17}$ at $z = 1$.

4. Cooling functions for the primordial gas

For cosmological problems, one generally adopts the analytical expressions for the H-H$_2$ cooling function given by Hollenbach & McKee (1979, 1989) or Lepp & Shull (1983). These formulae appear now to suffer considerably from the limitations and uncertainties associated with the collisional coefficients available at the time. At low temperatures, the uncertainty is associated with the choice of the interaction potential, since it is very difficult, even nowadays, to calculate the potential in the interaction region to the requisite level of accuracy ($\sim 10^{-3}$ hartrees). However, recent improvements in the calculations employing both fully quantum mechanical methods (Flower 1997, 1998) and classical trajectory methods (Boothroyd et al. 1996; Tiné et al. 1998) have greatly reduced the uncertainties in the cross-section and rate coefficients in the temperature interval 10^2 to 10^4 K (see Le Bourlot et al 1999; Flower, this volume).

Due to its small dipole moment, the lowest transition that can be excited in HD molecules ($J = 1 - 0$) has an energy of 128 K, four times less than H_2. Therefore, HD molecules are important to enhance the cooling function at low temperatures. Recently, Roueff & Flower (1999) have computed rate coefficients for rotational transitions induced by collisions of HD with H atoms using fully quantum mechanical methods and an updated potential surface. Their calculations include levels up to $J = 9$ of the vibrational ground state and cover the temperature range 100 to 1000 K.

Several H-H_2 cooling functions are shown in Fig. 3 for two values of the the density, $n(H) = 1$ cm^{-3} and $n(H) = 10^6$ cm^{-3}. Ortho- and para-H_2 were assumed in their equilibrium 3:1 ratio. The cooling functions given by Hollenbach & McKee (1989) and Lepp & Shull (1983), also shown in the figure, differ significantly from those recently computed by Forrey et al. (1997) and Le Bourlot et al. (1999), especially in the low-density limit, at both low and high temperatures. Of course, the differences disappear in the high density limit where the cooling function is computed in conditions close to LTE. The reduced cooling efficiency of H_2 molecules leaves its imprint on the determination of the characteristic mass scale of the first objects.

In considering the cooling by H_2, it is important not to neglect the effect of reactive collisions of H_2 with H or H$^+$ that cause ortho–para interchange and lead to a ratio out of the equilibrium value. In the conditions of the primordial gas with H$^+$/H$\gtrsim 10^{-4}$, the ortho–para exchange is dominated by collisions with protons: Pineau des Fôrets & Flower (2000) have presented self-consistent cooling calculations that follow the variation of the ortho-to-para ratio as a function of redshift.

5. Interaction of primordial molecules with the CBR

The discussion of the previous sections has neglected some physical processes resulting from the interaction of the photons of the CBR with the primordial gas. Here, we consider two effects: (a) the modification to the cooling function due to the CBR photons and (b) Thomson scattering of CBR photons on trace molecules.

5.1. Heating vs. cooling

The cooling functions described in Section 4 have been computed assuming that the temperature of the CBR is much smaller than the gas temperature. This approximation of course is valid at low redshifts, but becomes increasingly inaccurate at higher redshifts. Since most cosmological scenarios of structure formation begin at $z \simeq 100$, when $T_{\rm CBR} \simeq 300$ K, the level population of molecules is greatly affected by stimulated emission and absorption (see Capuzzo-Dolcetta et al. 1991).

For example, in the standard model of Sect. 2, the CBR temperature is always higher than, or equal to the matter temperature. Hence, the energy transfer function $(\Gamma - \Lambda)_{\rm mol}$ in eq. (4) becomes an effective heating source for the gas because the rate of collisional de-excitation of the rovibrational molecular levels is faster than their radiative decay (Khersonski 1986; Puy et al. 1993). As an illustration of the effect of accounting for the calculation of the level population in the presence of the CBR, we show in Figure 4 the

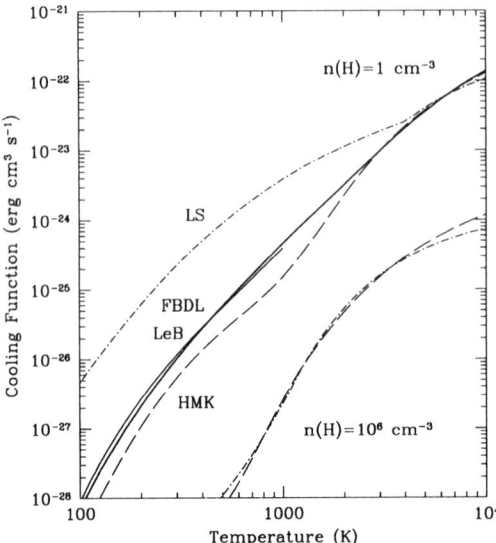

FIGURE 3. Cooling function of H_2 for H-H_2 collisions for two values of the density. The thick solid curve shows the results of Le Bourlot et al. (1999; LeBPF), while the thin solid line limited to $T < 1000$ K is from Forrey et al. (1997; FBDL). The other curves give the cooling functions computed by Lepp & Shull (1983; LS) and Hollenbach & McKee (1989; HMK). (Adapted from Galli & Palla 1998)

net heat transfer function $(\Gamma - \Lambda)_{H_2}$ computed at three selected redshifts. Note that for $T_{\rm gas} > T_{\rm CBR}$ the cooling function is significantly decreased from the value computed for $T_{\rm CBR} = 3$ K, because of the radiative depopulation of excited states. The sudden drop signals the condition when the two temperatures are equal. Finally, for $T_{\rm gas} < T_{\rm CBR}$ the function changes sign and becomes a heating term for the gas. Naturally, in cosmological simulations the heat transfer should be computed self-consistently at each redshift.

5.2. Thomson scattering due to molecules

In addition to thermal absorption and emission, molecules of the primordial gas can also interact with the CBR via Thomson scattering. As pointed out by Zel'dovich et al. (1968), the case for scattering is interesting since the cross-section for of a harmonic oscillator is many orders of magnitudes larger than for free electrons. Although scattering occurs only in narrow spectral lines, the spectrum of molecular transitions contains several bands and the cumulative effect can give rise to appreciable continuum absorption.

Spatial anisotropies in the CBR can be produced by Thomson scattering of photons on molecules (or electrons) located in protoclouds moving with a peculiar radial velocity component $v_{\rm pec}$ (e.g. Maoli et al. 1994). This process can create a curtain that blurs primordial CBR anisotropies (Dubrovich 1993). The level of secondary anisotropies in the CBR temperature at a frequency ν is simply

$$\frac{\Delta T}{T} = \frac{v_{\rm pec}}{c}\left(1 - e^{-\tau_{\rm cl}}\right) \qquad (5.1)$$

where $\tau_{\rm cl}$ is the optical depth through the cloud. The actual calculation of the resulting $\Delta T/T$ requires the solution of the radiative transfer equation in the moving cloud (see Bougleux & Galli 1997 for details).

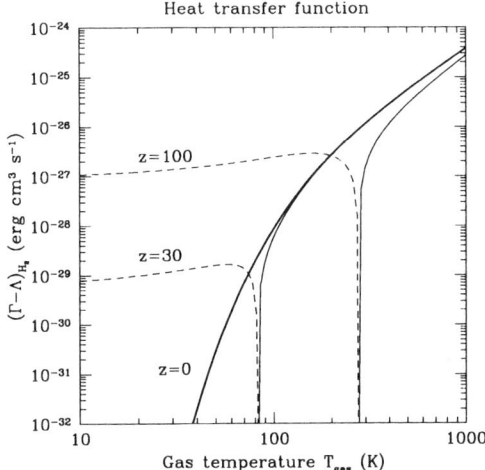

FIGURE 4. The heat transfer function $(\Gamma-\Lambda)_{H_2}$ versus the gas temperature at selected redshifts. The solid line is computed ignoring the effects of the CBR ($z=0$). When $T_{\rm gas} > T_{\rm CBR}$ the heat exchange is a cooling term (solid portion of the two other curves). In the opposite case, the transfer becomes a heating source for the gas (dashed curves).

In general, one can compute the total optical depth due to resonant Thomson scattering for all the relevant molecules, using the chemical evolution described in Sect. 2. This quantity is given by

$$\tau_{\rm tot} = \sum_i \int \sigma \, x_i \, n_b \, dl, \qquad (5.2)$$

where σ is the Thomson cross section, normalized to the linewidth $\Delta\nu_D$ due to Doppler broadening, x_i is the fractional molecular abundance, and dl is the photon path length in the redshift interval dz. In order to evaluate eq. (5.2), one must compute the level population of each molecule.

LiH and H_2D^+ are promising molecules because of their large dipole moments (Dubrovich 1993; Maoli et al. 1994). However, their tiny final abundances (LiH/H$\simeq$H$_2$D$^+$/H$\simeq$ 10^{-19}) yield negligible optical depths, $\tau_{\rm LiH} \simeq 10^{-10}$ at $\nu \simeq 30$ GHz and $\tau_{\rm H_2D^+} \simeq 10^{-13}$ at $\nu \simeq 40$ GHz. The only trace molecule of some interest turns out to be HeH$^+$ which, despite a freeze-out abundance of HeH$^+$/H$\simeq 10^{-12}$, has a maximum optical depth of $\tau_{\rm HeH^+} \simeq 10^{-5}$ at $\nu \simeq 50$ GHz due to the excitation of rotational transitions. The variation of the total optical depth due to Thomson scattering on HeH$^+$ is shown in Figure 5. From eq. (5.1), we see that even for $v_{\rm pec} \sim 300$ km s^{-1}, the resulting spatial variations of the CBR temperature are only $\Delta T/T \sim 10^{-8}$. On the other hand, the expected spectral distortion of the CBR should affect all angular scales smaller than the angular diameter of the horizon by an amount $(1 - e^{-\tau_\nu}) \simeq 10^{-5}$. Since ongoing CBR anisotropy experiments in the 10-50 GHz frequency range reach a sensitivity of $\sim 10 \, \mu$K, and the fluctuations measured on degree scales are of order $\sim 100 \, \mu$K, it is not completely out of the question that one may eventually be able to search for HeH$^+$ at high redshifts.

6. The Fragmentation and Collapse of primordial clouds

The discussion of the previous section has shown that the present knowledge of both the molecular abundances and the cooling properties of the primordial gas has greatly

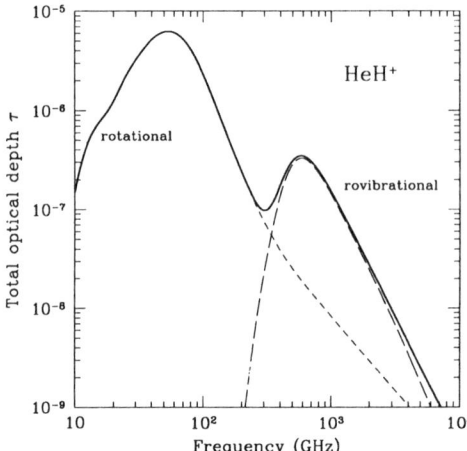

FIGURE 5. Optical depth of the universe generated by Thomson scattering of the CBR photons on HeH$^+$ molecules. The contribution of rotational and rovibrational transitions to the total (*solid* line) is shown by the *dashed* curves.

improved in recent years. These quantities provide the initial input for the actual calculations of the fragmentation and collapse processes that lead to star formation.

6.1. *Initial results*

Early models have greatly simplified the dynamics of collapse by using simple geometries and neglecting the presence of rotation and magnetic fields. The basic thermal and chemical properties of the collapse were worked out in detail, assuming that the collapse begins at an epoch when the photons of the CBR no longer prevent the formation of H$_2$ molecules via the H$^-$ ions. All calculations started with an initial temperature of the gas of several hundred degrees, and an ionization level corresponding to the freeze-out value. Bound clouds satisfying the condition of gravitational instability have typical masses $M_{\rm cl} > M_{\rm Jeans} \sim 10^5$–$10^6$ M$_\odot$, and experience continued fragmentation if the gas can radiate away the compressional heat of collapse.

The main features of the evolution consist in an initial adiabatic phase during which the temperatures increases to ~ 1000 K and the H$_2$ abundance to $\sim 10^{-3}$. Then, the temperature is kept almost constant by the efficient cooling provided by collisionally excited rotational transitions. The fraction of H$_2$ remains nearly constant since the H$^-$ abundance decreases slowly with time. At densities in excess of $\sim 10^8$ cm^{-3}, a new route for the formation of H$_2$ molecules is activated via the three-body reactions

$$H + H + H \longrightarrow H_2 + H \tag{6.3}$$
$$H + H + H_2 \longrightarrow H_2 + 2H \tag{6.4}$$

The effect is dramatic and all the atomic hydrogen is converted to molecules (Palla et al. 1983). The greatly enhanced cooling efficiency, due now to both rotational and vibrational transitions, allows the temperature to remain below ~ 2000 K for a wide density range. As a consequence, the instantaneous Jeans mass can drop to very low values, and fragmentation proceeds down to the smallest scales, M$_{\rm min} \sim 0.1$ M$_\odot$ (Palla et al. 1983; Lepp & Shull 1984). These values are similar to those found in present-day clouds, but the physical conditions are vastly different because of the much higher densities ($\sim 10^{14}$ cm^{-3} vs. $\gtrsim 10^5$ cm^{-3}) and temperatures ($\sim 10^3$ K vs. ~ 10 K).

The rôle of H$_2$ formation and cooling was investigated also in conditions not as quiescent as those of free fall collapse. It is likely that as the cloud collapses, shocks will form which will reheat and reionize the gas, destroying the molecular hydrogen. However, H$_2$ can reform efficiently behind shocks by the same reactions discussed in Section 2 although the chemistry proceeds in a highly non-equilibrium condition (Shapiro & Kang 1987; Palla & Zinnecker 1987). Detailed models of molecule formation behind shocks showed that H$_2$ can reach abundances of 10^{-3} in a wide variety of model assumptions and the gas could reach temperatures $\ll 10^4$ K (Mac Low & Shull 1986; Kang & Shapiro 1992). Similarly, HD formation is enhanced in these conditions, providing additional cooling that can drive the temperature down to $\lesssim 100$ K.

6.2. How Small Were the First Cosmological Objects?

The early studies described above have provided a first glimpse to the question of first structure formation. However, the approximations used both for the cosmological and numerical models were by necessity too crude to give complete answers. The problem has been recently treated realistically by numerical multidimensional hydrodynamical calculations that include radiative cooling, non-equilibrium chemistry and external radiation fields (Anninos et al. 1997; Gnedin & Ostriker 1997; Abel et al. 1998; Nishi & Susa 1999).

The basic question is: how much molecular cooling is needed for a gas cloud to be able to cool in a Hubble time? Haiman et al. (1997) and Tegmark et al. (1997) have provided an answer to the question of the formation of H$_2$ using an approximate method for the halo density profile and following the growth of an isolated density peak that is spherically symmetric. The critical H$_2$ fraction satisfies the condition that the cooling and the Hubble timescales must be equal. This fraction is given by

$$f_{H_2} \approx 2 \times 10^{-4} \left(\frac{h\Omega_b}{0.03}\right)^{-1} z_{100}^{-3/2} \left(1 + \frac{10\, T_3^{7/2}}{60 + T_3^4}\right) e^{512/T}, \tag{6.5}$$

where the last two terms on the left hand side come from an approximation to the cooling function obtained by the Hollenbach & McKee (1979). The H$_2$ fraction required exceeds typical initial abundances ($\sim 10^{-4}$) for all redshifts $z < 200$ when the temperature is less than $\sim 10^4$ K. Therefore, the low-mass high-redshift clouds can cool and collapse only if additional H$_2$ is produced. This can happen during the virialization phase of an overdense region that grows and goes nonlinear.

As a rule of thumb, Tegmark et al. (1997) find that if the virial temperature is high enough to produce an H$_2$ fraction of order 5×10^{-4}, then the cloud will collapse. Computing the density, temperature and H$_2$ abundance in virialized lumps as a function of $z_{\rm vir}$, they find that the minimum baryonic mass, $M_b = \Omega_b M_c$, is strongly redshift dependent, dropping from 10^6 M$_\odot$ at $z \sim 15$ to 5×10^3 M$_\odot$ at $z \sim 100$. The resulting minimum mass for collapse is shown in Figure 6. The solid line has been computed using the cooling function of Le Bourlot et al. (1999), but the qualitative behavior is the same for all functions. For $z > 100$, M_c decreases less rapidly with redshift as the CMB photons inhibit H$_2$ formation through the H$^-$ channel. For $z > 200$, the H$_2^+$ channel for H$_2$ formation becomes effective, driving M_c down toward $M_c \sim 10^3$ M$_\odot$. With a standard CDM power spectrum with $\sigma_8 = 0.7$, this implies that a fraction 10^{-3} of all baryons may have formed luminous objects by $z = 30$, which could be sufficient to reheat the universe.

These initial estimates have been confirmed and extended by full 3-D models. These

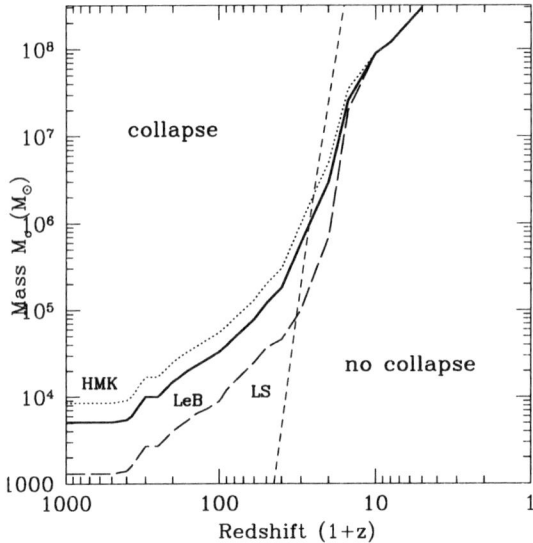

FIGURE 6. The minimum mass needed to collapse. The function M_c is plotted as a function of virialization redshift for standard CDM ($\Omega = 1$, $\Omega_b = 0.06$, $h = 0.5$). Only lumps whose parameters (z_{vir}, M_c) lie above the shaded area can collapse and form luminous objects. The three curves are for different H_2 cooling functions, labeled as in Fig. 3. The dashed line corresponds to 3σ peaks in standard CDM, normalized to $\sigma_8 = 0.7$, so such objects with baryonic mass $\Omega_b \times 2 \times 10^6 M_\odot \sim 10^5 M_\odot$ can form at $z = 30$. (Adapted from From Tegmark et al. 1997 and Abel et al. 1998)

simulations can explore large density contrasts and achieve very high spatial resolution by means of adaptive mesh refinement calculations that allow solution of self-gravitational problems that would be prohibitively expensive with fixed resolution (e.g. Abel et al. 1998; also Bromm et al. 1999 for SPH models). The resulting history of first structure formation indicates that a complex system of voids, filaments, sheets, and spherical knots form at the intersection of filaments. It is only within the knots that enough H_2 is formed ($f_{H_2} \gtrsim 5 \times 10^{-4}$) to cool the gas appreciably. However, only a small fraction of the total baryonic mass ($\lesssim 10\%$) can participate in the process, the rest remaining at rather high temperatures ($\gtrsim 10^3$ K).

Recently, Norman et al. (2000) have presented models where comoving scales from 128 kpc cab be followed with high accuracy down to 0.06 pc, i.e. the scale of *primordial cloud cores*! An important result is the existence of a contracting hydrostatic core of $\sim 100 M_\odot$. Within this core the gas density increases to values higher than 10^8 cm^{-3} where three-body reactions form H_2 very efficiently and the gas start cooling again. Once the gas is fully molecular, however, its opacity becomes very large and the cooling function changes substantially. The task of following the further evolution of this fragment is, however, extremely difficult, owing to the complexity of the radiative transfer of optically thick lines in collapsing clouds. Nevertheless, it is from the solution of this problem that one can hope to obtain the final answer to the characteristic mass of primordial stars. Because of the rather large value of M_c and because the cooling time at this stage becomes longer than the dynamical scale, Norman et al. estimate that these clouds are no longer subject to further fragmentation and that, therefore, the first stars were massive objects.

It is a remarkable achievement of the numerical simulations to have identified the typical scale for fragmentation to be the Bonnor-Ebert scale, similar to what happens in present-day molecular clouds. If indeed these are the typical initial conditions for gravitational collapse, what should be the outcome of the actual proto-stellar accretion phase? The combination of a high density regime ($n \gtrsim 10^{13}$ cm^{-3}) and relatively high temperatures (T$\gtrsim 10^3$ K) tends to suggest that proto-stellar collapse will be characterized by very short timescales. On the other hand, the large increase of the cooling rate following the onset of the three-body reaction will also lead to a dramatic density enhancement of the core. Silk (1983) has argued that such phase can lead to further fragmentation of the cloud, until individual fragments are opacity limited. If this is the case, then the formation of low-mass stars would be a natural outcome.

Whether the actual mass spectrum of the first stars is weighted toward massive stars or not is still an open question. However, the tremendous improvements described in this review both in the cosmological simulations of large-scale structure formation and in the multidimensional models of fragmentation and collapse have transformed the study of the first stars into a mature field that should be soon capable to make specific and testable predictions about their nature.

We would like to acknowledge useful comments and exchange of information on various aspects of the research described here with T. Abel, V. Bromm, A. Dalgarno, D. Flower, Z. Haiman and D. Puy. Special thanks to Francoise Combes for the invitation to express these ideas at the topical conference on H$_2$ molecules.

REFERENCES

Abel, T., Anninos, P., Norman, M.L., Zhang, Y. 1998, ApJ, 508, 518

Anninos, P., Zhang, Y., Abel, T., Norman, M.L. 1997, New Astron, 2, 209

Beers, T.C. 1999, in The Galactic Halo, eds. B.K. Gibson, T.S. Axelrod & M.E. Putnam, ASP Con. Ser. Vol. 165, 202

Black, J.H. 1991, in Molecular Astrophysics, ed. T.W. Hartquist (Cambridge: Cambridge Univ. Press), 473

Boothroyd, A.I., Keogh, W.J., Martin, P.G., Peterson, M.R. 1996, J Chem Phys, 104, 7139

Boschán, P., Biltzinger, P. 1998, AA, 336, 1

Bougleux, E., Galli, D. 1997, MNRAS, 288, 638

Bromm, V., Coppi, P.S., Larson, R.B. 1999, ApJ, 527, L5

Capuzzo-Dolcetta, R., Di Fazio, A., Palla, F. 1991, AAS, 88, 451

Cen, R., Ostriker, J.P., Peebles, P.J.E. 1993, ApJ, 415, 423

Couchman, H.M.P., Rees, M.J. 1986, MNRAS, 221, 53

Dalgarno, A., Lepp, S. 1987, in IAU Symp. 118 Astrochemistry, eds. M.S. Vardya & S.P. Tarafdar (Dordrecht: Reidel), 109

Dalgarno, A., Stancil, P.C., Lepp, S. 1998, in Stellar Evolution, Stellar Explosions and Galactic Chemical Evolution, ed. A. Mezzacappa (Inst. of Phys. Publ.), p.137

de Araújo, J.C.N., Opher, R. 1991, ApJ, 379, 461

Dubrovich, V.K. 1993, SvA Lett, 19, 53

Ferrara, A. 1998, ApJ, 499, L17

Flower, D.R. 1997, J Phys B, 30, 1

Flower, D.R. 1998, MNRAS, 297, 334

Galli, D., Palla, F. 1998, AA, 335, 403

Galli, D., Palla, F., Ferrini, F., Penco, U. 1995, ApJ, 443, 536

Gnedin, N.Y., Ostriker, J.P. 1997, ApJ, 486, 581

Grachev, S.I., Dubrovich, V.K. 1991, Astrophysics, 34, 124

Haiman, Z., Loeb, A. 1997, ApJ, 483, 21

Haiman, Z., Rees, M.J., Loeb, A. 1997, ApJ, 476, 458

Hollenbach, D., McKee, C.F. 1979, ApJS, 41, 555

Hollenbach, D., McKee, C.F. 1989, ApJ, 342, 306

Kang, H., Shapiro, P.R. 1992, ApJ, 386, 432

Khersonski, V.K. 1986, Astrophysics, 24, 114

Larson, R.B. 1998, MNRAS, 301, 569

Latter, W.B., Black, J.H. 1991, ApJ, 372, 161

Le Bourlot, J., Pineau des Forêts, G., Flower, D.R. 1999, MNRAS, 305, 802

Lepp, S., Shull, J.M. 1983, ApJ, 270, 578

Lepp, S., Shull, J.M. 1984, ApJ, 280, 465

MacLow, M.M., Shull, J.M. 1986, ApJ, 302, 585

Maoli, R., Melchiorri, F., Tosti, D. 1994, ApJ, 425, 372

Nishi, R., Susa, H. 1999, ApJ, 523, L103

Norman, M.L., Abel, T., Bryan, G.L. 2000, in The First Stars, eds. A. Weiss, T. Abel & V. Hill (Berlin: Springer), in press

Palla, F. 1988, in Galactic and Extragalactic Star Formation, eds. R.L. Pudritz & M. Fich (Dordrecht: Kluwer), 519

Palla, F., Zinnecker, H. 1987, in Starbursts and Galaxy Evolution, eds. T.X. Thuan, T. Montmerle & J. Tran Thanh Van (Gif sur Yvette: Editions Frontières), 533

Palla, F., Salpeter, E.E., Stahler, S.W. 1983, ApJ, 271, 632

Peebles, P.J.E. 1968, ApJ, 153, 1

Peebles, P.J.E. 1993, Principles of Physical Cosmology (Princeton Univ. Press: Princeton)

Peebles, P.J.E., Dicke, R.H. 1968, ApJ, 154, 891

Pineau des Forêts, G., Flower, D. 2000, MNRAS, in press

Puy, D., Alecian, G., Le Bourlot, J., Léorat, J., Pineau des Forêts, G. 1993, AA, 267, 337

Rawlings, J.M.C., Drew, J.E., Barlow, M.J. 1993, MNRAS, 265, 968

Rees, M.J. 1999, in AIP Conf. Proc. 470, After the Dark Ages: When Galaxies were Young, ed. S.S. Holt & E. Smith (Woodbury: AIP), 13

Roueff, E., Flower, D.R. 1999, MNRAS, 305, 353

Sasaki, S., Takahara, F. 1993, PASJ, 45, 655

Saslaw, W.C., Zipoy, D. 1967, Nature, 216, 967

Seager, S., Sasselov, D.D., Scott, D. 1999, ApJ, in press

Shapiro, P.R., Kang, H. 1987, ApJ, 318, 32

Silk, J. 1983, MNRAS, 205, 705

Stancil, P.C., Dalgarno, A. 1997a, ApJ, 479, 543

Stancil, P.C., Dalgarno, A. 1997b, ApJ, 490, 76

Stancil, P.C., Lepp, S., Dalgarno, A. 1998, ApJ, 509, 1

Tegmark, M., Silk, J., Rees, M.J., Blanchard, A., Abel, T., Palla, F. 1997, ApJ, 474, 1

Tiné, S., Lepp, S., Dalgarno, A. 1998, in Molecular Hydrogen in the Early Universe, eds. E. Corbelli, D. Galli & F. Palla, Mem Soc Astr It, 69, 345

Zel'dovich, Ya.B., Kurt, V.G., Sunyaev, R.A. 1968, Zh. Eksp. Teoret. Fiz., 55, 278. English transl., 1969, Sov Phys JETP, 28, 146

Evolution of primordial H_2 for different cosmological models

By Denis PUY

Paul Scherrer Institute, Laboratory for Astrophysics, 5232 Villigen (Switzerland)
Institute of Theoretical Physics, University of Zurich, 8057 Zurich (Switzerland)

Primordial chemistry began in the recombination epoch when the adiabatic expansion caused the temperature of the radiation to fall below 4000 K. The chemistry of the early Universe involves the elements hydrogen, its isotope deuterium, helium with its isotopic forms and lithium.
In this talk I will present results on the evolution of the primordial H_2 abundance for different cosmological models and the influence on the thermal decoupling.

1. Introduction

At early times the Universe was filled up with an extremely dense and hot gas. Due to the expansion it cooled below the binding energies of hydrogen, deuterium, helium, lithium, and thus one can expect the formation of these nuclei. As soon as neutrons and protons leave the equilibrium, the formation of deuterium followed by fast reactions lead finally to the formation of tritium and helium. Thus deuterium is the first stone of the nucleosynthesis but also the *passage obligé* for heavier elements such as lithium, beryllium and boron. The basic conclusions of the big bang nucleosynthesis on the baryon density Ω_ρ are

$$0.01 < \Omega_\rho h^2 < 0.025 \qquad (1.1)$$

See Sarkar (1996), Olive (1999) and references in Signore & Puy (1999).

2. Post-recombination chemistry

The study of chemistry in the post recombination epoch has grown considerably in recent years. From the pioneer works of Saslaw & Zipoy (1967), Shchekinov & Entél (1983), Lepp & Shull (1987), Dalgarno & Lepp (1987) and Black (1988), many authors have developed studies of primordial chemistry in different contexts. Latter & Black (1991), Puy et al. (1993), Stancil et al. (1996) for the chemical network and the thermal balance, Palla, Galli & Silk (1995), Puy & Signore (1996, 1997, 1998a, 1998b), Abel et al. (1997) and Galli & Palla (1998) for the study of the initial conditions of the formation of the first objects.

2.1. *History*

From the recombination phase, the electron density decreases which leads to the decoupling between temperature of the matter and temperature of radiation. Chemistry of the early Universe (i.e. $z < 2000$) is the gaseous chemistry of the hydrogen, helium, lithium and electrons species. The efficiencies of the molecular formation processes is controlled by collisions, matter temperature and temperature of the cosmic microwave background radiation (CMBR). In the cosmological context we have metal-free gas, and thus the formation of H_2 is not similar to the formation in the interstellar medium by adsorption on the surface of the interstellar grains.
The chemical composition of the primordial gas consists of a mixture of: H, H^+, H^-, D, D^+, He, He^+, He^{2+}, Li, Li^+, Li^-, H_2, H_2^+, HD, HD^+, HeH^+, LiH, LiH^+, H_3^+,

H_2D^+, e^-, and γ which leads to 90 reactions in the chemical network.

It is more convenient to reduce the reactions to only those that are essential to accurately model the chemistry and to reduce computer times. We adopt the concept of *minimal model* developed by Abel et al. (1997), and focus on the formation of molecular hydrogen. This way we can reduce the chemical network to 20 reactions.

From Saslaw & Zipoy (1967), and Shchekinov & Entél (1983) we know that the formation of primordial molecular hydrogen is due to the two main reactions:

$$H^- + H \longrightarrow H_2 + e^- \qquad (2.2)$$

$$H_2^+ + H \longrightarrow H_2 + H^+. \qquad (2.3)$$

These reactions are coupled with the photo-reactions, the associative, recombination and charge exchange reactions (Puy et al. 1993, and Galli & Palla 1998).

2.2. Equations of evolution

We consider here the chemical and thermal evolution in the framework of the Friedmann cosmological models. The relation between the time t and the redshift z is given by:

$$\frac{dz}{dt} = -H_o(1+z)\sqrt{1+\Omega_o z} \qquad (2.4)$$

where H_o is the Hubble constant and Ω_o the parameter of density of the Universe (open Universe $\Omega_o < 1$, flat Universe $\Omega_o = 1$ and closed Universe for $\Omega_o > 1$). The expansion is characterized by the adiabatic cooling:

$$\Lambda_{ad} = 3nkT_m H_o(1+z)\sqrt{1+\Omega_o z}, \qquad (2.5)$$

with the density of matter n and the temperature of matter T_m. The evolution of the matter density is given by:

$$\frac{dn}{dt} = -3nH_o(1+z)\sqrt{1+\Omega_o z}. \qquad (2.6)$$

and for temperature of radiation:

$$\frac{dT_m}{dt} = \frac{2}{3nk}\left[-\Lambda_{ad} + \Gamma_{compt} + \Psi_{molec}\right], \qquad (2.7)$$

where Γ_{compt} is the Compton scattering of CMBR photons on electrons. Below 4000 K only the rotational levels of H_2 can be excited (quadrupolar transitions). In the cosmological context Puy et al. (1993) have shown that the molecules heat the medium (due to the interactions between primordial molecules and the CMBR photons). The thermal molecular function Ψ_{molec} (heating minus cooling) is positive in this context.

All these equations are coupled with the set of chemical equations in order to calculate the evolution of the abundance of H_2.

2.3. Evolution of molecular hydrogen

We consider three sets of parameter for Ω_o which characterize the three particular Universe (open with $\Omega = 0.1$, flat with $\Omega_o = 1$ and closed with $\Omega_o = 2$). Moreover we consider two values for the baryonic fraction, the lower value obtained with the primordial nucleosynthesis $\Omega_\rho = 0.02$, and the other which characterizes a full baryonic Universe $\Omega_\rho = 1$.

In Fig. 1, we have plotted the different curves for the evolution of H_2. We see the classical two steps of H_2 formation (the first step correspond to the H_2^+ channel and the second one to the H^- channel). After this transient growth H_2 abundance becomes constant.

Primordial Molecular Hydrogen

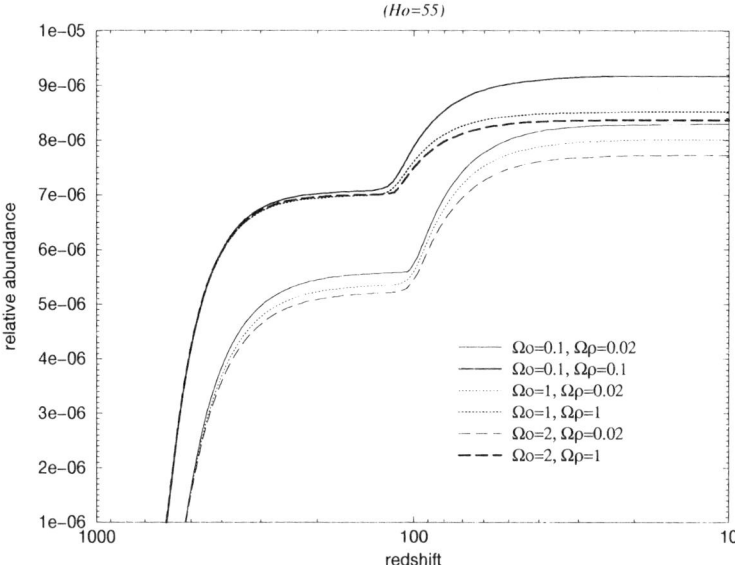

FIGURE 1. Evolution of the post-recombination abundance of primordial hydrogen for different cosmological models (Ω_o is the parameter of geometry and Ω_ρ the baryonic fraction. The upper curves correspond to the lower baryonic fraction ($\Omega_\rho = 0.02$), and vice versa, the lower curves correspond to the higher baryonic fraction ($\Omega_\rho = 1$).

3. Cosmological thermal decoupling

In Fig. 2, we have plotted the ratio between T_{molec} which is the temperature of matter with primordial molecules and T_{compt} the temperature of matter without molecules (we consider in this context only the Compton heating). The ratio is closed to unity for the lower value of baryonic fraction and close to 2.5 for the higher value. For $\Omega_b > 0.02$, we can expected that $T_{molec} \sim 1.25 \times T_{compt}$.

4. Outlook

The change of temperature, due to H_2, on the thermal decoupling could play a role during the transition between the linear regime and the non-linear regime (the turnaround point of gravitation collapse), (Puy & Signore 1996, Signore & Puy 1999). The temperature at the turn-around point is given by

$$T_{turn} \sim \left(\frac{3\pi}{4}\right)^{4/3} T_{compt}$$

where T_{compt} is the temperature of matter without the influence of molecules (Padmanabhan 1993). Taking into account the influence of the molecules, the temperature is 25 per cent more important than T_{compt}. Thus molecules could give other initial conditions for the dynamics of the gravitational collapse than ones predicted by the classical theory.

I would like to acknowledge Tom Abel, Lukas Grenacher, Philippe Jetzer and Monique Signore for valuable discussions. I thank Francoise Combes for organizing such a pleasant conference. This work has been supported by the *Dr Tomalla Foundation* and by the Swiss National Science Foundation.

Thermal Decoupling

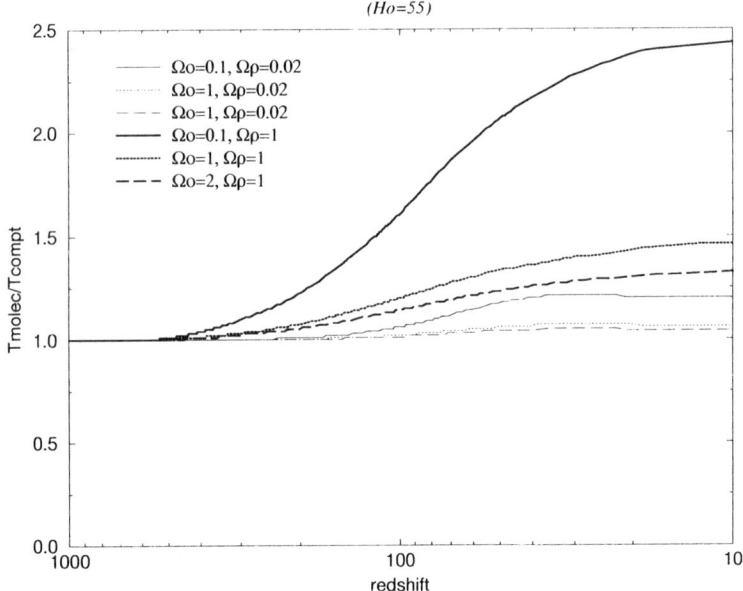

FIGURE 2. Evolution of the thermal decoupling for different cosmological models (Ω_o is the parameter of geometry and Ω_ρ the baryonic fraction).

REFERENCES

ABEL T., ANNINOS P., ZHANG Y., NORMAN M., 1997 *New Astr.* **2**, 181.
BLACK J., 1988, in: *Molecular Astrophysics*, Cambridge Univ. Press P. 473.
DALGARNO A., LEPP S., 1987, in: *Astrochemistry IAU Symp. 118*, p. 109.
GALLI D., PALLA F., 1998 *astro-ph/9803315*, 27 Mar 1998.
LATTER W., BLACK J., 1991 *ApJ* **371**, 161.
LEPP S., SHULL M., 1984 *ApJ.* **280**, 465.
OLIVE K.A., 1999 *astro-ph/9901231*, 18 Jan 1999.
PADMANABHAN T. 1993 in: *Structure formation in the Universe*, Cambridge University Press.
PALLA F., GALLI D., SILK J., 1995 *ApJ* **451**, 44.
PUY D., ALECIAN G., LEBOURLOT J., LEORAT J., PINEAU DES FORETS, 1993 *A&A* **267**, 337.
PUY D., SIGNORE M., 1996 *A&A* **305**, 371.
PUY D., SIGNORE M., 1997 *New Astr.*, **27**, 622.
PUY D., SIGNORE M., 1998a *New Astr.*, **3**, 27.
PUY D., SIGNORE M., 1998b *New Astr.*, **3**, 247.
PUY D., SIGNORE M., 1999 *New Astr. Rev.*, **43**, 223.
SARKAR S., 1996 *Rep. Prog. Phys.* **59**, 1493.
SASLAW W., ZIPOY D., 1967 *Nature* **216**, 967.
SHCHEKINOV Y.A., ENTÉL M.B., 1983 *Sov. Astr. Lett.* **27**, 622.
SIGNORE M., PUY D., 1999 *New Astr. Rev.*, **43**, 185.
STANCIL P., LEPP S., DALGARNO A., 1996 *ApJ* **458**, 401.

Dynamics of H_2 Cool Fronts in the Primordial Gas

By Miguel H. Ibáñez S., AND María C. Bessega L.

Centro de Astrofísica Teórica,
Facultad de Ciencias, Universidad de Los Andes.
Apartado Postal No 26, IPOSTEL, La Hechicera,
Mérida, Venezuela

Cool fronts originated by H_2 formation and supported by non saturated thermal conduction in the pregalactic gas, are analyzed. The pressure (p_2), number density (n_2), temperature (T_2) and flow velocity (v_2) behind the front are found as functions of the temperature ahead the cool front T_1 and the intake Mach number M_1. Compression behind the cool front occur for both, supersonic and subsonic intake flows providing that M_1 is larger than a threshold value, the exact value of which depends on T_1. But strongly compressed subsonic flows are left for larger values of M_1. Quasi-isobaric cool fronts ($p_2/p_1 \approx 1$) occur when the ratio n_1/n_2 is closed to the maximum value, where the compressional branch just emerges, beyond which the pressure of the flow behind the front increases when n_1/n_2 decreases, i.e. for denser subsonic flows behind the cool front. Implications of the above results on the formation of cool condensations in the primordial gas are outlined.

1. Introduction

Previous studies (Field 1965, Yoneyama 1973, Ibáñez & Parravano 1983, Fall & Rees 1985, Corbelli & Ferrara 1995, Puy et al. 1998) have showed that thermal instability can originate cool condensations in hot plasmas. Also it is believed that at large scales such cold structures are the precursors of the gravitational instability, because if a thermal instability is triggered, in cool regions the temperature decreases and the density increases, i.e. the Jeans mass ($\sim T^{3/2}\rho^{-1/2}$) could decrease below the value of the actual mass and therefore such regions should gravitationally collapse likely forming stars, globular clusters and galaxies.

On the other hand, from an analysis of the thermochemical equilibrium states of the primordial plasma (Rosenzweig et al. 1994) follows that when the number density of the gas is greater than about 1 *particle cm*$^{-3}$ the most efficient cooling agent is the hydrogen molecule. It also follows that when hydrogen molecules are the main cooling agent the plasma becomes thermally unstable. Therefore, it is expected that in a well advanced nonlinear state of the thermal instability, cool fronts propagate into the hot plasma regions (Doroshkevich & Zel'dovich 1981, Ibáñez & Bessega 2000, hereinafter reference IB). The present paper will be aimed to analyze, by a first approximation, the dynamics of cool fronts triggered by H_2 formation in the primordial gas.

2. Basic Equation

According to Landau and Lifshitz (1987), if one assumes that the interphase between a hot and cold region in a plasma has a characteristic length scale L much smaller than the scale length of the flow, one may consider the region where the temperature changes, from T_1 to T_2, as a short segment where the gas lost a net amount of heat per unit time Sq (S being a cross-sectional area). Under steady conditions, the mass, momentum and energy equations become

$$\rho_1 v_1 = \rho_2 v_2 = j , \qquad (2.1)$$

$$p_1 + j v_1 = p_2 + j v_2 , \qquad (2.2)$$

$$(w_1 + \frac{1}{2}v_1^2)j = (w_2 + \frac{1}{2}v_2^2)j + \bar{q} \qquad (2.3)$$

where ρ, v, p, w and $\bar{q}$ are mass density, normal velocity, enthalpy per unit mass and a mean energy loss flux to be defined below, respectively. Additionally, an ideal gas equation of state is assumed. The suffixes 1 and 2 refer to the two sides of the interphase, the cool front propagates into the hot region, i.e. the flow of gas being from the hot region 1 towards the cool region 2.

3. Cool Fronts

If $\Lambda(\rho, \xi, T)$ is the energy radiated by the hydrogen molecule H_2 per unit volume and time the averaged energy flux throughout the interphase can be written by a first approximation as

$$\bar{q} = \overline{\Lambda L} = \frac{1}{(\rho_2 - \rho_1)(T_1 - T_2)} \int_{\rho_1}^{\rho_2} \int_{T_2}^{T_1} \Lambda L d\rho dT , \qquad (3.4)$$

where L is an effective thickness of the cool front and

$$\Lambda = \rho^2 \phi_i(T, \xi) , \qquad (3.5)$$

$\phi_i(T)$ being the total vibrational/rotational cooling by H_2 (Hollenbach et al. 1979). Strictly speaking ϕ_i is a function ρ, however in the range of density under consideration such a dependence is weak and will be neglected by a first approximation. Additionally, the H_2 dissociation will be assumed to be frozen throughout the front with scale length L where the radiative losses are of the same order than the heat flux by thermal conduction assumed to be by neutral particles (Parker 1953). Introducing a slight change of notation, $\eta = n_1/n_2$, $z = p_2/p_1$, consequently $\eta z = T_2/T_1$, the averaged flux $\bar{q}$ can be written as

$$\bar{q} = \frac{n_1}{T_1}(\frac{1}{\eta} + 1)\frac{b_0 + b_1 \eta z + b_2 (\eta z)^2}{(1 - \eta z)} , \qquad (3.6)$$

where the coefficients b_0, b_1 and b_2 are obtained by numerical integration of the cooling function found by Hollenbach et al. i.e. $b_0 = 4.498 \times 10^{-5}$, $b_1 = 2.109 \times 10^{-5}$ and $b_2 = -6.366 \times 10^{-5}$. Equations (2.1)-(2.3) and (3.6) can be combined to obtain an expression for the ratio p_2/p_1 as a function of compressibility η. So, z are the roots of the equation

$$\alpha_2(\eta) z^2 + \alpha_1(\eta) z + \alpha_0(\eta) = 0 , \qquad (3.7)$$

where $\alpha_0(\eta)$, $\alpha_1(\eta)$ and $\alpha_2(\eta)$ are well defined coefficients (IB). For a given temperature T_1 ahead of the cool front equation (3.7) defines the branches $z(\eta)$ for any given Mach number ahead the front M_1.

If one assumes that the temperature T_1 is known, from equations (2.1)-(2.3) the ratio of densities, η ($n_1/n_2 = v_2/v_1$) and pressures, z, as well as the Mach number behind the cool front, are completely determined by the intake Mach number M_1, i.e.

$$z = 1 + \zeta , \qquad (3.8)$$

$$\eta = 1 - \frac{\zeta}{\gamma M_1^2} , \qquad (3.9)$$

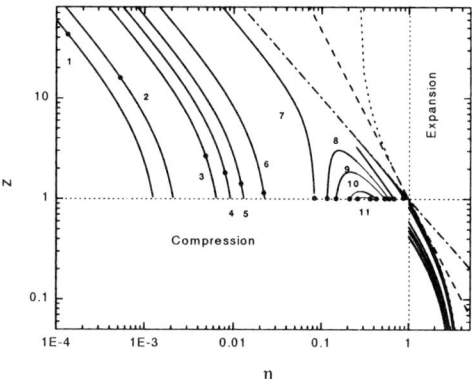

FIGURE 1. The ratio pressure z as function of the number density ratio η in a cool front for different values of M_1 and $\gamma = 5/3$, $T_1 = 10^6$ K . The ten curves $1-10$ correspond to $M_1 = 5$, 3, 1, 0.7, 0.5, 0.3, 0.1, 8×10^{-2}, 6×10^{-2}, 5.7×10^{-2}, respectively. Additionally, the isotherm (dashed point line), Poisson (dash line) and shock adiabatic(sa, dot line) are also shown.

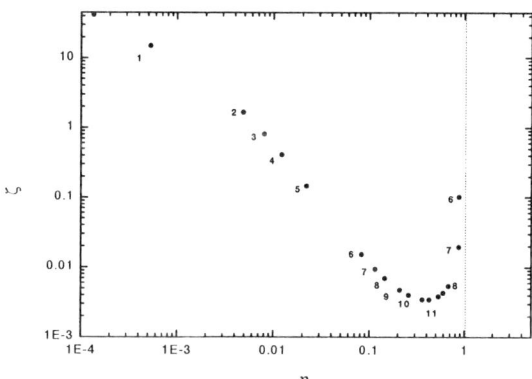

FIGURE 2. The roots ζ of equation (3.11) corresponding to the values of M_1 of Fig. 1.

$$M_2 = M_1 \sqrt{\frac{\eta}{1+\zeta}} \quad , \qquad (3.10)$$

ζ being the root of the polynomial

$$\beta_5 \zeta^5 + \beta_4 \zeta^4 + \beta_3 \zeta^3 + \beta_2 \zeta^2 + \beta_1 \zeta + \beta_0 = 0 \quad , \qquad (3.11)$$

where $\beta_0...\beta_5$ are known coefficients depending on M_1 and T_1 (see IB).

4. Results

Fig.1 shows the pressure ratio z through the cool front given by equation (3.7) as a function of compressibility η, for $\gamma = 5/3$, $T_1 = 5000$ K, and different values of M_1. As a reference, the isotherm (dashed dot line), the Poisson adiabatic (dash) and shock adiabatic (dot line, sa label) are also shown in the above Figure.

Compression behind the cool front occurs for both supersonic and subsonic intake flows

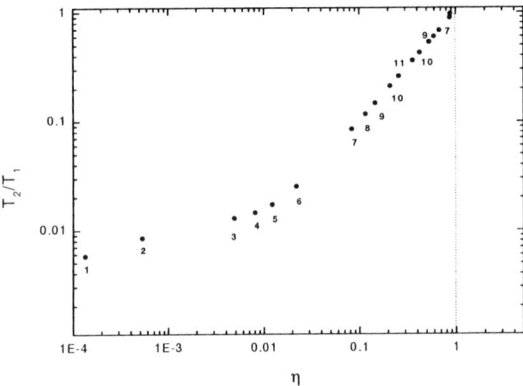

FIGURE 3. The ratio T_2/T_1 corresponding to the roots of equation (3.11) shown on Fig. 1.

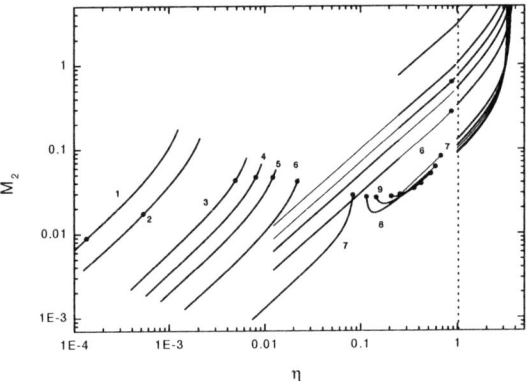

FIGURE 4. The Mach number M_2 as a function of η and the roots of equation (3.11) shown on Figs. 1-3.

providing that M_1 is larger than a threshold value M_t ($M_t \approx 5.62 \times 10^{-2}$, for $T_1 = 5000$ K). But strongly compressed subsonic flows are left for larger values of M_1. For values of $M_1 > M_t$, a strong compression branch appears for $\eta < (\gamma - 1)/(\gamma + 1)$ beyond the limiting value imposed by the condition $z > 1$ when $\eta < 1$. For any particular value of M_1, quasi-isobaric cool fronts ($z \approx 1$) occur at the value of η where the compressional branches just emerge beyond which, the pressure of the flow behind the front increases when η decreases, i.e. for denser subsonic flows behind the cool front.

The physically meaningful roots of equation (3.11), are shown in Fig 2 and the corresponding ratio of pressure are indicated by solid circles on Figures 1 - 4. Additionally, Fig. 3 shows the corresponding temperature ratio T_2/T_1 and Fig. 4 the Mach number behind the cool front.

For subsonic intake flows, there are two solutions ζ, those on the strong compression branch ($\eta < 1/4$) corresponding to higher pressure than those on the weak compression branch ($\eta > 1/4$), see Fig. 2. On the high compression branch ζ decreases, η and ratio of temperature T_2/T_1 (Fig 3) increases when the intake Mach number decreases, contrary to the weak compression branch where ζ, η and T_2/T_1 decrease when M_1 decreases. For subsonic intake flows, the Mach number behind the cool front M_2 is a decreasing function of M_1 for both compression branches, but to the solution on the higher compression branch corresponds a lower value of M_2 (Fig. 4). Supersonic flows leaves behind the cool front highly compressed flows, for instance, for $M_1 = 5$ (curve 1), $n_2 = 7.41 \times 10^3 n_1$ and $p_2 = 42.7 p_1$. However, if the velocity of the cool front is smaller than that of the shock wave, detachment of the two fronts occurs and the shock front leaves behind a subsonic intake flow for the cool front. Large density values behind a cool front is reached even at low subsonic intake velocities, say for instance, $n_2 = 8.20 \times 10^1 n_1$, $(p_2 = 1.41 p_1)$ and $n_2 = 12.06 n_1$ $(p_2 = 1.015 p_1)$, for $M_1 = 0.5$ (curve 5) and 0.1 (curve 7), respectively.

For subsonic and supersonic intake flows there are not any physical meaningful roots $\zeta < 0$, i.e. there are not any expanding flows behind strong cool fronts.

Comparing the results obtained here with those of the cool fronts in an optically thin hot plasma with solar abundances (see reference IB) one can conclude that the dynamics of the two cooling fronts is similar, i.e. in both cases one finds that for strong cooling fronts there are two compression branches: a very strong and a weak compression branch and in both cases there is a threshold value for M_1 for the formation of cooling fronts. However, there are quantitative differences between the two cases. In particular, the threshold value M_t for the ionized plasma (IB) is about two orders of magnitude less than that in the primordial gas. Similarly, for any particular value of M_1 the compression in the ionized plasma with solar abundances is about two order of magnitude larger than the corresponding one in the primordial plasma.

REFERENCES

CORBELLI, E. & FERRARA, A. 1995 *ApJ* **447**, 708.
DOROSHKEVICH, A.G. & ZEL'DOVICH, YA. B. 1981 *Sov. Phys. JETP* **53**, 405.
FALL, S. M. & REES, M. J. 1985 *ApJ* **298**, 18.
FIELD, G.B. 1983 *ApJ* **142**, 531.
HOLLENBACH, D. & MCKEE, C.F. 1979 *ApJ Supplement Series* **41**, 555.
IBÁÑEZ S., M.H. & BESSEGA L., M. C. 2000 *To appear in ApJ*
IBÁÑEZ S., M.H. & PARRAVANO, A. 1983 *ApJ* **275**, 181.
LANDAU, LD & LIFSHITZ 1987 *Fluid Mechanics* (London, Pergamon Press).
PARKER, E. N. 1953 *ApJ* **117**, 431.
PUY, D. & SIGNORE, M. 1998 *New Astronomy* **3**, 247.
ROSENZWEIG, P., PARRAVANO, A., IBÁÑEZ S., M.H. & IZOTOV, Y.I. 1994 *ApJ* **432**, 485.
YONEYAMA, T 1973 *PASJ* **25**, 349.

Yuri Shchekinov and Miguel Ibanez exchanging ideas about thermochemical instabilities and clump formation.

Is Reionization Regulated by H_2 in the Early Universe ?

By A. Ferrara[1], B. Ciardi [2] & P. Todini[2]

[1] Osservatorio Astrofisico di Arcetri, Firenze, Italy

[2] Dipartimento di Astronomia, Universitá di Firenze, Firenze, Italy

Molecular hydrogen is a key species for the formation of the first luminous objects in the early universe. It is therefore crucial to understand the various physical processes leading to its formation and destruction and the feedbacks regulating this chemical network. Here we review both the radiative and SN-induced feedbacks and we assess the role of the objects relying on H_2 for their collapse in the evolution of the reionization of the universe.

1. Introduction

At $z \approx 1100$ the intergalactic medium (IGM) is expected to recombine and remain neutral until the first sources of ionizing radiation form and reionize it. Until recently, QSOs were thought to be the main source of ionizing photons, but observational constraints suggest the existence of an early population of pregalactic objects (Pop III hereafter) which could have contributed to the reheating, reionization and metal enrichment of the IGM at high redshift. In order to virialize in the potential well of dark matter halos, the gas must have a mass greater than the Jeans mass ($M_b > M_J$), which, at $z \approx 20-30$ corresponds to very low virial temperatures ($T_{vir} < 10^4$ K). To have a further collapse and fragmentation of the gas, and to ignite star formation, additional cooling is required. It is well known that in these conditions the only efficient coolant for a plasma of primordial composition, is molecular hydrogen (Abel et al. 1997; Tegmark et al. 1997; Ciardi, Ferrara & Abel 2000 [CFA]). As stars form in Pop III objects, their photons in the energy range 11.26-13.6 eV are able to penetrate the gas and photo-dissociate H_2 molecules both in the IGM and in the nearest collapsing structures, if they can propagate that far from their source. Thus, the existence of UV radiation below the Lyman limit capable of dissociating the H_2, could deeply influence subsequent small structure formation. This is what in jargon is referred to as the radiative feedback and we will discuss it in the next section. However, dust is produced in the early universe, as a byproduct of the first supernova explosions, and it might open a new, efficient channel for H_2 formation and strongly influence the formation of early objects and, eventually, the reionization of the universe.

2. Radiative feedback

We model the spectrum impinging onto a collapsing objects as the one produced by a nearby PopIII one, calculated using Bruzual & Charlot spectrophotometric code. The spectra at different evolutionary times are characterized by the value of the parameter β, the ratio between photo-dissociating and ionizing photons; its value is in the range 1 to 50 depending on the evolutionary stage of the stellar cluster inside the PopIII object.

Initially, we assume that the gas in the collapsing object is homogeneous, with a mean density $\langle \rho_b \rangle$ 18 π^2 times higher than the background intergalactic medium. Starting from these initial conditions we have followed the chemical evolution up to a free-fall time. Further details of the calculations can be found in CFA. The conditions are such that

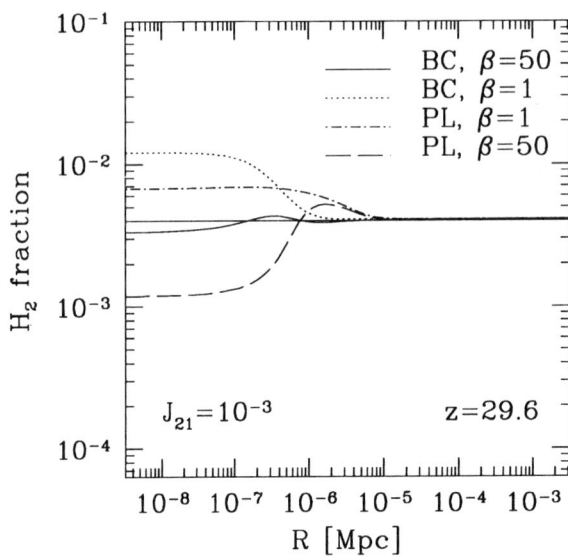

FIGURE 1. Molecular hydrogen fraction as a function of depth inside the PopIII object

in the absence of radiation the fraction of H_2 is high enough to allow for the collapse of the object on a time scale comparable to the free-fall time. The effect of the radiation field is to decrease this abundance in the external regions which therefore cannot cool. As an example we discuss the case of an object which virializes at $z = 29.6$, exposed to a radiation field with intensity $J_{s,0} = J_{21}10^{-21}$ erg s^{-1} cm^{-2} Hz^{-1} sr^{-1} at the Lyman limit. The critical mass for collapse, M_{crit}, corresponding to that redshift is $M_{crit} = 9 \times 10^5 \, M_\odot$. Fig. 1 shows the corresponding molecular hydrogen fraction, f_{H_2}, after a free-fall time for different values of β. For sake of comparison, we also show the cases relative to a power law spectrum (PL) with $\alpha=1.5$ and a cutoff energy of 40 keV.

In the external regions, the most important mechanisms for the H_2 destruction are direct photoionization by photons with energies above 15.4 eV and LW photons dissociation. A PL spectrum with the same value of β yields a larger H_2 destruction, while if the upper PL spectrum cutoff is increased the curves shift towards larger depths due to penetrating photons. Note that when $\beta=1$, H_2 is formed rather than destroyed due to the larger abundance of free electrons and paucity of LW photons available. Finally, the f_{H_2} smoothly approaches the initial value in the internal regions due to the loss of free electrons. A more global view of the radiative feedback effect is presented in Fig. 2. There we show the values of M_{sh}, the minimum mass required in order for an object to self-shield from the external flux, for different values of $J_{s,0}$ at various redshifts. We then conclude that the mass range in which negative feedback is important (solid portion) lies approximately in $10^6 - 10^8 M_\odot$, depending on redshift. In order for the negative feedback to be effective, fluxes of the order of $10^{-24} - 10^{-23}$erg s^{-1} cm^{-2} Hz^{-1} sr^{-1} are required.

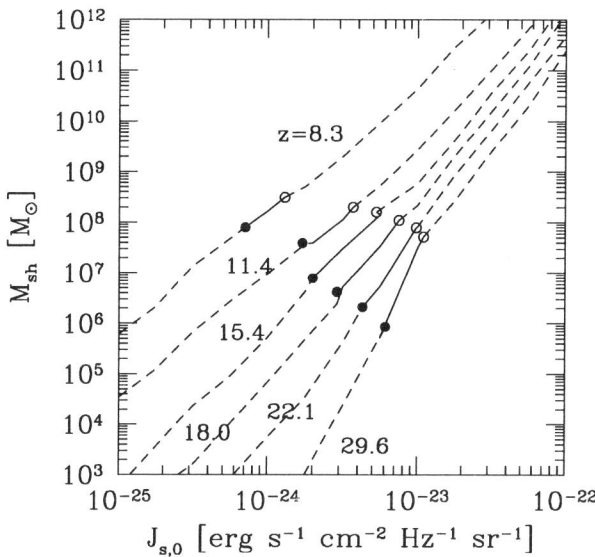

FIGURE 2. Minimum total mass for self-shielding from an external incident flux with intensity $J_{s,0}$ at the Lyman limit. The curves are for different redshift: from the top to the bottom z =8.3, 11.4, 15.4, 18.0, 22.1, 29.6. Radiative feedback works only in the solid portions of the curves.

3. SNe, dust and formation of H_2

Once the first stars have formed in the host protogalaxy, they can deeply influence the subsequent star formation process through the effects of mass and energy deposition due to winds and supernova explosions. While low mass objects may experience a blow-away, expelling their entire gas content into the IGM and quenching star formation, larger objects may instead be able to at least partially retain their baryons. However, even in this case the blowout induces a decrease of the star formation rate due to the global heating and loss of the galactic ISM. These two regimes are separated by a critical mass, M_{by}, explicitly calculated in CFGJ (see also MacLow & Ferrara 1999).

Pop III objects can produce regions of considerably high molecular hydrogen abundance behind blow-away shocks propagating in the IGM. Ferrara (1998) has shown that the H_2 thus produced can exceed the amount of relic H_2 destroyed inside the photodissociation region surrounding a given Pop III. This is due to the fact that the shock eventually becomes radiative and allows the swept gas to cool in a dense shell. This cooling transient is characterized by a strong non-equilibrium condition in which recombination lags behind the temperature decrease; this is a favorable condition for H_2 formation. The H_2 thus produced can exceed the amount of relic H_2 destroyed inside the photodissociation region surrounding a given Pop III and tends to balance the H_2 destruction due to the (negative) radiative feedback discussed in the previous Section.

SNe also initiate H_2 formation on dust grain surfaces, a more standard process in present day galactic environments. At high redshift Type II SNe are the only possible sources of dust, due to the short age of the universe and the long evolutionary timescales characterizing more conventional dust sources, as for example evolved stars. A detailed nucleation study of dust formation in the ejecta of Type II SNe whose initial chemical composition is primordial has been very recently carried on (Todini & Ferrara, in preparation). This case is suitable for the study of the very first SNe, as those originating in PopIII objects. From this study we have been able to derive several important properties characterizing the first solid particles produced in the universe, as for example the total dust mass produced, the dust composition and the grain size distribution. The preliminary results reported at this conferences show that in the range of SN masses considered, $M_{SN} = 12 - 40 M_\odot$, typically 0.1 $M_\odot$ of dust is produced. SNe towards the low end of the mass distribution mostly produce silicate grains, whereas the most massive ones predominantly produce graphite grains; this fact is easy to understand in terms of the C/O ratio of their ejecta. The size distribution is approximated by a power-law for about two decades in radius (from ≈ 1Å to 100Å) and shows a smooth cutoff beyond that grain radius. Thanks to these results we can now pose the following question: what is the minimum amount of dust required in order for the molecular hydrogen formation on grains to become competitive with the gas phase one ? An order-of-magnitude answer can be obtained by comparing the two formation rates. At the low densities relevant here, H_2 is formed in the gas phase mainly via the channel $H + e^- \rightarrow H^- + h\nu$, at rate k_8 (the rate coefficient k_8 is given in Abel et al. 1997); formation via the H_2^+ channel, when included, is found to be negligible in our case. Therefore the formation rate in the gas phase is $\mathcal{R} \simeq k_8 n_{H^-} n_H$. The formation rate on grain surfaces is instead given by $\mathcal{R}^d \simeq 0.5 \langle \gamma c_s \sigma \rangle n_d n_H$, where γ is the sticking coefficient, c_s is the sound speed in the gas, and σ is the grain cross section. The equality between the two rates can be cast into the following form: $\mathcal{D} = 0.1 \sqrt{T} x_e$, where $\mathcal{D}$ is the dust-to-gas ratio normalized to its Galactic value, T is the gas temperature and x_e the gas ionization fraction. For typical parameters of the PopIII objects, H_2 production on dust grains becomes dominant once $\mathcal{D}$ is larger than 5% of the local value. With the dust yields per SN calculated above we then conclude that only about 15 SNe are required to enrich in dust to this level a primordial object. Clearly, early dust formation might play a role in the formation of the first generation of objects.

4. Reionization History Driven by Feedbacks

In this final Section we summarize some of the results obtained by CFGJ on the evolution of inhomogeneous reionization, regulated by the feedback mechanisms touched above and regulated by the molecular hydrogen network. For a broader presentation the reader is advised to refer to CFGJ. We have determined the spatial distribution of the ionizing sources from high resolution numerical N-body simulations within a periodic cube of comoving length $L = 2.55 h^{-1}$ Mpc for a critical density cold dark matter model ($\Omega_0=1$, $h=0.5$ with $\sigma_8=0.6$ at $z=0$). This allows us to describe the topology of the ionized and dissociated regions at various cosmic epochs and derive the evolution of H, He, and H_2 filling factors, soft UV background, cosmic star formation rate and the final fate of ionizing objects. There are three free parameters in the computation: f_b, the fraction of virialized baryons that is able to cool and become available to form stars, $f_\star$ the star formation efficiency, and f_{esc} the photon escape fraction from the protogalaxy. The main results can be summarized as follows. For the reference case (Run A) presented by CFGJ, in which the values of the free parameters have been fixed at $f_b = 0.08, f_\star = 0.15, f_{esc} = 0.2$, galaxies are able to reionize the neutral atomic hydrogen by redshift $z \approx 10$, while molecular hydrogen is completely dissociated at very high redshift

($z\approx25$). The reionization is basically driven by objects collapsed through H line cooling, while small mass PopIII stars play only a minor role and even in the absence of a radiative negative feedback they would not be able to reionize the IGM. Note that only about 2% of the stars observed at $z = 0$ is required to reionize the universe completely. This corresponds to an average IGM metallicity (assuming a Salpeter IMF) at redshift $z\approx10$ equal to $\langle Z \rangle \approx 3 \times 10^{-4} Z_\odot$.

REFERENCES

ABEL, T., ANNINOS, P., ZHANG, Y. & NORMAN, M. L. 1997, *NewA*, **2**, 181.
CIARDI, B., FERRARA, A. & ABEL, T. 2000, *ApJ*, in press (CFA)
CIARDI, B., FERRARA, A., GOVERNATO, F. & JENKINS, A. 2000, *ApJ*, in press (CFGJ)
FERRARA, A. 1998, *ApJ*, **499**, L17.
MAC LOW, M.-M. & FERRARA, A. 1999, *ApJ*, **513**, 142.
TEGMARK, M. ET AL. 1997, *ApJ*, **474**, 1.

Harvey Liszt and Peter Kalberla discussing, with Craig Kulesa onlooking.

H_2 in Galaxies

By Françoise COMBES

DEMIRM, Observatoire de Paris, 61 Av. de l'Observatoire, F-75 014, Paris, France

The bulk of the molecular gas in spiral galaxies is under the form of cold H_2, that does not radiate and is only suspected through tracer molecules, such as CO. All tracers are biased, and in particular H_2 could be highly underestimated in low metallicity regions. Our knowledge is reviewed of the H_2 content of galaxies, according to their types, environment, or star-forming activities. The HI and CO components are generally well-mixed (spiral arms, vertical distribution), although their radial distributions are radically different, certainly due to radial abundance gradients. The hypothesis of H_2 as dark matter is discussed, as well as the implications on galaxy dynamics, or the best perspectives for observational tests.

1. How to observe H_2 in galaxies?

The bulk of molecular hydrogen in a galaxy is cold, around 10-20K, and therefore invisible. The first rotational level, accessible only through a quadrupolar transition, is more than 500 K above the fundamental. The presence of H_2 is inferred essentially from the CO tracer. The carbon monoxide is the most abundant molecule after H_2; its dipole moment is small (0.1 Debye) and therefore CO is easily excited, the emission of CO(1–0) at 2.6mm (first level at 5.52K) is ubiquitous in the Galaxy.

1.1. The H_2/CO conversion ratio

To calibrate the H_2/CO ratio, the most direct and natural is to compare the UV absorption lines of CO and H_2 along the same line of sight (Copernicus, e.g. Spitzer & Jenkins 1975; ORFEUS, cf Richter et al., this conference). However, only very low column densities are accessible, in order to see the background source, and therefore these observations sample only the diffuse gas, which is not representative of the global molecular component. It is well known now that the conversion ratio might vary considerably from diffuse to dense gas (see below). The CO molecule is excited by H_2 collisions, and should be a good tracer; but its main rotational lines are most of the times optically thick. One can then think of observing its isotopic substitutes ^{13}CO or $C^{18}O$, but these are poor tracers since they are selectively photo-dissociated, and trace only the dense cores.

The main justification to use an H_2/CO conversion ratio is the Virial hypothesis: in fact, the CO profiles do not yield the column densities, but they give the velocity width ΔV of molecular clouds. Once the latter are mapped, and their size R known, the virial mass can be derived, proportional to $\Delta V^2 R$. There exists a good relation between the CO luminosity and the virial mass, as shown in Figure 1. The relation has a power-law shape, but with a slope different from 1. Both are not proportional, and the conversion ratio should vary by more than a factor 10 from small to Giant Molecular Clouds (GMC). In external galaxies, the observations provide only an average over many clouds, and it has been hoped that the clouds are of the same nature from galaxy to galaxy. If T_b is the brightness temperature of the average cloud, the conversion ratio X should vary as $n^{1/2}/T_b$, where n is the average density of the cloud. This does not take into account the influence of the gas metallicity. And the CO luminosity varies with the metallicity Z, sometimes more than linearly. In the Magellanic Clouds, LMC or SMC (Rubio et al 1993), the conversion ratio X might be 10 times higher than the "standard" ratio. The ratio can be known for local group galaxies, since individual clouds can be resolved, and virial masses computed (Wilson 1995).

1.2. Dust as a tracer

At millimetric wavelengths, in the Rayleigh-Jeans domain, dust emission depends linearly on temperature, and its great advantage is its optical thinness. In some galaxies, CO and dust emission fall similarly with radius, like in NGC 891 (Guélin et al. 1993). In other, such as NGC 4565 (Neininger et al. 1996), the dust emission falls more slowly than CO, although more rapidly than HI emission. This can be interpreted by the exponential decrease of metallicity with radius. The dust/HI ratio follows this dependency, while CO/HI is decreasing more rapidly (either due to metallicity, or excitation problems).

1.3. Gamma-rays

Their emission is proportional to the product of the Cosmic Ray density and the gas density. But both densities, and their radial profiles are not known independently. The γ-ray radial distribution is however much more extended than that of the supernovae remnants, the source of cosmic ray acceleration (e.g. Bloemen 1989). Recently: EGRET on board GRO has observed an excess of gamma-rays in the halo of our Galaxy (see below).

1.4. Direct H_2 observations

Of course, H_2 can also be observed directly when it is warm. Starbursts and mergers reveal strong 2.2 μm emission, like in NGC 6240 (DePoy et al 1986). The source of excitation has long been debated (X-ray heating, UV fluorescence, shocks...) and it was recently concluded that global shocks were responsible (van der Werf et al. 1993, Sugai et al. 1997). Pure rotational lines have been observed with ISO. In Arp220, as much as 10% of the ISM could be in the warm phase, i.e. $3\ 10^9\ M_\odot$ (Sturm et al. 1996) while CO observations conclude to a total $M(H_2) = 3.5\ 10^{10}\ M_\odot$ (Scoville et al. 1991). In normal galaxies, the warm H_2 could be less abundant (Valentijn et al. 1996). At least, the warm CO component does not affect the H_2/CO ratio.

2. CO and H_2 content of galaxies

From a CO survey of more than 300 galaxies, it has been concluded that the average molecular content was comparable to the atomic content: $M(H_2)/M(HI) \sim 1$ (Young & Knezek 1989; Young & Scoville, 1991). But most of galaxies in this survey were selected from their IRAS flux, and this could introduce a bias. A recent survey by Casoli et al (1998) near the Coma cluster has shown an average $M(H_2)/M(HI) \sim 0.2$.

2.1. Variation with morphological type

It is well established that the HI component is proportionally more abundant relative to the total mass in late-type galaxies. The opposite trend is observed for the H_2, at least as traced by the CO emission. $M(H_2)/M(HI)$ is therefore smaller for late-types, by a factor ~ 10. However, this could be entirely a metallicity effect. Since the metallicity is increasing with the mass of the galaxy, a test is to select the most massive galaxies of late-type. For these high-mass galaxies, there is no trend of decreasing H_2 fraction with type (Casoli et al. 1998).

2.2. Dwarf and LSB galaxies

The strong dependency of the H_2/CO conversion ratio on metallicity Z is also the main problem in the observations of dwarf and Low Surface Brightness (LSB) galaxies. Both

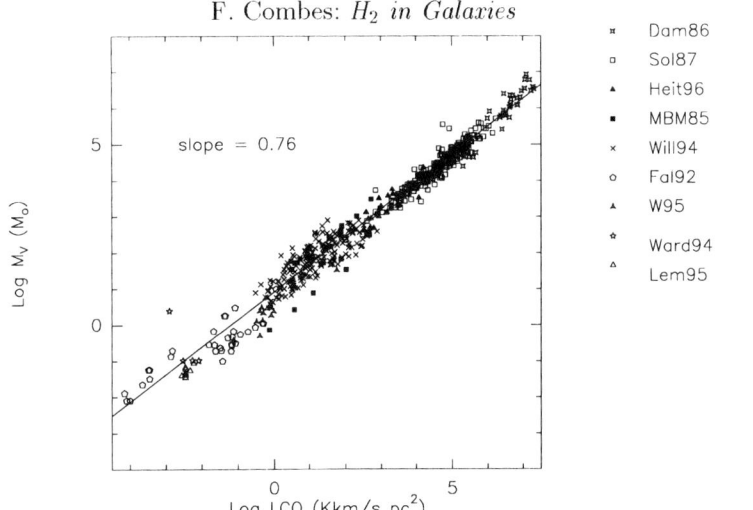

FIGURE 1. Virial mass versus CO luminosity for molecular clouds in the Milky Way. The fit corresponds to $M_V \propto L_{CO}^{0.76}$. The data are from Dam86: Dame et al. (1986); Sol87: Solomon et al. (1987); Heit96: Heithausen (1996); MBM85: Magnani et al. (1985); Will94: Williams et al. (1994); Fal92: Falgarone et al. (1992); W95: Wang et al. (1995); Ward94: Ward-Thompson et al. (1994); Lem95: Lemme et al. (1995).

have low metallicity. It appears that the conversion factor X can vary linearly and even more with metallicity, as predicted by Maloney & Black (1988). Not only, the low abundance of C and O lowers the abundance of CO, but also the dust is less abundant, and therefore the UV light is less absorbed, and spread all over the galaxy, photo-dissociating the CO molecules. When the dust is depleted by a factor 20, there should be only 10% less H_2, but 95% less CO (Maloney & Black 1988).

In dwarf galaxies, CO emission is very low, and it is difficult to know the H_2 content. If the HI/H_2 ratio is assumed constant from galaxy to galaxy, then X varies with $Z^{-2.2}$ (Arnault et al 1988). Recent results by Barone et al (1998), Gondhalekar et al (1998) and Taylor et al (1998) confirm this strong dependency on metallicity, increasing sharply below 1/10th of solar metallicity.

Low-surface-brightness galaxies have large characteristic radii, large gas fraction and are in general dark matter dominated; they are quite un-evolved objects. Their total gas content is similar to that of normal galaxies (McGaugh & de Blok 1997). But CO is not detected in LSB (de Blok & van der Hulst 1998). Due to their low surface density, below the threshold for star formation, these galaxies have a very low efficiency of star formation (Van Zee et al 1997). The cause could be the absence of companions, since LSB live in poor environments (Zaritsky & Lorrimer 1993). It is well known that galaxy interactions, by driving in a high amount of gas, trigger star formation.

2.3. Ultra-luminous IRAS galaxies

At the opposite, there exists a class of galaxies, characterized by their bursts of star formation; these are ultra-luminous in far-infrared, because of the emission of dust heated by the new stars. These objects possess large amounts of gas, particularly condensed in the inner parts, certainly due to interactions and mergers. CO emission is highly enhanced in these starbursting galaxies, and large H_2 masses are deduced, even with a modified (lower than standard) conversion ratio (Solomon et al 1997). The prototype of these objects is the nearby Arp220: new CO interferometer data show that CO is

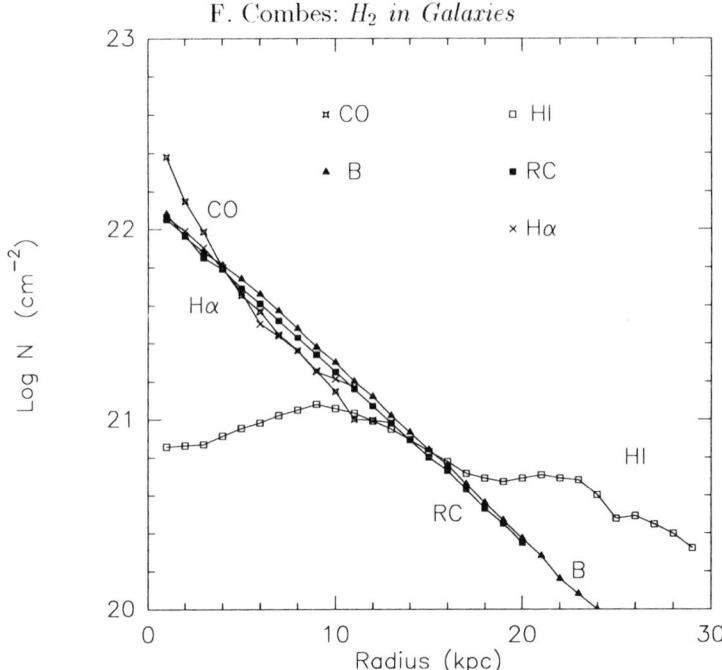

FIGURE 2. Radial distributions of various surface densities in a typical spiral galaxy NGC 6946: H_2(CO) and HI column densities, Blue, Radio-continuum and Hα surface densities (adapted from Tacconi & Young, 1986).

in rotating nuclear disks (Downes & Solomon 1998), where the surface density of gas is about 30% of the total surface density.

3. Spatial distribution

3.1. Radial Distributions

The differences between HI and H_2 (or CO) radial distributions in galaxies is striking (cf figure 2). While all components related to star formation, the blue luminosity from stars, the Hα (gas ionised by young stars), the radio-continuum (synchrotron related to supernovae), and even the CO distribution, follow an exponential distribution, the HI gas alone is extending much beyond the "optical" disk, sometimes in average by a factor 2 to 4 (R_{HI} = 2-4 R_{opt}). The HI gas has very often a small deficiency in the center. Would this mean that the atomic gas is transformed in molecular phase in the denser central parts? This is possible in some galaxies, where the HI and CO distribution appear complementary, but it is not the general case, all possibilities have been observed, including a central gaseous hole, both in CO and HI (like in M31, for example).

3.2. Large-scale Structure

Within the optical disk, where CO is observed easily, there is a very good large-scale correlation between both gas components (see e.g. Neininger et al. 1998). They appear well mixed, and follow the spiral arms with large contrast. This is also true for the ionised gas (HII regions). Of course, this is only at 100pc-1kpc scale; at very small scale the various components can be anti-correlated, the HI gas being found more at the envelopes of dense molecular clouds, the ionised gas being also anti-correlated with the neutral gas.

3.3. Vertical Structure

In our own Galaxy, and in external galaxies seen edge-on, the galaxy disks appear much narrower in CO emission than in HI. This suggests that the molecular gas is more confined to the plane, and that its vertical dynamical oscillations are of less amplitude than for the atomic gas. The consequence should be a vertical velocity dispersion much lower for the molecular gas, since for a given restoring force due to the stellar disk, the maximum height above the plane is proportional to the z-velocity dispersion. Surprisingly, this is not the case: in face-on galaxies both CO (Combes & Becquaert 1997) and HI (Kamphuis 1992) velocity dispersions are observed of similar values ($\sigma_v \sim 6$ km/s), and remarkably constant with radius. This is not a saturation effect of the CO lines, since the ^{13}CO spectra show the same. A possible interpretation is that both gas are well mixed, in fact it is the same dynamical component, which changes phase along its vertical oscillations. It is possible that the H_2 gas follows the HI, but the CO is photo-dissociated at high altitudes, or not excited. Or even the H_2 could disappear, since the chemistry time-scale ($\sim 10^5$ yr) is much smaller than the dynamical z-time-scale ($\sim 10^8$ yr).

3.4. Small scale structure of clouds

The molecular component is also characterized by its remarkable self-similar structure, a hierarchical system of clouds, tightly related to a fractal structure. It can be quantified by power-law relations between cloud size and linewidth, or size and mass (Larson, 1981). These relations are observed, whatever the radial distance to the center, and in the HI component as well. The fact that the same structure is observed outside of the star-forming regions is puzzling: the HI gas outside the optical disk displays a very clumpy structure, implying that it is unstable at all scales (spiral arms, self-similar structure of clumps). The fact that these gravitational instabilities do not trigger star formation must be explained.

4. H_2 as a dark matter candidate

One of the driver to propose cold H_2 as a dark matter candidate is our increasing knowledge about evolution of galaxies along the Hubble sequence (e.g. Pfenniger, Combes & Martinet, 1994). Because of spiral waves and bars, galaxies progressively concentrate their mass towards the center, and the late-type galaxies evolve to early-types in the sequence. Besides, HI observations of rotation curves have shown that the fraction of dark matter in the total mass is larger in late-type galaxies: therefore, some of the dark matter must be transformed into stars during evolution (cf Pfenniger, this conference).

4.1. Baryonic mass fraction

The quantity of baryons in the Universe (and more precisely the fraction of the critical density in baryons Ω_b) is constrained by the primordial nucleosynthesis to be $\Omega_b = 0.015$ h^{-2}, with h = $H_0/(100$ km/s/Mpc) is the reduced Hubble constant. With h = 0.4, Ω_b is 0.09, and more generally Ω_b is between 0.01 and 0.09 (Walker et al. 1991, Smith et al. 1993), while the visible matter corresponds to $\Omega_* \sim 0.003$ (M/L/5) h^{-1} (+ 0.006 h$^{-1.5}$ for hot gas). Therefore, most of the baryons (90%) are dark.

In rich clusters of galaxies, the baryons are more visible, under the form of hot gas, they constitute $\sim$ 30% of the total mass (White et al. 1993). Since clusters must be representative of the baryonic fraction of the Universe, this implies that the total mass cannot be larger than 3 times the baryons mass (or $\Omega_m < 0.3$).

4.2. The smallest fragments

The existence of a large number of gas clumpuscules (of ~ 10 AU in size) in the Galaxy has already been invoked to explain the observed ESE (Extreme Scattering Events) in front of quasars, by Fiedler et al. (1987, 1994). About 300 QSOs were observed during a few years, 150 over 12 yr. More than 10 ESE events were detected, due to diffraction or refraction by a region of high electronic density (n_e). From the duration of the events, sizes of ~ 10 AU are derived, and from their frequency, the number of clumpuscules in the Galaxy must be about 10^3 the number of stars. The neutral density of these objects is still a matter of debate. Their stability is best explained in the hypothesis that they are self-gravitating. The mass of one clumpuscule is then of the order of 10^{-3} $M_\odot$. Walker & Wardle (1998) have recently built models of self-gravitating clouds, with envelopes ionised by the interstellar radiation field: they found for the electronic density the right order of magnitude to account for the observed ESE. This hypothesis is supported by direct observations through HI VLBI in absorption in front of remote radio sources (Diamond et al. 1989, Davis et al. 1998, Faison et al. 1998); large column densities ($\sim 10^{21}$ cm^{-2}) are observed with sizes of ~ 10 AU, leading to HI densities of 10^6 cm^{-3} or more.

4.3. Gamma-rays

Dixon et al. (1998) from EGRET observations have recently detected an excess of diffuse γ-ray emission in the galactic halo. This could be interpreted in several ways: either coming from un-resolved sources associated to the Galaxy; or being due to high-latitude inverse Compton emission; or finally to extra molecular gas in the halo, through cosmic ray/nucleon reaction giving π_0 then γ-rays. This has been developed by de Paolis et al. (1999) and Kalberla et al. (1999), see also Shchekinov (this conference). Cosmic rays are stopped by thick clumpuscules, that have enough column density to be opaque for both cosmic rays and gamma rays. Sciama (1999) proposes that cosmic rays are fragmented in clouds, heat the clouds, and are responsible for the their FIR emission (Sciama 1999). However, the absorbed energy is non negligible, and since clouds in these halo models are assumed to move through the optical plane (in their z-oscillations), sweep up high-metallicity gas, and therefore contain CO molecules, they should be visible through CO emission.

4.4. Various models of H_2 as dark matter

The first model proposes to prolong the visible gaseous disk towards large radii, with thickening and flaring, following the HI disk. The cold and dark H_2 component is supported by rotation, exists only outside the optical disk, where it is required by rotation curves (Pfenniger et al 1994, Pfenniger & Combes 1994). The gas is stabilised through a constantly evolving fractal structure, experiencing Jeans instabilities at all scales, in thermal equilibrium with the cosmic background radiation at T = 2.7 (1+z) K.

Other models distribute the dark molecular gas in a spherical or flattened halo, with no hole within the optical disk. The molecular gas is not so cold, and is associated with clusters of brown dwarfs or MACHOS (de Paolis et al. 1995, Gerhard & Silk 1996, Shchekinov, this conference).

In the clumpuscule model, the HI gas can be considered as a tracer, the interface between the molecular clumps and the extra-galactic radiation field. Beyond the HI disk, there could be an ionization front, and the interface might become ionized hydrogen. In this context, there should exist a distribution correlation between the dark matter and the HI gas. This is indeed the case, as already remarked by Bosma (1981), Broeils (1992) or Freeman (1993): there is a constant ratio between the surface density of dark matter, as deduced from the rotation curves, and the HI surface density, $\Sigma_{DM}/\Sigma_{HI} =$

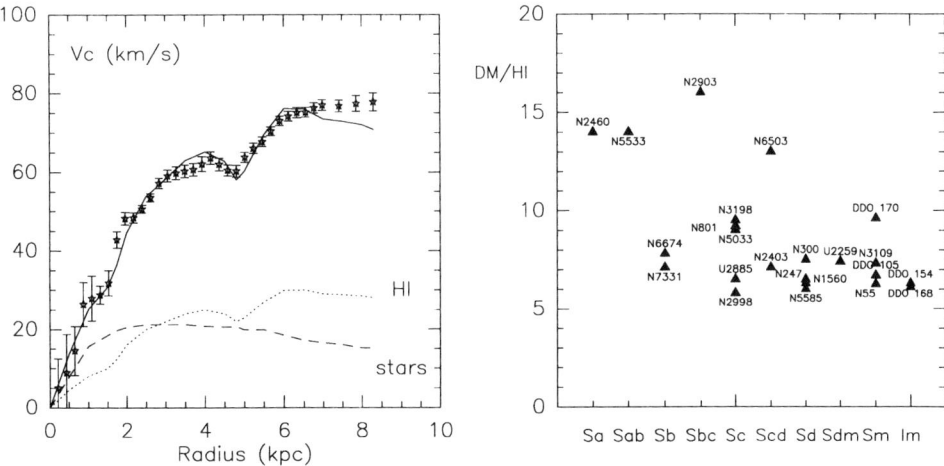

FIGURE 3. *Left*: HI rotation curve of NGC 1560 (dots +error-bars), with the rotation curve due to the HI itself (dotted line) and the stellar component (dash). The full line is the resulting expected rotation curve, when the HI mass has been multiplied by 6.2. *Right*: The ratio of surface densities of dark matter to HI required to explain the rotation curve of galaxies, as a function of type. Data from 23 galaxies have been taken from Broeils (1992) and references therein

7-10 (cf figure 3, and a recent work by Hoekstra et al. 1999). This ratio is constant with radius in a given galaxy, and varies slightly from galaxy to galaxy, being larger in early-types. However, the dark matter does not dominate the mass in the latter, and therefore the estimate of its contribution is more uncertain. The correlation is the most striking in dwarf galaxies, which are dominated by dark matter. The observed velocity curve is almost exactly proportional to the velocity curve expected from the HI component alone. Figure 3 shows the example of NGC 1560, from Broeils (1992). Let us note that dwarf galaxies represent a hard test for all models of dark matter, since the stellar component does not dominate the mass. They rule out cold dark matter (CDM) profiles (Burkert & Silk 1997), and hot dark matter (HDM) models are also unable to concentrate as much as is observed (Lake 1989, 1990, Moore 1996). Baryonic dark matter is thus required.

4.5. *Detection possibilities*

If there exists a transition region where the cold H_2 is mixed in part with evolved gas with enough metallicity and dust, it might be possible to detect cold dust emission. Encouraging results have been found by the COBE satellite, concluding to the existence of a cold (4-7K) component with column densities 10 times that of the warm component (at 18K), and more confined to the outer Galaxy (Reach et al. 1998).

Another promising tool for detection is H_2 absorption in the UV electronic lines. The main problem is the low expected surface filling factor of the cold gas ($\sim$ 1%). H_2 has already been detected in front of QSO through intervening galaxies (Foltz et al. 1988, Ge et al 1997); but these observations suffer from severe confusion problem in the Lyα forest. With the Hubble Space telescope, it was not possible to observe the fundamental lines at zero redshift, but it will be possible with FUSE. Let us remark that heavy lines of sight will be impossible to observe, due to obscuration of the background source (e.g. Combes & Pfenniger 1997). Finally, observations of the lowest pure rotational lines of H_2 have suggested some clues for the existence of large quantities of H_2 in galaxies (Valentijn & van der Werf 1999) and should be pursued in external galaxies, at much further radius than was possible with ISO.

5. Conclusions

The bulk of the H_2 mass in galaxies is cold and furtive. The main tracer is the CO molecule, but the H_2/CO conversion ratio is very variable, according to the physical conditions in molecular clouds (density and temperature), but mainly with the metallicity.

Probably because of radial abundance gradients, the molecular gas traced by the CO emission is only observed to extend over the optical disk, while the atomic gas component is prolonged much farther out in radius. Nevertheless, the HI and H_2 at the same radius are tightly correlated, at large-scale (kpc scale). They trace the same spiral structure for example. In the vertical direction, the two components are also well mixed: both reveal a constant vertical velocity dispersion with radius, of comparable amplitude. This suggests that the HI could serve as the tracer of the H_2 component, that would then also extends far out in radius. The constant ratio between dark matter and HI surface densities in galaxies support this hypothesis. If cold molecular gas is a good candidate of baryonic dark matter, observational tests should be pursued: H_2 UV absorption lines may be the best probe, and data from the FUSE satellite will certainly make big advances on the subject.

REFERENCES

Arnault P., Kunth D., Casoli F., Combes F. 1988 A&A 205, 41

Barone L.T., Heithausen A., Fritz T., Klein U. 1999, in The Physics and Chemistry of the Interstellar Medium, Zermatt Proceedings, September 22-25, 1998, ed. V. Ossenkopf

Bloemen, H.: 1989, ARAA 27, 469

Bosma A.: 1981, AJ 86, 1971

Broeils A.: 1992, PhD thesis, Groningen University.

Burkert A., Silk J., 1997, ApJ 488, L55

Depoy, D. L., Becklin, E. E., Wynn-Williams, C. G.: 1986, ApJ 307, 116

Casoli F., Sauty S., Gerin M. et al. 1998, A&A 331, 451

Combes F., Becquaert J-F.: 1997, A&A 326, 554

Combes F., Pfenniger D.: 1997, A&A 327, 453

Dame T.M., Elmegreen B.G., Cohen R.S., Thaddeus P.: 1986, ApJ 305, 982

Davis R.J., Diamond P.J., Goss W.M.: 1996, MNRAS 283, 1105

de Blok W.J.G., van der Hulst J.M. 1998, A&A 336, 49

De Paolis F., Ingrosso G., Jetzer P. et al.: 1995, A&A 299, 647

De Paolis, F., Ingrosso, G., Jetzer, Ph., Roncadelli, M.: 1999, ApJ 510, L103

Diamond P.J., Goss W.M., Romney J.D. et al: 1989, ApJ 347, 302

Dixon, D.D., Hartmann, D.H., Kolaczyk, E.D. et al.: 1998, New A. 3, 539

Downes D., Solomon P.M. 1998 ApJ 507, 615

Faison M.D., Goss W.M., Diamond P.J., Taylor G.B., 1998, AJ 116, 2916

Falgarone, E., Puget J-L., Perault M., 1992, A&A 257, 715

Fiedler R.L., Dennison B., Johnston K., Hewish A.: 1987, Nature 326, 675

Fiedler R.L., Pauls T., Johnston K., Dennison B.: 1994, ApJ 430, 595

Foltz, C. B., Chaffee, F. H., Jr., Wolfe, A. M.: 1988, ApJ 335, 35

Freeman K.: 1993, in "Physics of nearby galaxies: Nature or Nurture?" ed. T.X. Thuan, C. Balkowski, J.T.T. Van, Ed. Frontieres, p. 201

Ge J., Bechtold J.: 1997, ApJ 477, L73

Gerhard O., Silk, J.: 1996, ApJ 472, 34

Gondhalekar P.M., Johansson L.E.B., Brosch N., Glass I.S., Brinks E.: 1998 A&A 335, 152

Guélin M., Zylka R., Mezger P.G. et al. 1993, A&A 279, L37
Heithausen A.: 1996 A&A 314, 251
Hoekstra H., van Albada T.S., Sancisi R.: 1999, MNRAS, preprint
Kalberla P.M.W., Shchekinov Y.A., Dettmar R-J.: 1999, A&A preprint (astro-ph/9909068)
Kamphuis J.: 1992, PhD thesis, Groningen
Lake G.: 1989, AJ 98, 1253
Lake G.: 1990, MNRAS 244, 701
Larson R.B.: 1981, MNRAS 194, 809
Lemme C., Walmsley C.M, Wilson T.L., Muders D., 1995, A&A 302, 509
Magnani L., Blitz L., Mundy L. 1985, ApJ 295, 402
Maloney P., Black J.H. 1988, ApJ 325, 389
McGaugh S., de Blok W.J.G 1997 ApJ 481, 689
Moore B.: 1996, ApJ 461, L13
Neininger N., Guelin M., Ungerechts H. et al. 1998, Nature 395, 871
Neininger N., Guelin M., Garcia-Burillo S. et al. 1996, A&A 310, 725
Pfenniger D., Combes F., Martinet L.: 1994, A&A 285, 79
Pfenniger D., Combes F.: 1994, A&A 285, 94
Reach, W. T., Dwek, E., Fixsen, D. J., et al.: 1995, ApJ 451, 188
Richter, P., de Boer K.S.: 1999, this conference
Rubio M., Lequeux J., Boulanger F. 1993, A&A 271, 9
Sciama, D.: 1999, MNRAS, in press (astro-ph/9906159, 9909226)
Scoville N.Z., Sargent, A. I., Sanders, D. B., Soifer, B. T.: 1991, ApJ 366, L5
Shchekinov Y.A..: 1999, Astron. Rep. in press, (astro-ph/9811434)
Solomon P.M., Downes D., Radford S.J.E., Barrett J.W. 1997, ApJ 478, 144
Solomon P.M., Rivolo A.R., Barrett J.W., Yahil A.: 1987, ApJ 319, 730
Smith M.S., Kawano L.H., Malaney R.A.: 1993, ApJS 85, 219
Spitzer L., Jenkins E.B.: 1975, ARAA 13, 133
Sturm, E., Lutz, D., Genzel, R., et al.: 1996, A&A 315, L133
Sugai, H., Malkan, M. A., Ward, M. J., Davies, R. I., Mclean, I. S.: 1997, ApJ 481, 186
Tacconi L., Young J.S.: 1986, ApJ 308, 600
Taylor C.L., Kobulnicky H.A., Skillman E.D. 1998, AJ 116, 2746
Valentijn E. A., van der Werf, P.P., de Graauw, T., de Jong, T.: 1996, A&A 315, L145
Valentijn E. A., van der Werf, P.: 1999, ApJ 522, L29
van der Werf P., Genzel, R., Krabbe, A. et al.: 1993, ApJ 405, 522
van Zee L., Haynes M.P., Salzer J.J. 1997 AJ 114, 2497
Walker, T. P., Steigman, G., Kang, H-S., Schramm, D. M., Olive, K. A.: 1991, ApJ 376, 51
Walker M., Wardle M., 1998, ApJ 498, L125
Wang Y., Evans N.J. II, Zhou S., Clemens D.P., 1995, ApJ 454, 217
Ward-Thompson D., Scott P.F., Hills R.E., André P., 1994, MNRAS 268, 276
White S.D.M., Navarro J.F., Evrard A.E., Frenk C.S.: 1993, Nature 366, 429
Williams J.P., de Geus E.J., Blitz L., 1994, ApJ 428, 693
Wilson C.D. 1995, ApJ 448, L97
Young, J., Knezek, M. 1989, ApJ, 347, L55
Young, J., Scoville N.Z. 1991, A.R.A.A. 29, 581
Zaritsky D., Lorrimer S.J. 1993 in The Evolution of Galaxies and Their Environment, Proceedings NASA. Ames Research Center, p. 82-83

Pierre Cox, Françoise Combes, Jacques Le Bourlot and Guillaume Pineau des Forêts forming a molecular cycle.

Transformation of Galaxies Within the Hubble Sequence

By Daniel Pfenniger

Geneva Observatory, University of Geneva, Switzerland

Currently there are three quite different views about galaxy evolution, each one improving the previous state of knowledge:
(1) The older one ("ELS") in which galaxies form by collapse early, quickly, and synchronously (during the "galaxy formation epoch"), ending the dynamically active period; subsequent galaxy evolution is merely a matter of stellar formation processes in a rigid potential.
(2) An alternative one ("SZ") in which disks are viewed as forming inside out over an extended period of time. Galaxy evolution occurs without important internal dynamical instabilities.
(3) The slowly emerging picture, after 40 years of N-body simulations and the obvious evidences from recent high-z observations: galaxies evolve both dynamically and chemically over most of the Hubble time in a widely asynchronous way at different speeds, depending on the environment. The Hubble sequence, from late to early types, appears to represent a broad description of the general aging process.

Thus galaxies appear now as evolving structures over typical time-scales of order of 1 Gyr. A fundamental aspect of the micro-physics in galaxies is star formation and gas processes in which the H_2 molecule must play a key role: indeed interstellar gas must first form H_2 before being able to form stars, so star forming regions do trace molecules, although CO might not have been detected.

1. Introduction

In the following we explain why galaxies can no longer be viewed as rigid or stable structures over many Gyr, an idea still very widespread. To a large part stimulated by N-body simulations, and reinforced during the last years by very deep universe observations, galaxies are increasingly perceived as secularly evolving structures with the main visible function of transforming gas into stars, and light elements into heavy elements.

A direct by-product of the secular transformation of galaxies is a constraint on the composition of galaxies at different stages of their evolution. Comparing the detected components and the undetected part, the dark matter, at different stages of evolution allows to draw constraints, or at least indications, about the nature of the dark matter (Pfenniger, Combes & Martinet 1994).

As the star formation process is understood, stars form inside molecular clouds over time-scales much shorter than the galaxy proper time-scales. Thus actively forming star regions can be viewed as witnessing the quasi-simultaneous presence of molecules (de Geus & Phillips 1996). On the other hand, old stellar populations trace the cumulated stellar formation history, yielding a lower bound on the cumulated amount of formed molecules.

Therefore, the comparison of star formation regions and regions where molecules are directly detected suggests, since the correspondence is not one-to-one, that not all the molecules are traced by direct observations, generally the CO ones. Along a similar line of arguments, the galactic far-infrared emission traces well cold dust, i.e., grains made for a substantial part of large molecules containing a cocktail of light and heavy elements. In such physico-chemical conditions, it is hard to imagine that no free floating molecules are associated with cold dust. Thus the presence of many molecules should be expected when dust is observed, even when molecules have not been directly detected (e.g., cirrus clouds, far-infrared emission in the outer galactic disks).

2. Evolution of the Ideas about Galaxy Evolution

Decades of research have slowly improved about our physical understanding on galaxies and their evolution. For a long time the Hubble sequence has appeared just as a curious collection of objects with smooth variations of properties (correlations), but lacking of a deeper meaning given by fundamental physical laws. Presently, the Hubble sequence turns out to take more signification because we can both interpret it in much more detail with the increasingly powerful computers, and because modern instruments allow us to check back directly most of the history of galaxies.

2.1. The ELS Scenario

The mental picture of galaxy evolution, still widespread among a large part of astronomers and particularly physicists, is close to the seminal ELS scenario for the formation of the Milky Way (Eggen, Lynden-Bell, & Sandage 1962), based on the following reasoning: since galaxies form presumably by collapse induced by the gravitational instability in a homogeneous medium, which has a well defined, universal free-fall time $\tau_{\rm ff} \approx 1/\sqrt{G\rho_0}$, one supposes that galaxies form by billions quasi synchronously over an epoch lasting a few 10^8 yr. After all, shouldn't the collapse phase be of the order of the free-fall time?

At this point usually ad hoc star formation mechanisms in spherical halos are invoked to explain the old stellar halos and globular cluster systems. Eventually condensing galaxies settle later around stable states in which dynamical factors are balanced. Thus the subsequent changes to discuss are only coming from the slow transformation of the leftover gas into stars. This occurs then in a rigid potential, dynamics has been absorbed by a steady and symmetric state, and thus in practice its most fundamental characteristics are ignored. Below we will see by examples how this well impregnated picture needs to be seriously updated.

2.2. First Crack

As any scenario, the ELS scenario needs to be confirmed by more detailed works, particularly because in the early 60's computer simulations were just beginning. They had a profound influence by opening in nearly all sectors of sciences the field of unstable, transitory and chaotic phenomena, neglected up then due to the almost exclusive interest of scientists for the stable, steady, and symmetric states of nature.

So the first point should be to check whether the collapse picture of the ELS scenario has been confirmed by later works. In fact, starting in the 60's (e.g. Lin, Mestel, & Shu 1965; Larson 1969; Zel'dovich 1970) it became more and more clear than the spherical homogeneous collapse associated to a unique collapse time-scale is far from being *generic*: in a spherical collapse a slight residual pressure produces rapidly inhomogeneous profiles where the outer parts delay collapse indefinitely, and an initial slight ellipsoidal deformation leads to pancakes and filaments, at least two contracting directions, but possibly one dilating direction. Filamentary condensations in pure N-body simulations are nice examples of possible mass concentrations not requiring any radiative cooling to reach dense and cool states, since the whole simulated process is purely gravitational. In the more realistic situations simulated in cosmological models the systematic results are highly inhomogeneous, with regions having widely different densities ("clusters", filaments, sheets, and voids) in which by necessity subsequent sub-structure formation must occur at very different speeds.

The basic assumption of a synchronous evolution of galaxies is challenged because such irregular matter distributions imply not only that mergers must be significant in dense environments, but also that the long lasting harassment caused by galaxy interactions and mergers must be highly variable from regions to regions. The very concept of "galaxy formation epoch" appears then irrelevant, because time-scales and length-scales are not clearly separated during the whole galaxy formation process.

2.3. Further Cracks

Assuming nevertheless that in some cases regular collapses can lead to isolated galactic disks, the next point to check is whether a really stable state follows. Indeed the respective energies in interaction are quite different: for dynamics it is of the order of $v_{\rm rot}^2 \approx 500\,{\rm eV/nucleon}$, which makes a large energy reservoir in comparison with other energy reservoirs such gas pressure, cosmic rays, or magnetic fields (each about $1\,{\rm eV/nucleon}$). The developments in galactic dynamics following the ELS scenario showed amply that violent instabilities in spirals must be considered as normal, the most important one being the bar instability. The spirals themselves appeared as transient structures that require regeneration mechanisms. The N-body simulations in the 60's showed early that the bar phenomenon occurs easily; a first hint that Hubble types were not necessarily permanent was given (the transformation SA $\to$ SB was found).

Yet the acceptance of the community to this surprise was slow. Instead of taking dynamics at face, dark halos were invented to suppress the easy transformation to SB, considered then as a disease. In fact it turned out that dark halos were unable to prevent the bar instability, and given the observed large fraction of barred spirals one must on the contrary wonder about their high frequency (Debattista & Sellwood 1998) if a substantial amount of dark matter exists within the optical disks.

2.4. The SZ Improvement and the Merger Subversion

Beside taking observational constraints on Galactic stellar clusters, the SZ scenario (Searle & Zinn 1978) at least takes into account the theoretical finding of Larson (1969) that more realistic spherical collapses proceed quicker in the central regions than in the outer ones. The center collapses well within a free-fall time, but the outskirts, still keeping a substantial part of the mass, take an indefinite long time (a multiple of $\tau_{\rm ff}$) due to the non-zero pressure *gradients* with the surrounding medium. Therefore the gradual building of a galactic disk as described by SZ must be seen as more realistic, although not complete since building disks requires to consider the angular momentum budget, as well as internal and external dynamical factors that turned out to remain of major importance.

The main galactic dynamics topic of debate in the 70's was the hypothesis pushed forward by Toomre & Toomre (1972), that ellipticals may result from the merging of spirals and other galaxies (the transformations S + S $\to$ E, E + S $\to$ E and E + E $\to$ E). This simple reasonable idea met extraordinary resistance precisely because it challenged directly the quasi-dogma that galaxy morphologies are fixed by the "birth" conditions. In a well developed review Schweizer (1998) shows in great detail all the arguments, up to the latest HST results, which were required to convince the most skeptical opinions that mergers represent a significant aspect of the building and evolution of galaxies.

Two arguments did slow the acceptance of the merger hypothesis. One was the conservation of phase space volumes (Liouville's theorem) as predicted assuming collisionless fluids: mixing collisionless systems can only dilute the apparent phase space density.

The observed higher phase space density in ellipticals than in spirals thus appears to contradict the merging hypothesis. Later in turned out that Liouville's theorem is very sensitive to dissipative perturbations, indeed a small fraction of gas can condense in the core of galaxies and increases its central phase space density to very high values.

The other invoked counter-argument was the higher number of globular clusters per unit light in ellipticals than in spirals, while merging, by eventually triggering star formation in the gas parts of spirals, should decrease rather than increase the globular light fraction. The answer to this objection appears now to lie in the well documented formation of a large number of globulars and dwarf galaxies in the tidal tails following strong mergers (see, e.g., Schweizer 1998).

3. Further Cases of Morphological Transformations

Additional cases of galaxy type transformations have shown via computer simulations during the 80's and 90's to be possible and likely. Over the years the general picture of galaxy existences has become more and more dynamical, and many observational findings (a few examples are the Milky Way bar, the Sagittarius dwarf, the double nucleus in M31, the numerous irregular galaxies in the Hubble deep field, the complex kinematics in ellipticals and spirals) just reinforce such a view.

3.1. Internal Transformations

Not so many degrees of freedom are left when simulating realistic spirals. The known high flattening of disks constrains the velocity dispersion. As a consequence, realistic spiral disks are never very far from a global instability. This is why almost all galaxy simulators, starting with Lindblad (1960), Miller & Prendergast (1968), and many followers, found that the bar instability appears nearly always. Further, taking into account gas dissipation tends to lower the global velocity dispersion, and therefore to push the disks toward global instability. The bar instability and the easy generation of spirals appear then natural.

The average global state of such disks competing between dissipative effects and dynamical heating is a marginal stability state. Spiral disks are therefore able to react to perturbation by strongly amplifying the effect of disturbances. There are many ways to trigger instabilities in such a situation. For example, even weak tidal interactions induce the formation of an internal bar which persists well after the perturber passage (e.g. Noguchi 1988).

Beside dynamical heating, a global gravitational instability in a gaseous disk can also induce star formation, which, via the nuclear energy freed by massive stars, may produce a substantial heating slowing down the general gas cooling. Such feed-back mechanisms between dynamics and star formation have been discussed many times (e.g. Quirk 1972; Kennicutt 1998). If such a feed-back mechanism is active, it means that the perhaps slight disturbances to equilibrium regulating the galaxy are nevertheless essential to understand properly the galaxy evolution over several Gyr.

In non star forming disks, such as in low-surface brightness galaxies, if star formation is unable to compensate the gas cooling (which must act since this gas had surely to dissipate and condense once to form a disk!), global dynamical heating must damp gravitational instability. In this case dynamics and gas cooling are the major agents to understand well in order to describe the long term evolution of the galaxy.

A bar is itself subject to instabilities of various kinds. As shown by Combes et al. (1990), Pfenniger & Friedli (1991), Raha et al. (1991) and others, a bar is often subject to a bending instability transverse to the galaxy plane. The end-evolution of the bending instability is a peanut- or box-shaped bulge much resembling the Milky Way bulge-bar. Therefore we see here a case of a minor morphology transformation not only explaining a way to obtain the observed peanut-shaped bulges, but also showing how to inflate a disk into a bulge much after the disk has formed via vertical (transverse) dynamical resonances.

Strongly non-axisymmetric perturbations such as bars typically exchange angular momentum globally much faster than any other known process. Bars are therefore ideal to regulate the excess of angular momentum that dissipating inflowing gas carries. The gas within a bar tends either to flow toward its inner Lindblad resonance, or outside its corotation resonance. The inflowing gas at some point may become self-gravitating and form a partially decoupled secondary bar inside the larger one. For some rotation periods the galaxy may appear to possess two embedded bars, which are indeed observed in many spirals. The mutual torques must on the long run destroy the structure. From N-body simulations (Friedli & Martinet 1993) the end-result is usually a three-dimensional spheroidal structure much resembling a bulge. This is a nice example how a finer study of dynamical processes bring to the inescapable conclusion of fast morphological changes induced by dynamics coupled to dissipative gas.

In fact any increase of the inner mass of a bar by a few percents is sufficient to destroy the whole bar into a three-dimensional spheroidal bulge as large as the original bar (e.g. Pfenniger & Norman 1990; Friedli & Pfenniger 1991; Hasan, Pfenniger, & Norman 1993). For example the accretion of a small galaxy weighting 10% the main galaxy is sufficient to destroy its initial bar with a surprisingly short time-scale, $20-30\,\mathrm{Myr}$ (Pfenniger 1991). The destruction of bars is therefore a process SB $\rightarrow$ SA contributing to explain why not all spirals are barred!

Another effect of bars on stars is to let a substantial amount of them to diffuse from the center to several times the corotation radius. The formation of strong bars destabilizes most of the stars originally circling near the corotation radius. The chaotic diffusion of stellar trajectories described in Pfenniger (1985) has been examined in more details recently (Ollé & Pfenniger 1998) and leads to predict that metal-rich stars around the Sun are diffusing from the inner Milky Way bulge-bar up to substantially more than the solar orbit and back. This mixing process is important regarding the abundance gradients. In summary, a bar may enhance an abundance gradient when triggering star formation at some particular place, but also at the same time it mixes the abundances when diffusing stars and transporting gas.

3.2. External Perturbations

When dissolving, a bar is able to make a bulge as big as itself, but not much bigger. We discussed a few year ago a process to make big bulges (Pfenniger 1992, 1993). When several small and uncorrelated galaxy satellites merge into a large galaxy, they bring not much mass but much more kinetic energy. It suffices to accrete $5-10\%$ of mass this way to cause a global heating of the whole disk to lenticular morphologies much resembling galaxies such as the Sombrero galaxy. The heating can be secular (over several Gyr) or rapid (over several Myr) depending on the number, mass and arrival times of the satellites. Such processes thus transform $S \rightarrow S0$, and can be seen as the weak version of strong mergers transforming $S \rightarrow E$.

An interesting idea of Zwicky (1956) of morphological changes and galaxy formation at intermediate times was nicely illustrated by Barnes (1994): some dwarf galaxies must be produced following strong mergers in the escaping tidal tails. These tails escape or follow elongated paths around one of the once interacting bigger galaxies, and recondense under their own gravity. Thus they may form small galaxies with already a rich stellar evolution history in the parent galaxy. Repeated burst of star formation in dwarf galaxies are hard to understand otherwise because any starburst in such small galaxies should either eject and/or consume the gas, preventing future star formation periods.

3.3. *The Milky Way*

For obvious reasons the Milky Way is the reference galaxy, but apparently its proximity has not much helped to grasp that the Milky Way is subject to complex dynamical events too. Still today many studies assume axisymmetric, and time-independent mass models of the Milky Way. Only in the recent years it became obvious, in particular with the COBE data, that the Milky Way bulge is indistinguishable from a thick, peanut-shaped bar, which already suggests an eventful past.

In his Ph.D. thesis, Fux (1997) has studied how to reconcile the H I, CO and IR observations with completely self-consistent 3D N-body simulations with unprecedented numerical resolution. The essential results of Fux's work are not only to strengthen the existence of a bar in the Milky-Way to certainty, but also to show that to interpret correctly the available data, such as the H I longitude-velocity maps, even when considering only their gross features, the degree of detail and asymmetries is such that kpc wide structures are changing over short time-scales, due to dynamics. The usual assumption of equilibrium allowing to forget about dynamics is invalidated. The real Milky Way, as most spirals, shows a number of asymmetries such as a bar and several spiral arms that must be properly included into the models, including their time-dependence, to reflect correctly their state. Thus evaluations of the long term Milky Way evolution must include its time-dependent dynamical effects before hoping to describe well its full history. Apparently we begin to discover that the Milky Way "weather" is as rich, when properly scaled, as the Earth atmosphere one.

3.4. *Harassment in Clusters*

One could discuss in more detail the dynamics of galaxy clusters, where the morphology-density relationship shows in which way environment effects modify galaxies: the more likely the galaxy interactions are, in denser cluster parts, the earlier in average the Hubble types are. To reach this conclusion one must not forget that galaxies *move* across a cluster, so the position is not reflecting any "birthplace". Therefore the galaxy type gradients must reflect a fast transformation of Hubble types from late to early when spirals cross the dense parts of a cluster. Recent estimates of large fraction of intergalactic stars and large fraction of hot gas in clusters suggest that much mass is lost during tidal interactions with a proportion of gas much larger that the one of stars.

4. Modernizing the Hubble Sequence

Dynamics is able to reproduce and explain a lot of properties and detailed features in galaxies, provided that some old assumptions are dropped, such as the invariance of Hubble types, the synchronous evolution of galaxies, and well defined ages and populations. Instead of the old views on a regular evolution of galaxies, the picture which emerges after ≈ 40 yr of modeling is a chaotic one. Both secular and erratic brief episodes of dynamical evolution lead to the transformation of galaxy morphologies, in general toward earlier Hubble types, at widely different speeds. The bad news are that chaotic processes mean more difficulty to learn about the past, particularly in systems with complex histories, i.e. the early types. Complex histories of galaxies appear inescapable: if one accepts Newtonian dynamics for understanding some aspects of galaxies (e.g. their equilibrium), to be consistent *all* its consequences must be adopted, not just the easy ones.

So the whole assumption that dynamics can be neglected for most of the age of a galaxy appears today as requiring a serious revision. But if the morphologies are not fixed "at birth", then morphologies are due to other factors and may evolve, and the whole Hubble sequence might trace evolutive processes and not only initial conditions of formation. The notion that Hubble types may change becomes then an essential feature of galaxies. The Hubble sequence and all the associated correlations must be ordered according to known dynamical and physical secular processes. It is clear that once stars are formed, galaxies subject to morphological transformations must remain within the Hubble sequence, because the considered transformations do not destroy stars. In the other sense, since stars are made from cold molecular gas, the number of galaxies may change over the Gyr. This number may increase if gaseous disks form stars, or if tidal interactions produce dwarfs, but the number of galaxies can also decrease by mergers. Therefore galaxy counts according to redshift are not revealing well galaxy evolution, as well as galaxy counts selected by Hubble types.

On a more physical side, the known irreversible processes in galaxies show clearly that evolution must proceed from late to early types, indeed the correlations along the Hubble sequence are consistent with such an ordering measuring *aging*:

(*a*) A measure of energy dissipation in proportion of the depth of the potential is just given by the rotation curve and/or the velocity dispersion, giving the energy per unit mass that had to be released to reach the bound state.

(*b*) The fraction of bulge to disk measures also a stellar dynamical irreversible mixing, stellar disks can be heated to bulges, but not the contrary.

(*c*) The winding of spiral arms and the degree of symmetry measure the amount of quiet rotation periods.

(*d*) The ratio of gas to stars measures the amount of star formation history.

(*e*) The ratio of heavy to light elements measures the amount of nucleosynthesis.

All these criteria are varying monotonously along the Hubble sequence in the sense that late type galaxies trace less aging than early type galaxies. From this the sense of evolution is clear, and galaxies can be ordered roughly by aging with their Hubble type from I–Sm–Sd to Sa–S0–E.

In principle, it is possible to rejuvenate galaxies by massive infall of fresh gas. Yet this must occur with some precautions because even light mass infall lead to strong dynamical heating, and shocked gas lead to starbursts which accelerate aging. The most favorable accretion route is to have gentle ∼corotating gas accretion in the outer galactic disks which would lead to a slow growth of the outer gaseous spiral disks "à la SZ". But it is hard to imagine that such a process would decrease by more than a factor 10 the metallicity of a galaxy, therefore the variations of metallicity from galaxy to galaxy are unlikely to be entirely due to gas infall. Further when later mass accretion exceeds 100% the very meaning of galaxy age becomes questionable.

5. Conclusions

Galaxies can be ordered along a sequence of aging measuring the number of "events" participating to the overall transformations from late to early types. The gas to star transformation is perhaps one of the most characteristic one of galaxies, and its characteristics is to imply the formation of at least correspondingly large amounts of H_2 molecules. The fact that so few H_2 molecules are presently inferred from the CO observations even in star formation regions (particularly those in outer galactic disks) should be viewed as an indication that not all the H_2 molecules are correctly inferred.

We are grateful to the organizers for this stimulating conference. This work is supported by the Swiss Foundation for Scientific Research.

REFERENCES

BARNES, J.E. 1994 in *The Formation and Evolution of Galaxies*, Proceedings of the 5th Canary Islands Winter School of Astrophysics, eds. C. Muñoz-Tuñón, & F. Sanchez (Cambridge: Cambridge University Press), 399

COMBES, F., DEBBASCH, F., FRIEDLI, D., & PFENNIGER, D. 1990 *Astron. Astrophys.*, **233**, 82

DEBATTISTA, V.P., & SELLWOOD, J.A. 1998 *Astrophys. J.*, **493**, 5

DE GEUS, E. J., & PHILLIPS, J. A. 1996 in *Unsolved problems of the Milky Way*, Proceedings of the 169th Symposium of the IAU, Dordrecht Kluwer, L. Blitz & P. Teuben (eds.), 575

EGGEN, O.J., LYNDEN-BELL, D., & SANDAGE, A.R. 1962 *Astrophys. J.*, **136**, 748 (ELS)

FRIEDLI, D., & MARTINET, L. 1993 *Astron. Astrophys.*, **227**, 27

FRIEDLI, D., & PFENNIGER, D. 1991 in *Dynamics of Galaxies and their Molecular Clouds Distribution*, IAU Symp. 146, eds. F. Casoli, & F. Combes, (Dordrecht: Reidel), 362

FUX, R. 1997 Ph.D. thesis, Université de Genève, Switzerland

HASAN, H., PFENNIGER, D., & NORMAN, C.A. 1993 *Astrophys. J.*, **409**, 91

KENNICUTT, R.C. 1998 in *Galaxies: Interactions and Induced Star Formation*, Saas-Fee Advanced Course 26, eds. D. Friedli, L. Martinet, & D. Pfenniger (Berlin: Springer), 1

LARSON, R.B. 1969 *Monthly Not. Roy. Soc.*, **145**, 405

LIN, C.C., MESTEL, L., & SHU, F.H. 1965 *Astrophys. J.*, **142**, 1431

LINDBLAD, P.O. 1960 *Stockholm Obs. Ann.*, **21**, No. 3-4

MILLER, R.H., & PRENDERGAST, K.H. 1968 *Astrophys. J.*, **151**, 699

NOGUCHI, M. 1988 *Astron. Astrophys.*, **203**, 259

OLLÉ, M., & PFENNIGER, D. 1998 *Astron. Astrophys.*, **150**, 112

PFENNIGER, D. 1985 *Astron. Astrophys.*, **150**, 112

PFENNIGER, D. 1991 in *Dynamics of Disc Galaxies*, ed. B. Sundelius (Göteborg: Göteborg University), 191

PFENNIGER, D., 1992 in *Physics of Nearby Galaxies: Nature or Nurture?*, XIIth Moriond Astr. Meeting, eds. T.X. Thuan, C. Balkowski, & J. Trân Thanh Vân (Gif-sur-Yvette: Editions Frontières), 519

PFENNIGER, D. 1993 in *Galactic Bulges*, IAU Symp. 153, eds. H. Dejonghe, & H.J. Habing (Dordrecht: Kluwer), 387

PFENNIGER, D., COMBES, F., & MARTINET, L. 1994 *Astron. Astrophys.*, **285**, 79

PFENNIGER, D., & FRIEDLI, D. 1991 *Astron. Astrophys.*, **252**, 75

PFENNIGER, D., & NORMAN, C.A. 1990 *Astrophys. J.*, **363**, 391

QUIRK, W.J. 1972 *Astrophys. J.*, **176**, L9

RAHA, N., SELLWOOD, J.A., JAMES, R.A., & KAHN, F.D. 1991 *Nature*, **353**, 411

SCHWEIZER, F. 1998 in *Galaxies: Interactions and Induced Star Formation*, Saas-Fee Advanced Course 26, eds. D. Friedli, L. Martinet, & D. Pfenniger (Berlin: Springer), 105

SEARLE, L., & ZINN, R. 1978 *Astrophys. J.*, **225**, 357 (SZ)

TOOMRE, A., & TOOMRE, J. 1972 *Astrophys. J.*, **178**, 623

ZEL'DOVICH, Y.B. 1970 *Astron. Astrophys.*, **5**, 84

ZWICKY, F. 1956 *Ergeb. Exakten Naturwiss.* **29**, 344

Extragalactic H$_2$ and its variable relation to CO

By F.P. Israel

Sterrewacht Leiden, Postbus 9513, 2300 RA Leiden, the Netherlands

We derive and discuss the strong dependence on metallicity of the CO to H$_2$ conversion factor $X = N(\text{H}_2)/I_{CO} = 12.2 - 2.5\,log\,[O]/[H]$ appropriate to extragalactic objects, as well as the weaker dependence found for such objects from interferometer measurements.

1. Introduction

The difficulty of directly observing molecular hydrogen (H$_2$), the major constituent of the interstellar medium in galaxies, and ways of doing so indirectly are reviewed elsewhere in this volume (Combes 2000). Usually, H$_2$ cloud properties are derived by extrapolation from more easily conducted CO observations. For instance, observed CO cloud sizes and velocity widths yield total molecular gas masses under the assumption of virial equilibrium. However, in extragalactic systems especially, this method is beset by pitfalls (see Israel, 1997, hereafter Is97) and requires high linear resolutions (i.e. use of interferometer arrays). More seriously, *the fundamental assumption of virialization appears to be false.* As individual components ('clumps') have velocities of only a few km s^{-1} and CO complex sizes are 50–100 pc, crossing times are comparable to CO complex *lifetimes* of only a few times 10^7 years or less (Leisawitz et al. 1989; Fukui et al. 2000; see also Elmegreen 2000). As equilibrium cannot be reached in a single crossing time or less, the virial theorem is not applicable to such complexes. Indeed, the elongated and interconnected filamentary appearance of many large CO cloud complexes do not suggest virialized systems (see also Maloney 1990).

The observed CO intensity is the weighted product of CO brightness temperature and emitting surface area; actual CO column densities are completely hidden by high optical depths. However, in large beams CO cloud ensembles may be assumed to be statistically identical so that CO intensities scale with CO mass within the beam, i.e. beam-averaged CO column density. If we can determine the proportionality, the H$_2$-to-CO conversion factor X, subsequent CO measurements can be used to find the appropriate H$_2$ column density and mass. In the Milky Way, the calibration of X is controversial by a factor of about two (cf. Combes 2000), and frequently based on application of the virial theorem (but see preceding paragraph ...).

In other extragalactic environments, the assumption of statistical CO cloud ensemble similarity becomes questionable. Very clumpy, even fractal molecular clouds are very sensitive to e.g. variations in radiation field intensity and metallicity. As H$_2$ and CO, supposedly tracing H$_2$, react differently to such variations, X is also very sensitive to them (Maloney & Black 1988). The determination of the dependence of X on metallicity and radiation field intensity, needed to correctly estimate amounts of H$_2$ in environments (dwarf galaxies, galaxy centers) different from the Solar Neighbourhood thus requires H$_2$ mass determinations independent of CO.

2. H$_2$ determinations from dust continuum emission

Fortunately, H$_2$ and HI column densities are traced by optically thin continuum emission from associated dust particles. Unfortunately, dust emissivities depend strongly on temperature, dust particle properties are not accurately known and dust-to-gas ratios are frequently uncertain. The effect of these uncertainties are minimized if we can avoid the need for determining absolute values of the dust column density and the dust-to-gas ratio. Far-infrared/submillimeter continuum fluxes and HI intensities from spatially nearby positions, preferably in dwarf galaxies that lack strong temperature or metallicity gradients, can be used to obtain reasonably accurate H_2 column densities (Is97). The ratio of dust continuum emission to HI column density at locations lacking substantial molecular gas provides *a measure* for the dust-to-gas column density ratio. Without requiring its absolute value, we can apply this measure to a nearby location rich in molecular gas to find the total hydrogen column density and, after subtraction of HI, the H_2 column density. Division by the CO intensity yields the local value of X in absolute units with an accuracy better than a factor of two (Is97).

Individual molecular cloud complexes in the nearby Magellanic Clouds were used by Is97 to determine the effects of radiation field intensity (as sampled by far-infrared surface brightness) on X. Over a large range of intensities, X is linearly proportional to the radiative energy available per nucleon (σ). Quiescent regions in the LMC yield X values close to those of the Solar Neighbourhood, whereas a value 40 times higher is obtained for the radiation-saturated 30 Doradus region. The more metal-poor SMC exhibits higher X values, but again linearly proportional to σ.

3. Dependence of X on metallicity

To further study the relation between X and metallicity, we have added several recent results to the database given by Is97. These include data for NGC 7331 (3 points; Israel & Baas 1999), the Milky Way center and centers of NGC 253 (both from Dahmen et al. 1998), NGC 891 (Guélin et al. 1993; Israel et al. 1999), NGC 3079 (Braine et al. 1997; Israel et al. 1998a) as well as IC 10 (Madden et al. 1997) and D 478 in M 31 (Israel et al. 1998b). Although they were obtained somewhat differently from those in Is97, they are quite comparable (Figs. 1 and 2).

In Fig. 1, radiation-corrected values $X' = X/\sigma$ are plotted against metallicity [O]/[H]. In Fig. 2, values of X are plotted in in its usual form. Figs. 1 and 2 yield the relations:

$$\log X' = \log X/\sigma = -4 \log ([O]/[H]) + 33.9 \quad (3.1)$$

and

$$\log X = -2.5 \log ([O]/[H]) + 12.2 \quad (3.2)$$

With a larger sample size, these results differ only slightly from those published by Is97. The points representing high-metallicity regions in NGC 7331 extend rather well along the relation defined by the low-metallicity dwarfs, as do those representing the galaxy centers with a larger scatter. Both correlations are highly significant. Thus, *(3.2) should in general be used to convert CO intensities observed in large beams to obtain H_2 column densities* within a factor of about two. Note that the result may greatly differ from that obtained by applying 'standard' Milky Way conversion factors (i.e. lower by a factor of 4–10 for high-metallicity galaxy centers and higher by a factor of 10–100 for low-metallicity irregular dwarf galaxies).

In Fig. 2, we have also included X values derived by virial theorem application to CO clouds mapped with interferometer arrays taken from Wilson (1995 – replacing her M 31 and M 33 metallicities by those from Garnett et al. 1999), Taylor & Wilson (1998) and Taylor et al. (1999). These points define a different dependence of X on metallicity, with a much shallower slope of only -1.0. Generally, these X values are much lower than those in Is97.

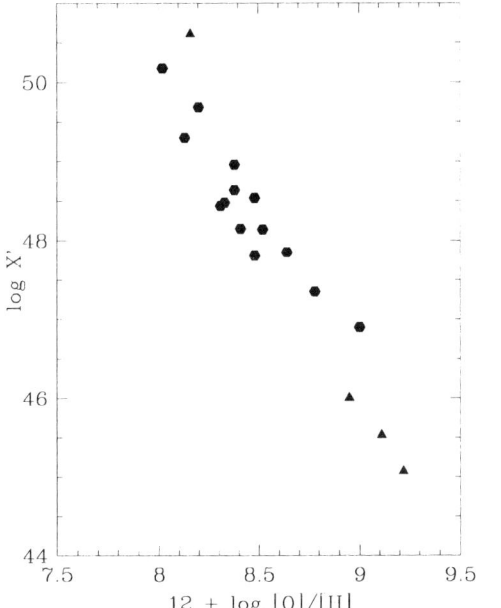 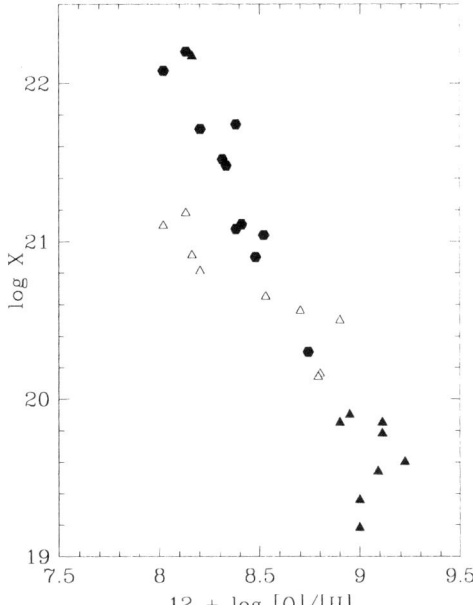

FIGURE 1. Dependence of X' on metallicity. For definition and units of X', see Is97. Filled hexagons: points taken from Is97; filled triangles: additional points (see text).

FIGURE 2. Dependence of conversion factor X (in mol H_2 cm^{-2}/ K km s^{-1}) on metallicity. Open triangles: points from millimeter array observations (see text); otherwise as Fig.1.

4. Discussion

A *steep* dependence of X on metallicity can be understood within the context of photon-dominated regions (PDRs). In weak radiation fields and at high metallicities, neither H_2 or CO suffers much from photo-dissociation, and the CO volume will fill practically the whole H_2 volume. However, when radiation fields become intense, CO photo-dissociates more rapidly than H_2 because it is less strongly self-shielding. Thus, the projected CO emitting projected area will shrink and no longer fill that of H_2. The observed CO intensity, proportional to the shrinking emitting area, then requires use of a *higher* X factor to obtain the correct, essentially unchanged H_2 mass. We have found that at constant metallicity, X must be increased linearly with radiation field intensity.

We may somewhat quantitatively estimate the effects of metallicity on CO (self)shielding and thereby on X. From Garnett et al. (1999) we find that over the range covered by Figs. 1 and 2, log [C]/[H] $\propto$ 1.7 log [O]/[H]. Thus, CO abundances drop significantly more rapidly than metallicity [O]/[H], as do dust abundances given by $M_{dust}/M_{gas} \propto$ 2 log [O]/[H] (Lisenfeld & Ferrara 1998). Thus, a ten times lower metallicity (cf. Figs 1 and 2) implies a [CO]/[H_2] ratio lower by a factor of 50, and less CO shielding by a factor of 5000! The precise effect on X depends on the nature of the cloud ensemble, but at lower metallicities PDR effects very quickly increase in magnitude. In a standard H_2 column, there is less CO to begin with, and this smaller amount is even less capable of resisting further erosion by photodissociation. With decreasing metallicity, CO is losing both its self-shielding and its dust-shielding, so that even modestly strong radiation fields completely dissociate extended but relatively low-density diffuse CO gas, leaving only embedded smaller higher-density CO clumps intact. As CO intensities primarily sample emitting surface area, the loss of extended diffuse CO strongly reduces them, even when actual CO mass loss is still modest. Further metallicity decreases cause further erosion and molecular clumps of ever higher column density lose their CO gas. CO is thus occupying an ever-smaller fraction of the H_2 cloud which still fills most of the PDR. Its destruction releases a large amount of atomic carbon which is ionized and forms a large and bright cloud of CII filling the entire PDR. This and the expected anti-correlation between CO and CII intensities is indeed observed in the Magellanic Clouds and in IC 10 (Is97; Israel et al. 1996; Madden et al. 1997; Bolatto et al. 2000). As the strongly self-shielding H_2 is still filling most of the PDR (cf. Maloney & Black 1988), the appropriate value of X becomes ever higher. In the extreme case of total CO dissociation, any amount of H_2 left defines an infinitely large value of X!

In contrast, use of e.g. interferometer maps to find resolved CO clouds for the determination of X introduces a strong bias in low-metallicity environments. In the PDR, only those subregions are selected which have most successfully resisted CO erosion, with CO still filling a relatively large fraction of the local H_2 volume. The relatively low X values thus derived, although appropriate for the selected PDR subregions, are not at all valid for the remaining PDR volume where CO has been much weakened or has disappeared; the PDR has a much higher overall X value than the selected subregion. It is because of this bias that the array-derived points in Fig. 2 are much lower than the large-beam points and exhibit a much weaker dependence on metallicity. Incidentally, it also explains the suggested dependence of X on observing beamsize (Rubio et al. 1993).

REFERENCES

BOLATTO, A.D., JACKSON J.M., ISRAEL, F.P, ZHANG, X. & KIM, S. 2000 *Ap. J.* in prep.
BRAINE, J., ET AL. 1997 *A&A* **326**, 963–975.
COMBES, F. 2000 *These Proceedings*, ***_***
DAHMEN G., ET AL. 1998 *A&A* **351**, 959–976.
ELMEGREEN, B.G. 2000 *Ap. J.*, in press
FUKUI, Y., ET AL. 2000 *P. A. .S. J.* in press
GARNETT, D.R., ET AL. 1999 *Ap. J.* **513**, 168–179.
GUÉLIN, M. ET AL. 1993 *A&A* **279**, L37–L40.
ISRAEL, F.P. 1997 *A&A* **328**, 471–482, (Is97).
ISRAEL, F.P. ET AL. 1996 *Ap. J.* **465**, 738–747.
ISRAEL, F.P., ET AL. 1998a *A&A* **336**, 433–444.
ISRAEL, F.P., TILANUS, R.P.J. & BAAS, F. 1998b *A&A* **339**, 398–404.
ISRAEL, F.P., VAN DER WERF, P.P. & TILANUS, R.P.J. 1999 *A&A* **344**, l83–L86.
ISRAEL, F.P. & BAAS, F. 1999 *A&A* **351**, 10–20.
LEISAWITZ, D., BASH F.N. & THADDEUS, P. 1989 *Ap. J. Suppl.* **70**, 731–812
LISENFELD, U. & FERRARA, A. 1998 *Ap. J.* **496**, 145–154.
MADDEN, S.C., ET AL. 1997 *Ap. J.* **483**, 200–209.
MALONEY, P.R. 1990 *Ap. J. L.* **348**, L9–L12
MALONEY, P.R. & BLACK J.H. 1988 *Ap. J.* **325**, 389–401.
RUBIO, M., LEQUEUX, J. & BOULANGER, F. 1993 *A&A* **271**, 9–17.
TAYLOR, C.L. & WILSON, C.D. 1998 *Ap. J.* **494**, 581–586.
TAYLOR, C.L., HÜTTEMEISTER S., KLEIN, U. & GREVE, A. 1999 *A&A* **349**, 424–435.
WILSON C.D. 1995 *Ap. J.* **448**, L97–L100.

The Galactic dark matter halo: is it H_2?

By P.M.W. Kalberla[1], J. Kerp[1] AND U. Haud[2]

[1]Radioastronomisches Institut der Universität Bonn, Auf dem Hügel 71, 53121 Bonn, Germany

[2]Tartu Observatory, 61602 Toravere, Estonia

We discuss several recent observational results according to which a significant fraction of the Galactic dark matter halo may exist in the form of dense self-gravitating clumps of H_2.

We model the large scale mass distribution in the Milky Way and derive in a self-consistent way the associated gravitational potential. The basic constraint in our model is that the shape of the gaseous Galactic halo, observable from synchrotron radiation, γ-rays, soft X-ray background and from H I lines with high-velocity dispersion needs to be explained by the mass distribution.

The resulting mass model has a flat rotation curve. In the solar vicinity surface density, volume density and gravitational acceleration K_z are consistent with all known constraints. We find that the distributions of H I gas and dark matter are closely related to each other. Furthermore, the mass distribution implies a co-rotation of disk and halo for radii $R > 5$ kpc.

Our analysis supports strongly the hypothesis, that the gaseous halo of the Milky Way traces dark matter in the form of dense self-gravitating H_2 condensations as indicated from an analysis of the γ-ray background observed with EGRET (Shchekinov et al., these proceedings). In addition we find evidence for an extended non-baryonic dark matter halo, which co-exists with the baryonic component. We derive for the Milky Way a baryonic mass fraction of 12 % in close agreement with cosmological predictions.

1. Introduction

Since many decades, the origin of dark matter is one of the major unsolved problems in astrophysics. Here we discuss the proposal, that the Galactic baryonic dark matter may be in the form of self-gravitating molecular gas clumps. On large scales ($R < 350$ kpc) we find evidence for an extended non-baryonic component in the Milky Way.

The hypothesis, that the Galactic dark matter may be in the form of molecular gas was originally proposed by Lequeux et al. (1993) and worked out in detail by Pfenniger & Combes (1994), de Paolis et al. (1994), and by Gerhard & Silk (1996). Recently, Walker & Wardle (1998) argued that the so-called extreme scattering events (ESE), discovered by Fiedler et al. (1987) might be caused by gas clumps with masses of $\sim 10^{-3} M_\odot$ and sizes of a few AU.

A baryonic dark matter component in the form of molecular gas has to be traceable in γ-ray emission due to the interaction of high energy cosmic rays with the gas (Salati et al, 1996, De Paolis et al. 1999, Sciama 1999). The process $p + p \to p + p + \pi^0 \to p + p + 2\gamma$ should lead to a significant γ-ray emission. Such an emission was deduced from the analysis of EGRET data (Dixon et al. 1998), but much weaker as expected. A faint emission, as discussed by Shchekinov et al. (these proceedings) and Kalberla et al. (1999) is consistent with dense molecular clumps, having radii of a few AU, as needed to explain the extreme scattering events. Such molecular dark matter clumps are optically thick with respect to penetrating cosmic rays and γ-ray emission.

Concerning the shape of the dark matter halo, Kalberla et al. (1999) claim that a flattened halo with an axial ratio $q \sim 0.3$ fits the γ-ray data best. A gaseous halo with such a flattening ratio was found recently to be associated with soft X-ray emission (Pietz et al. 1998) and faint H I emission lines with high velocity dispersion (Kalberla et al. 1998). Highly ionized atoms emit UV-lines which appear to share the turbulent

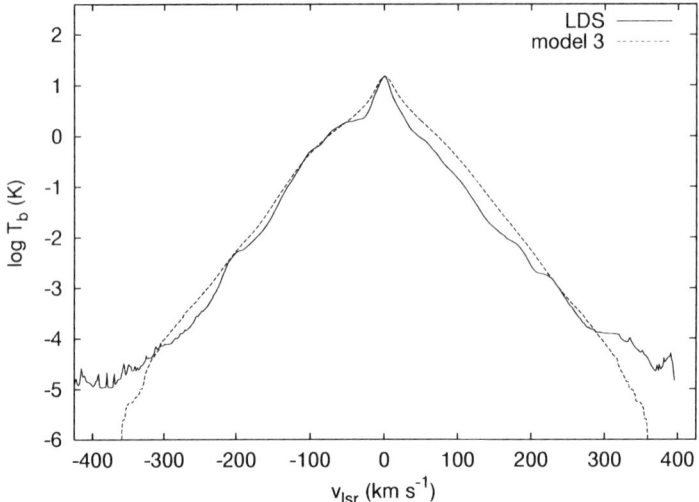

FIGURE 1. Total H I emission for latitudes $b \geq 0°$ derived from the Leiden/Dwingeloo survey (LDS) after Gaussian decomposition (solid line) compared with the emission profile according to the mass model 3 (dashed line). Note the logarithmic scale and the excellent agreement over 5 orders of magnitude up to velocities $|v| \sim 300$ km s^{-1}.

properties of the H I halo component as well as the scale height perpendicular to the disk (Savage et al. 1997). For a gaseous halo with a radial scale length of $h_R \sim 15$ kpc and a scale height of $h_z \sim 4.4$ kpc in hydrostatic equilibrium with the gravitational potential, Kalberla & Kerp (1998) derived pressure equilibrium between gas, magnetic fields and cosmic rays. Accordingly, the major part of the Galactic synchrotron radiation appears to originate from the same volume which is occupied by a low density multiphase gas, but also by dense clumps of molecular gas which apparently produce the extended γ-ray emission. The aim of this contribution is, to discuss, whether these clumps of molecular gas may host the dark matter which is needed to explain the flat rotation curve in the Milky Way.

2. Galactic mass models and gravitational forces

We assume that the dark matter clumps are distributed as an isothermal component and supplement this mass distribution with gas and stellar components in the disk. We calculate for such a mass distribution in a self-consistent way the gravitational acceleration K_R and K_z and derive the rotation curve using the constraint that the rotational velocity at the position of the Sun ($R_\odot = 8.5$ kpc) should be $v_{rot} = 220$ km s^{-1}, as recommended by the IAU. We demand that the parameters for the dark matter distribution and the resulting gravitational potential have to be consistent with the observed density distribution of the low density gaseous halo. These conditions are sufficient to determine uniquely the local density of the baryonic dark matter component (Kalberla et al. 2000).

For our calculations we use 3 different mass models for the stellar distribution: 1) bulge and stellar disk as parameterized by Kuijken & Gilmore (1989), 2) a stellar disk with a lower surface density as determined by Gould et al. (1997) from HST data, and 3) including a central bar as indicated from NIR COBE data. In all cases, we derive a flat rotation curve for galacto-centric radii $R < 25$ kpc.

Stars and molecular gas in the disk are concentrated towards the Galactic center with

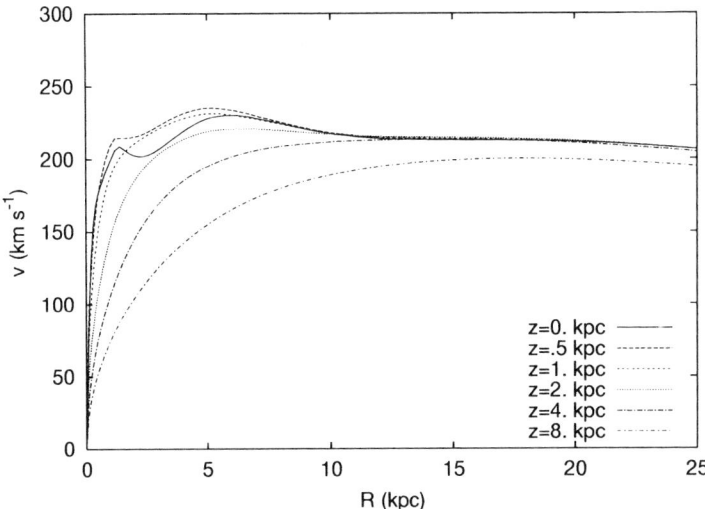

FIGURE 2. Rotation curves derived from model 3 at z distances of 0, 0.5, 1, 2, 4, and 8 kpc above the Galactic plane. The halo co-rotates with the disk, except for layers beyond $z > 8$ kpc, or within the inner Galaxy, $R < 8.5$ kpc. Such a co-rotation was observed previously for the gaseous halo (Savage et al. 1997 and Kalberla et al. 1998).

a radial scale lengths $h_R < 4.5$ kpc. We determine a significantly larger scale length of the H I gas in the disk, $h_R \sim 15$ kpc, *identical* with the radial distribution of the dark matter component. Fig. 1 shows the H I distribution from our model in comparison with the observations. Extragalactic astronomers, observing our Galaxy in H I and measuring the rotation curve will find a "conspiracy" between H I gas and dark matter.

We find another remarkable relation between gas and dark matter. The dynamics between H I disk, halo gas, and mass distribution are closely coupled to each other. The mass distribution obtained by us leads to gravitational forces which cause that most of the halo matter is co-rotating with the disk. In Fig. 2 we display for model 3 (which includes a central bar) the rotation curve (solid line) and in addition curves for positions which are offset by $\Delta z = 0.5, 1, 2, 4$, and 8 kpc from the galactic plane. These rotation curves show that for $R > 5$ kpc the halo is essentially co-rotating with the disk. This velocity field is consistent with the analysis by Savage et al. (1997) and the finding of Kalberla et al. (1998). Contrary, in the inner Galaxy layers with prograde as well as retrograde rotation do exist. A similar behavior was observed by Rand (1997) who positioned the slit of his spectrometer at a distance of 5 kpc perpendicular to the disk of NGC 891.

From our mass model we derive a total mass density in the vicinity of the Sun $\rho_0 \sim 0.09$ $M_\odot pc^{-3}$, which is in good agreement with recent determinations from HIPPARCOS (e.g. Holmberg & Flynn 1999). Also the local surface density for $|z| < 1.1$ kpc is consistent with $\Sigma_{1.1} = 71\ (\pm 6) M_\odot\ pc^{-2}$ as determined by Kuijken & Gilmore (1989). The local gravitational acceleration, K_z, as derived by us, agrees with previous results by Kuijken & Gilmore (1989) and Bienaymé et al. (1987). In summary, our mass model, based on a gaseous isothermal dark matter component, passes all tests regarding observational parameters of the Milky Way. The derived gravitational potential is consistent with the distribution of the low density halo tracer phase as well as the γ-ray distribution as discussed by Shchekinov et al. (these proceedings) and by Kalberla et al. (1999).

3. Summary and conclusion

We have modeled the large scale mass distribution in the Milky Way assuming that the dark matter is represented by an isothermal component associated with a low density gaseous halo. We determined in a self-consistent way mass distribution as well as gravitational potential, considering several different stellar mass distributions and analyzing in detail the distribution of the H I gas in the disk. Our model satisfies all known observational constraints like rotation curve, local gravitational potential, local mass density and scale heights h_z for the tracer gas in the halo. In addition we find a close relation between dark matter halo and H I gas: the radial distributions are identical and the dark matter co-rotates with the gaseous disk. Our conclusion, that gas and dark matter are closely related to each other, supports the model as presented by Shchekinov et al. (these proceedings) according to which the dark matter may be in the form of dense, self-gravitating molecular clumps containing predominantly H_2. We determine the velocity dispersion of these clumps to $\sigma_{DM} = 110 \pm 10$ km s^{-1}.

We derive a total mass for the Milky Way $\sim 2.5 \cdot 10^{11} M_\odot$ within a radius $R \sim 35$ kpc. 71% of this matter is dark. This baryonic mass is however insufficient to explain the large scale mass distribution in the Milky Way. Considering a radius of 350 kpc, half the distance to M31, it can be shown that the dark matter component at large distances from the Galactic plane is best described by a spherical distribution (Kalberla et al. 2000). If we associate this extended spherical component with non-baryonic cold dark matter we are led to the conclusion that the baryonic mass fraction for the Milky Way is 12%. Cosmology determines the average baryonic mass fraction to 13%±1.5% (e.g. Turner 1999).

REFERENCES

BIENAYMÉ O., ROBIN A.C., CRÉZÉ M. 1987 *A&A* **180**, 94
DE PAOLIS, F., INGROSSO, G., JETZER, PH. & RONCADELLI, M. 1994 *Phys. Rev. Lett.* **74**, 14
DE PAOLIS, F., INGROSSO, G., JETZER, PH. & RONCADELLI, M. 1999 *ApJ* **510** L103
DIXON, D. D. ET AL. 1998 *New Astronomy* **3**, 539–561
FIEDLER, R. L., DENNISON, B., JOHNSON, K. J. & HEWISH, A. 1987 *Nat* **326**, 675
GERHARD, O. & SILK, J. 1996 *ApJ* **472**, 34
GOULD A., BAHCALL J.N., FLYNN C. 1997 *ApJ* **482**, 913
HOLMBERG J., FLYNN C. 1999 *astro-ph/9906159*
KALBERLA, P. M. W. & KERP, J. 1998 *A&A* **339**, 745
KALBERLA, P. M. W., KERP, J. & HAUD, U. 2000 *A&A* submitted
KALBERLA, P. M. W. WESTPHALEN G., MEBOLD U., ET AL. 1998 *A&A* **332**, L61
KALBERLA, P. M. W., SHCHEKINOV, YU. A. & DETTMAR, R.-J. 1999 *A&A* **350**, L9
KUIJKEN, K., GILMORE, G. 1989 *MNRAS* **239**, 571
LEQUEUX, J., ALLEN, R.J., GUILLOTEAU, S. 1993 *A&A* **280**, L23
PFENNIGER, D. & COMBES, F. 1994 *A&A* **285**, 94
PIETZ J., KERP J., KALBERLA P.M.W., ET AL. 1998 *A&A* **332**, 55
RAND R.J. 1997 *ApJ* **474**, 129
SALATI, P., ET AL. 1996 *A&A* **313**, 1
SAVAGE B.D., SEMBACH K.R., LU L. 1997 *AJ* **113**, 2158
SCIAMA, D. W. 1999 *astro-ph/9906159*
WALKER M. & WARDLE, M. 1998 *ApJ* **498**, L125
TURNER M. 1999 *PASP* **111**, 264

Observations of H2 in Quasar Absorbers

By Jill BECHTOLD [1]

[1] Steward Observatory, University of Arizona, Tucson, AZ 85721, USA

The ultraviolet Lyman and Werner absorption lines of H_2 have been searched for in a number of high redshift quasar spectra, and detected unambiguously in at least 3 systems at redshifts $z \sim 2$. The lack of detectable H_2 in most absorbers results from the strong selection in quasar studies against lines-of-sight with significant dust extinction. At high redshift, the ultraviolet radiation field is inferred to be higher than that observed in the local solar neighborhood, suggesting that vigorous star-formation is underway in these galaxies.

1. Introduction

Recent observations of the high redshift Universe, interpreted in the context of a new generation of computer simulated model Universes, are providing a clear picture of how large galaxies like the Milky Way formed. A number of different observations suggest that large galaxies were assembled from what appear at $z = 2 - 3$ to be several star-forming proto-galactic fragments (PGF's), widely distributed in space (Windhorst et al. 1994, Pascarelle et al. 1996ab, 1998; Steidel et al. 1996ab, Bechtold et al. 1998). Computer simulations suggest that initially small clumps of material collapsed at the intersection of sheets and filaments in the intergalactic medium, and began forming stars, and that eventually these clumps merged to form large galaxies (Haehnelt, Steinmetz & Rauch 1998, Steinmetz 1998 and references therein). Searches for the galaxies associated with damped Ly-α quasar absorbers show that at $z \sim 2$ they are the same population of objects seen in the Hubble Deep Field faint galaxies and the Lyman dropout galaxies (Steidel et al. 1996ab; Bechtold et al. 1998). They appear to be multiple PGF's, with several clumps of star-formation spread over many megaparsecs.

Since high redshift quasar absorbers are pointers to gas-rich concentrations with active star-formation, they undoubtedly also contain molecular gas. Warm H_2 can be detected through IR emission, but most H_2 is probably contained in a cold phase, measurable by ultraviolet Lyman and Werner absorption (Black & Dalgarno 1976). Shortly after the detection of H_2 absorption in local interstellar clouds with Copernicus in the 1970's (Spitzer et al. 1973), searches for H_2 in quasar absorbers began (Aaronson, Black & McKee 1974; Carlson 1974). However, evidence for molecular gas or dust in quasar absorbers has been scant (Bechtold 1996 and references therein). The inclusion of quasars for absorption studies that are bright and blue selects against lines-of-sight which contain dust, and hence detectable molecular absorption. Note that searches for molecular absorption in the millimeter (reviewed by Combes & Wiklind 1999 and Wiklind & Combes 1999) have been successful only in heavily reddened objects, whose optical/UV continuum is too faint for absorption line spectroscopy.

2. Observations

Table 1 summarizes the searches for H_2 in damped Ly-α quasar absorption systems (from Ge & Bechtold 1999) It includes ground-based spectroscopy, as well as archival data from the *Faint Object Spectrograph* aboard the *Hubble Space Telescope*. Figure 1 shows one of the detections, for the $z(abs) = 2.34$ absorber of the $z(em) = 2.567$ quasar Q1232+0815 obtained at the Multiple Mirror Telescope (MMT). The data and analysis

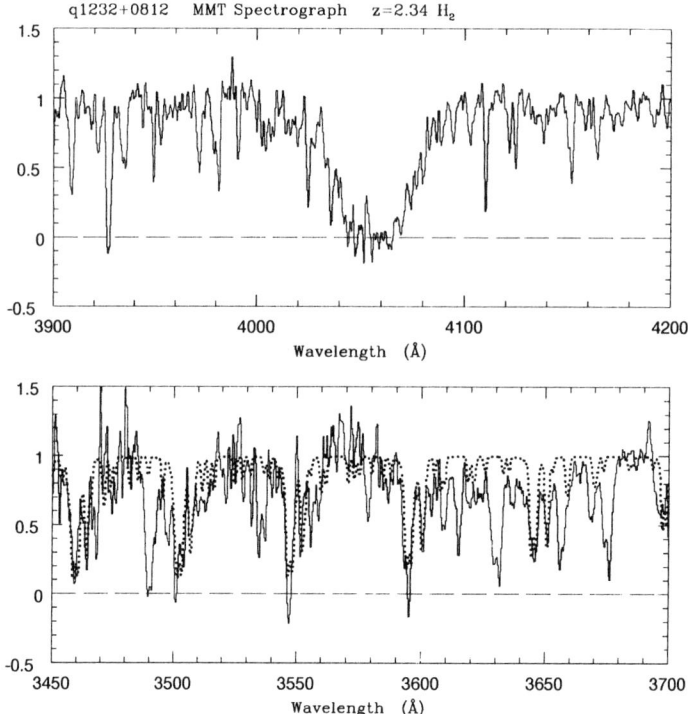

FIGURE 1. Spectrum of QSO 1232+08 obtained with the MMT Blue Spectrograph. Top Panel: Ly-α absorption from intervening system at z=2.34. Bottom Panel: H_2 absorption in the Ly-α forest; dotted lines show model H_2 spectrum. From GBK.

are described in Ge, Bechtold & Kulkarni 2000, hereafter GBK. The top panel shows the Ly-α absorption line, and the bottom panel shows a portion of the molecular hydrogen spectrum, in the Ly-α forest. A cross-correlation analysis with synthetic Ly-α forest spectra shows that the probability that the molecular hydrogen spectrum is produced by the chance superposition of Ly-α forest lines is less than 10^{-4}.

Table 1 gives the H_2 fraction, defined as $f = 2N(H_2)/[2N(H_2) + N(HI)]$ for the quasar absorbers observed to date. The values appear bimodal, with $f > 0.05$ for two cases, and $f \ll 10^{-4}$ for the rest. This is similar to what was observed in the Copernicus survey of 76 stars in the local solar neighborhood by Savage et al. (1977). They found that f undergoes a transition from $f < 0.01$ to $f > 0.01$ at E(B-V)$\sim$0.08 and N(H I + H_2)$\sim 5\times 10^{20}$ cm^{-2}. It isn't possible to derive E(B-V) for quasars in the way one does for stars, but the relative over-abundance of the easily depleted element Cr compared to undepleted Zn, suggests a low dust-to-gas ratio (Table 1). The bimodality of f is understood as the result of self-shielding. Once H_2 begins to form, it shields itself from subsequent photodissociation by ultraviolet radiation, and f increases.

Low extinction is probably not the whole story, however. The lack of molecules in some absorbers may also be the result of an enhanced ultraviolet radiation field. For example, the well-studied absorber of PHL 957 (Q0100+13) has N(HI) = 2.5$\times 10^{21}$ cm^{-3}, the column density is high enough that H_2 should have been detected (Black, Chaffee & Foltz 1987). They suggested that the lack of H_2 in that system resulted from photodissociation by the ultraviolet radiation field produced by the intense star-formation in the galaxy within which the absorption features arise.

Table 1. **Summary of H_2 from Quasar Absorbers**

QSO	z_{ab}	$N(HI)(cm^{-2})$	[Zn/H]	[Cr/Zn]	$f_{H_2}^a$	Ref
1328+30	0.69	2.0×10^{21}	-1.2	-0.6	$< 5 \times 10^{-6}$	1,2
0454+03	0.86	6.0×10^{20}	-0.7	-0.1	$< 8 \times 10^{-5}$	1,3
0935+41	1.37	2.5×10^{20}	-1.1	-0.1	$< 2 \times 10^{-4}$	1,4
0013−00	1.97	6.4×10^{20}	-0.8	-1.2	0.22	1,5
0458−02	2.04	8.0×10^{21}	-1.1	-0.7	$< 2 \times 10^{-6}$	1,5
0100+13	2.30	2.5×10^{21}	-1.55	-0.24	$< 4 \times 10^{-6}$	7,8
1232+08	2.34	8.0×10^{20}	-0.86	-1.04	0.07	1
0112+03	2.42	1.0×10^{21}	-1.0	-0.9	$< 2 \times 10^{-5}$	1,5
1223+17	2.47	3.0×10^{21}	-1.5	-0.6	$< 9 \times 10^{-7}$	1,5
1337+11	2.79	8.0×10^{20}	< -1.0	?	$< 1.3 \times 10^{-4}$	5,9
0528−25	2.81	2.0×10^{21}	-0.76	-0.47	5×10^{-5}	10,11
1946+76	2.84	2.0×10^{20}	-2.5	?	$< 9 \times 10^{-6}$	1,6
0000−26	3.39	2.0×10^{21}	< -1.9	< -0.3	$< 3 \times 10^{-6}$	5,12

[a] Upper limits based on $T_{ex} = 1000$ K, b = 5 km s^{-1}.

REFERENCES: 1 - Ge & Bechtold 1999; 2 - Meyer et al. 1992; 3 - Steidel et al. 1995; 4 - Meyer et al. 1995; 5 - Pettini et al. 1994; 6 - Lu et al. 1995 7 - Wolfe et al. 1994; 8 - Black et al. 1987; 9 - Lanzetta et al. 1989; 10 - Sriannand & Petitjean 1998; 11 - Lu et al. 1996; 12 - Levshakov et al. 1992.

High resolution spectroscopy of the H_2 lines give estimates of excitation temperatures, volume densities (assuming some of the excitation is collisional), and the ultraviolet radiation field which can excite or photodissociate the molecules. Details of these quantities for the quasar absorbers with detected H_2 are given in Sriannand & Petijean (1998) and GBK. Here we discuss the quantity β_0, which is the photodissociation rate of molecules in the Lyman and Werner bands. The local solar neighborhood value is $\beta_0 = 5 \times 10^{-10}$ s^{-1} (Jura 1975ab; Savage et al. 1977). Note that the metagalactic H I ionizing radiation field, which determines the ionization of the Ly-α forest (Bechtold 1994), is irrelevant: for $J_\nu(912$ Å$) = -21$, $\beta_0 = 2 \times 10^{-12}$ s^{-1} (Sriannand and Petitjean 1998).

In the case of Q 0528-25, the absorber redshift is close to the quasar redshift, and the radiation field is very high, possibly due to proximity to the quasar: $\beta_0 = 3 \times 10^{-7}$ s^{-1}. In both cases where the absorber is intervening, the inferred value of β_0 is also high: for q0013-004, $\beta_0 = 6.7 \times 10^{-9}$ s^{-1} (Ge & Bechtold 1999), and for q1232+0815, $\beta_0 = 7 \times 10^{-9}$ s^{-1} (GBK). These last results must be regarded with caution, however, since they are based on moderate resolution data, and β_0 would be systematically *overestimated* if, for example, more than one velocity component is present. If confirmed with high resolution spectra, however, these high values of β_0 compared to the solar neighborhood values suggest that star-formation rates are higher in PGFs at z$\sim$2 than in the present-day Milky Way.

3. Future Prospects

Although all known damped Ly-α absorbers appear to be enriched with heavy elements (Pettini et al. 1999 and references therein), as are the quasars used to find them

(Ferland et al. 1996, Hamann 1999), someday one may be able to find UV bright objects at high enough redshift to probe truly primordial H_2. The chemistry of primordial molecule formation is important for models of the thermal history of the early intergalactic medium since H_2 cooling controls the collapse and fragmentation of the first PGFs. Observations of primordial H_2 would allow interesting measures of the ultraviolet radiation background, which may suppress the formation of low-mass objects (Couchman & Rees 1986; Efstathiou 1992; Haiman, Rees & Loeb 1996).

In the shorter term, we note that to date, the observations of H_2 in damped Ly-α absorbers from the ground have been carried out on 4m-class telescopes which can only reach the brightest, and least reddened objects. With 8m class telescopes, higher dispersion spectra can be obtained of the Lyman and Werner bands, as well as the weaker vibrationally excited H_2 transitions which suffer less from saturation and blending with the Ly-α forest (e.g. Federman et al. 1995). Future very large telescopes, currently being planned, will be able to observe fainter objects, and probe heavily extincted and truly molecular lines-of-sight. The reddened background quasars may be found in all sky surveys in the near-IR and radio, currently underway.

The recent launch of FUSE and the installation of STIS on HST will greatly extend the observations of H_2 in local galaxies and low redshift quasar absorbers. As this review was written, observations were underway with these instruments, but results were not yet available. Spectra of a few lines-of-sight in the LMC and SMC observed with Orpheus II suggest that future observations will provide a rich source of information on H_2 in the interstellar medium in nearby galaxies (Richter et al. 1998, de Boer et al. 1998).

Future missions being studied by NASA, particularly SUVO, a large ultroviolet-optimized telescope with an echelle spectrometer (see Shull 1999), would be ideal for detailed studies of molecular hydrogen in quasar absorbers.

I am indebted to Jian Ge and John H. Black, long time collaborators on this work, for their contributions. I am grateful for encouragement from D. Meyer and A. Stopeck. Support was provided by NSF grant AST-9617060. Observations reported in this paper were obtained at the Multiple Mirror Telescope Observatory, a facility operated jointly by the University of Arizona and the Smithsonian Institution. Support for this work was provided by NASA through grant number AR-05785.01-94A from STScI, which is operated by AURA, Inc., under NASA contract NAS 5-26555.

REFERENCES

AARONSON, M., BLACK, J. H. AND MCKEE, C. F. 1974 *Ap.J.* **191**, L53-L56.

BECHTOLD, J. 1994 *Ap.J.Supp.* **94**, 1.

BECHTOLD, J. 1996 in *Molecules in Astrophysics: Probes and Processes.* ed. E. van Dishoeck, IAU symposium 178, p. 525.

BECHTOLD, J., ELSTON, R., YEE, H. K. C., ELLINGSON, E., AND CUTRI, R. M. 1998 in *The Young Universe: Galaxy Formation and Evolution at Intermediate and High Redshift.* S. D'Odorico, A. Fontana, and E. Giallongo, eds. ASP Conf. Ser. **146**, p.241

BLACK, J. H. AND DALGARNO, A. 1976 *Ap.J.* **203**, 132-142.

BLACK, J. H., CHAFFEE, F. H. AND FOLTZ, C. B. 1987 *Ap.J.* **317**, 442-449.

CARLSON, R. W. 1974 *Ap.J.* **190**, L99-L100.

COMBES, F. AND WIKLIND, T. 1999 in *Highly Redshifted Radio Lines*, ASP Conf. Series **156**, C. L. Carilli, S. J. E. Radford, K. M. Menten, and G. I. Langston, eds. p. 210.

COUCHMAN, H. M. P. AND REES, M. J. 1986 *MNRAS* **221**, 53.

DE BOER, K. S., RICHTER, P., BOMANS, D. J., HEITHAUSEN, A., AND KOORNNEEF, J. 1998 *Astron. Astrophys.* **338**, L5-L8.

EFSTATHIOU, G. 1992 *MNRAS* **256**, P43.

FEDERMAN, S. R., CARDELLI, J. A., VAN DISHOECK, E. F., LAMBERT, D. L. AND BLACK, J. H. 1995 *Ap.J.* **445**, 325-329.

FERLAND, G. J. ET AL. 1996 *Ap.J.* **461**, 683-697.

FOLTZ, C. B., CHAFFEE, F. H. AND BLACK, J. H. 1988 *Ap.J.* **324**, 267-278.

GE, J. AND BECHTOLD, J. 1999 in *Highly Redshifted Radio Lines*, Ibid, p. 121.

GE, J., BECHTOLD, J. AND BLACK, J. H. 1997 *Ap.J.* **477**, L73.

GE, J., BECHTOLD, J. AND KULKARNI, V. 2000 *Ap.J.*, submitted. (GBK)

HAEHNELT, M. G., STEINMETZ, M. AND RAUCH, M. 1998 *Ap.J.* **495**, 647.

HAIMAN, Z., REES, M. J. AND LOEB, A. 1996 *Ap.J.* **476**, 458.

HAMANN, F. 1999 in *Structure and Kinematics of Quasar Broad Line Regions* ASP Conference Series, **175** Ed. C. M. Gaskell, W. N. Brandt, M. Dietrich, D. Dultzin-Hacyan, and M. Eracleous, p.33.

JURA, M. 1975a *Ap.J.* **197**, 575-580.

JURA, M. 1975b *Ap.J.* **197**, 581-586.

LANZETTA, K. M. ET AL. 1989 *Ap.J.* **344**, 277.

LEVSHAKOV, S. A. ET AL. 1992 *Astron. Ap.* **262**, 385.

LU, L. SAVAGE, B. D., TRIPP, T. D. & MEYER, D. M. 1995 *Ap.J.* **447**, 597.

LU, L. ET AL. 1996 *Ap.J.Supp.* **107**, 475.

MEYER, D. M. ET AL. 1992 *Ap.J.* **399**, L121.

MEYER, D. M., LANZETTA, K. M. AND WOLFE, A. M. 1995 *Ap.J.* **451**, L13.

PASCARELLE, S. M., WINDHORST, R. A., DRIVER, S. P., OSTRANDER, E. J. AND KEEL, W. C. 1996a *Ap. J.* **456**, L21-L24.

PASCARELLE, S. M., WINDHORST, R. A., KEEL, W. C. AND ODEWAHN, S.C. 1996b *Nature* **383**, 45-50.

PASCARELLE, S. M., WINDHORST, R. A. AND KEEL, W. C. 1998 *A. J.* **116**, 2659-2666.

PETTINI, M. ET AL. 1994 *Ap. J.* **46**, 79.

PETTINI, M., ELLISON, S. L., STEIDEL, C. C., BOWEN, D. V. 1999 *Ap. J.* **510**, 576-589.

RICHTER, P. ET AL. 1998 *Astron. Astrophys.* **338**, L9-L12.

SAVAGE, B. D., BOHLIN, R. C., DRAKE, J. F. AND BUDICH, W. 1977 *Ap.J.* **216**, 291-307.

SHULL, J. M. 1999 in *Ultraviolet-Optical Space Astronomy Beyond HST*, ASP Conf. Ser. **164**, J. A. Morse, J. M. Shull and A. L. Kinney, eds., 229-233.

SPITZER, L., DRAKE, J. F., JENKINS, E. B., MORTON, D. C., ROGERSON, J. B. AND YORK, D. G. 1973 *Ap.J.* **181**, L116-L121.

SRIANNAND, R. AND PETITJEAN, P. 1998 *Astron. Astrophys.* **335** 33-40.

STEIDEL, C. C. ET AL. 1995 *Ap. J.* **440**, L45.

STEIDEL, C. C., GIAVALISCO, M., DICKINSON, M., AND ADELBERGER, K. L. 1996a *A. J.* **112**, 352-359.

STEIDEL, C. C., GIAVALISCO, M., PETTINI, M., DICKINSON, M., AND ADELBERGER, K. L. 1996b *Ap.J.* **462** L17-L21.

STEINMETZ, M. 1998 in *Structure et Evolution du Milieu Inter-Galactique Revele par Raies D'Absorption dans le Spectre des Quasars*, 13th Colloque d'Astrophysique de l'Institut d'Astrophysique de Paris, Ed. Patrick Petitjean and Stephane Charlot, p.281.

WIKLIND, T. AND COMBES, F. 1999 in *Highly Redshifted Radio Lines*, Ibid.,p. 202.

WINDHORST, R. A. ET AL. 1994 *Ap. J.* **435**, 577-598.

WOLFE, A. M. ET AL. 1994 *Ap. J.* **435**, 101.

The participants preparing for the group photo.

H_2 emission as a diagnostic of physical processes in starforming galaxies

By Paul P. VAN DER WERF[1]

[1]Leiden Observatory, P.O. Box 9513, NL - 2300 RA Leiden, The Netherlands

Observations and interpretation of extragalactic rotational and rovibrational H_2 emission are reviewed. Direct observations of H_2 lines do not trace bulk H_2 mass, but excitation rate. As such, the H_2 lines are unique diagnostics, if the excitation mechanism can be determined, which generally requires high-quality spectroscopy and suitable additional data. The diagnostic power of the H_2 lines is illustrated by two cases studies: H_2 purely rotational line emission from the disk of the nearby spiral galaxy NGC 891 and high resolution imaging and spectroscopy of H_2 vibrational line emission from the luminous merger NGC 6240.

1. Introduction

Direct observations of H_2 emission from external galaxies have become standard practice in the past decade through the revolution in ground-based near-infrared instrumentation. As a result, the near-infrared H_2 rovibrational lines are now readily detectable throughout the local universe (e.g., Moorwood & Oliva 1988, 1990; Puxley *et al.* 1988, 1990; Goldader *et al.* 1995, 1997; Vanzi *et al.* 1998). More recently, the Short Wavelength Spectrograph (SWS) on the Infrared Space Observatory (ISO) has for the first time allowed detection of the purely rotational H_2 lines in the mid-infrared spectral regime. For instance, the first detection (outside the solar system) of the H_2 S(0) line at 28.21 μm was reported by Valentijn *et al.* (1996) from the star forming nucleus of the nearby spiral galaxy NGC 6946.

The interpretation of these data is, however, far from trivial. At typical molecular cloud temperatures ($T \sim 20$ K) the upper levels of even the lowest H_2 transitions are essentially un-populated, and hence H_2 emission is, unlike for instance CO $J = 1{\to}0$ emission, not a tracer of bulk molecular gas mass. Instead, an excitation mechanism capable of populating these energy levels is required. Furthermore, the excitation rate needs to be high enough to maintain an excited level population sufficient for producing detectable emission. Hence the H_2 line luminosities measure the *rate of excitation*; as such they provide unique and highly diagnostic information, that cannot be obtained in any other way. In addition, since the H_2 lines are forbidden quadrupole transitions with very small Einstein A values, the lines are optically thin; hence there is, in contrast to the situation with most other molecular lines, no need to solve a complicated, geometry-dependent radiative transfer problem for a physical interpretation of the H_2 lines. The near-infrared H_2 lines will of course suffer from extinction by dust, but this effect can usually be quantified adequately by using suitable ratios of hydrogen recombination lines, [Fe II] lines or H_2 rovibrational lines in the same spectral range, with accurately known intrinsic flux ratios.

The situation is complicated, however, by the fact that H_2 excitation can be brought about by a variety of mechanisms, which may all play a role. For instance, in a starburst galaxy, H_2 emission may be generated by UV-pumping in starforming regions or by shocks due to supernova remnants or outflows; moreover, these processes are expected to occur together in the same small volume occupied by the starburst, and thus a combination of excitation mechanisms may be expected at every position. Generally, this combination will be difficult to separate (e.g., NGC 253 observations: Forbes *et al.* 1993; Engelbracht *et al.* 1998). In addition, shocks due to large-scale streaming motions in spiral arms, bars (favoured for the barred Seyfert 2 NGC 1068 by Tacconi *et al.* 1994) or merger-driven flows (e.g., in NGC 6240, Van der Werf *et al.* 1993; see also § 3) may play a role. Furthermore, X-ray excitation may be produced by multiple supernova remnants, or, if present, by an active galactic nucleus (proposed for e.g., Cyg A by Ward *et al.* 1991 and for Cen A by Bryant & Hunstead 1999) or a cooling flow (Jaffe & Bremer 1997). In the absence of sufficient spatial resolution for separating the various excitation mechanisms, spectral diagnostics must be used. Models for UV-pumped, shocked or X-ray excited H_2 emission have reached considerable predictive power; however, the densities of the emitting regions are often sufficiently high to thermalize the relevant level populations, quenching the typical signatures of non-thermal excitation processes. The only accessible diagnostics are then fluxes of lines with very high critical densities. However, these lines are intrinsically faint, and accurate fluxes for such lines require long integrations with large telescopes. The complexity of the problem is well illustrated by the H_2 emission of the merging system NGC 6240, which has been attributed to X-ray excitation (Draine & Woods 1990), UV-pumping (Tanaka *et al.* 1991), shocks (Van der Werf *et al.* 1993) and formation pumping (Mouri & Taniguchi 1995). However, all of the non-thermal excitation processes relied on poorly measured fluxes of faint lines. More recent spectroscopy of NGC 6240 by Sugai *et al.* (1997) showed that only shock excitation can account for the H_2 rovibrational spectrum (see also § 3).

Generally a combination of accurate multi-line spectroscopy and suitable additional information such as high resolution spatial information is needed for a proper analysis of the dominant excitation mechanism and hence a physical analysis of the H_2 emission. A complicating factor when combining H_2 rovibrational and purely rotational lines is the fact that the dominant excitation mechanisms of these lines may be different, because of the different excitation requirements of these transitions: while the lowest rotational lines require $T > 100$ K and are excited at moderate densities ($n_{H_2} > 10^2$ cm^{-3}), the rovibrational lines require $T > 2000$ K and $n_{H_2} > 10^4$ cm^{-3}. The following sections will therefore present two case studies. First an analysis of the extended purely rotational H_2 emission from the nearby edge-on spiral galaxy NGC 891 is given (§ 2). Then, the more extreme conditions in the merger NGC 6240 are discussed, using new high resolution near-infrared data. The general implications of these two cases are discussed in § 4.

2. Extended H_2 rotational line emission in the disk of NGC 891

Valentijn & Van der Werf (1999a) have observed seven positions in the disk of the nearby (distance $D = 9.5$ Mpc) edge-on spiral galaxy NGC 891 with the ISO SWS. The positions observed include the nucleus and positions spaced along the disk to galactocentric distances R of 8 kpc south of the nucleus and 11 kpc north of the nucleus. At the 11 kpc north position, the ^{12}CO $J = 1 \rightarrow 0$ (Garcia-Burillo *et al.* 1992; Scoville *et al.* 1993; Sakamoto *et al.* 1997), and dust emission (Israel *et al.* 1999) are barely detected. With ISO, the H_2 S(0) (28.21 μm) and S(1) (17.03 μm) lines were detected at all positions. These are the first detections of these lines outside starburst or active nuclei, with the exception of the detection in the disk of NGC 6946 reported by Valentijn & Van der Werf (1999b).

The simplest analysis of these data would assume that the emission is fully thermalized (i.e., the high-density limit is assumed) and arises from an isothermal gas layer with an ortho-para ratio of three, in agreement with the statistical weights of the ortho- and para-varieties. Under these assumptions the S(1)/S(0) ratio yields the temperature of the emitting region, which can be combined with the observed fluxes to give the

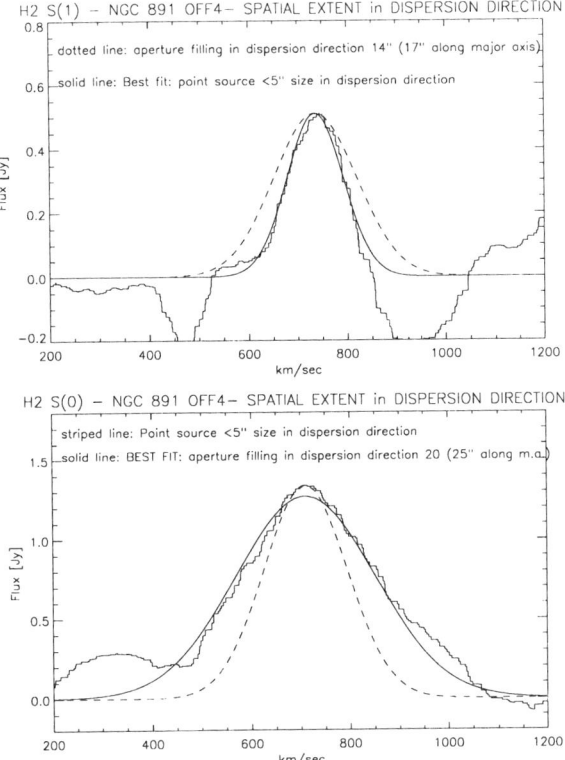

FIGURE 1. H$_2$ S(1) (upper panel) and S(0) (lower panel) lines from $R = 8$ kpc south in the disk of NGC 891, with fits for a point source and an aperture-filling source as indicated.

column density of emitting H$_2$, averaged over the SWS aperture. However, these simple assumptions lead to unacceptable results. For instance, at the position 8 kpc north (a typical "disk" position), the S(0) and S(1) data then imply $N(\mathrm{H}_2) = 2.7 \cdot 10^{22}$ at $T = 76$ K. This result is incompatible with the CO $J = 2{\rightarrow}1/J = 1{\rightarrow}0$ line ratio of 0.75, at this position (Garcia-Burillo et al. 1992), which strongly rules out a large mass of dense molecular gas at $T = 76$ K. This argument shows that the simple assumptions used above do not suffice.

The discrepancy can be removed by allowing lower densities, and therefore subthermal excitation of the H$_2$ $J=3$ level, giving rise to a higher implied temperature for a given S(1)/S(0) ratio. Because the temperatures involved are still much lower than the upper levels of the transitions involved, the implied H$_2$ mass will be extremely sensitive to the adopted temperature, and thus strongly decrease as lower densities are allowed. An additional advantage of this procedure is that the CO $J=2$ level will now also be subthermally populated, so that CO $J = 2{\rightarrow}1/J = 1{\rightarrow}0$ line ratios lower than the thermal value can be tolerated in the analysis. Indeed the multi-level multi-isotope CO data by Garcia-Burillo et al. (1992) indicate dominant densities of $n_{\mathrm{H}_2} \sim 10^3$ cm^{-3} in the nuclear region and $n_{\mathrm{H}_2} \sim 200$ cm^{-3} throughout the disk of NGC 891. Adopting these densities, the ISO data imply H$_2$ column densities varying from 10 to 30% of the value derived from CO, at temperatures of 120 to 130 K, throughout the inner disk ($R < 8$ kpc) of NGC 891.

At the outer disk positions ($R = 8 - 11$ kpc) however, this solution is not adequate, since the S(0) and S(1) line profiles at these positions are significantly different. The

SWS uses an aperture much larger than the diffraction beam and hence the observed line width depends on the extent of the emission region; the S(0) line shows the broad profile characteristic of aperture-filling emission, while the S(1) profile is narrow, indicating emission from a region much smaller than the aperture (Fig. 1). Thus the S(0) and S(1) lines in the outer disk of NGC 891 must originate in separate regions: a warm component ($T > 130$ K) dominating the S(1) emission and located in isolated regions in the disk and a separate more pervasive component, dominating the S(0) line. The latter component may contain very significant mass, depending on its temperature, and the implications of this possibility for the mass budget at the outer positions have been explored by Valentijn & Van der Werf (1999a). However, a solution where the component dominating the S(0) line is cool ($T \sim 90$ K) is problematic, since the heating required for maintaining a very large mass of H_2 at $T \sim 90$ K cannot be provided. In thermal equilibrium the heating rate should equal the cooling rate, which is dominated by [C II] 158 μm emission (which at $T \sim 90$ K is an extremely efficient cooler, since the upper level of the 158 μm transition is at 91 K), with possible contributions from H_2 rotational lines, CO rotational lines, and [C I] emission. The [C II] 158 μm emission from NGC 891 has been observed by Madden et al. (1994), and it is easily verified that in the outer disk the [C II] emission dominates the H_2 emission (SWS data discussed here), CO emission (barely detected in the outer disk) and [C I] emission (Gérin & Phillips 1997). If all available carbon (assumed to have solar abundance) is in the form of C^+, the [C II] emission allows a column density of at most $N(H_2) \sim 7 \cdot 10^{21}$ cm^{-2} at $T \sim 90$ K in the outer disk, averaged over the SWS aperture, while $N(H_2) \sim 10^{23}$ cm^{-2} would be required to produce the observed S(0) line flux.

This problem can be solved by relaxing the final assumption: that of an ortho-para of three. A lower ortho-para ratio raises the implied temperature and lowers the implied mass in the same way as a lower density does (as discussed above). Assuming an ortho-para ratio of unity, the implied mass is lowered sufficiently that the required heating can be accounted for.

This analysis thus favours an interpretation in which the H_2 rotational emission arises in low-density ($n_{H_2} \sim 200$ cm^{-3}), warm ($T \sim 120$ K) molecular gas with an ortho-para ratio of about unity, which pervades the disk of NGC 891 at least to the end of the detectable CO disk and dominates the S(0) emission; it contains 10 to 30% of the mass implied by CO observations (lower fractions will result if an ortho-para ratio below unity is adopted); concentrations of warmer gas (plausibly identified with active star forming regions) are ubiquitous in the inner disk and rarer in the outer disk and give rise to the S(1) emission. Given the density and temperature of the extended warm gas, this component can most likely be identified with extended low-density photon-dominated regions (PDRs) which form the warm envelopes of giant molecular clouds, heated by the local interstellar radiation field. Observations of Galactic molecular cloud edges in the 21 cm H I line reveal the presence of such warm envelopes by a "limb brightening" in H I (Wannier et al. 1983; Van der Werf et al. 1988, 1989). Modeling of molecular cloud envelopes by Andersson & Wannier (1993) shows that the temperatures and densities estimated here are typical for such regions, especially in the zone where the transition from molecular to atomic hydrogen takes place. This interpretation of the present H_2 data in terms of extended diffuse PDRs in cloud envelopes is corroborated by the excellent numerical agreement with column density estimates based on [C II] 158 μm, the principal coolant for such regions. In addition, the fact that the implied ortho-para ratio is significantly lower than three also points to a low-density PDR origin of this emission, since shocks or hot, high-density PDRs would produce an ortho-para ratio of three (the high temperature thermal value) by spin exchange reactions with H and H^+.

In summary, the H$_2$ rotational lines in the disk of NGC 891 arise in warm, extended molecular cloud envelopes; these envelopes provide a physical link between the giant molecular clouds and the atomic medium in which they are embedded (Chromey et al. 1989). The H$_2$ rotational lines provide unique diagnostics of these regions.

3. Vibrational H$_2$ emission in the nuclear region of the luminous merger NGC 6240

The vibrational H$_2$ emission of NGC 6240 has attracted attention because of its extraordinary luminosity: $7 \cdot 10^7$ L$_\odot$ in the H$_2$ $v = 1 \to 0$ S(1) line alone (for $H_0 = 75$ km s^{-1} Mpc^{-1} and with no correction for extinction). This line contains 0.012% of the bolometric luminosity of NGC 6240, which is considerably higher than any other galaxy (Van der Werf et al. 1993). Together the vibrational lines may account for 0.1% of the total bolometric luminosity.

Imaging of the H$_2$ $v = 1 \to 0$ S(1) emission from NGC 6240 has shown that the H$_2$ emission peaks *between* the two remnant nuclei of the merging system (Van der Werf et al. 1993). This morphology provides a unique constraint on the excitation mechanism, since it argues against any scenario where the excitation is dominated by the stellar component (e.g., UV-pumping, excitation by shocks or X-rays from supernova remnants). Instead, the favoured excitation mechanism is slow shocks in the nuclear gas component, which, as shown by recent high resolution interferometry in the CO $J = 2 \to 1$ line (Tacconi et al. 1999), also peaks between the nuclei of NGC 6240. A multi-line H$_2$ vibrational spectrum (Fig. 2) confirms this excitation mechanism (Van der Werf et al. 2000), in agreement with more limited spectroscopy by Sugai et al. (1997). The interpretation in terms of slow shocks also naturally accounts for the high ratio of H$_2$ line emission to infrared continuum emission, which is a characteristic of such shocks (e.g., Draine et al. 1983).

What is the role of these shocks? In the shocks mechanical energy is dissipated and radiated away, mostly in spectral lines (principally H$_2$, CO, H$_2$O and [O I] lines). This energy is radiated away at the expense of the orbital energy of the molecular clouds in the central potential well. Consequently, the dissipation of mechanical energy by the shocks will give rise to an infall of molecular gas to the centre of the potential well. Therefore, *the H$_2$ vibrational lines measure the rate of infall of molecular gas* into the central potential well in NGC 6240. This conclusion can be quantified by writing

$$L_{\rm rad} = L_{\rm dis}, \tag{3.1}$$

where $L_{\rm rad}$ is the total luminosity radiated by the shocks and $L_{\rm dis}$ the dissipation rate of mechanical energy, giving rise to a molecular gas infall rate $\dot{M}_{\rm H_2}$ given by

$$L_{\rm dis} = \frac{1}{2} \dot{M}_{\rm H_2} v^2, \tag{3.2}$$

where v is the circular orbital velocity at the position where the shock occurs.

Using a K-band extinction of $0\overset{m}{.}15$ (Van der Werf et al. 1993), the total luminosity of H$_2$ vibrational lines from NGC 6240 becomes $7.2 \cdot 10^8$ L$_\odot$; inclusion of the purely rotational lines observed with the ISO SWS approximately doubles this number, so that $L_{\rm rad} = 1.5 \cdot 10^9$ L$_\odot$.

In order to use this number to estimate $\dot{M}_{\rm H_2}$, it is necessary to establish more accurately the fraction of the H$_2$ emission that is due to infalling gas. Observations with NICMOS on the Hubble Space Telescope (HST) by Van der Werf et al. (2000) provide the required information (Fig. 3). The NICMOS image shows that the emission consists of a number of tails (presumably related to the superwind also observed in Hα emission),

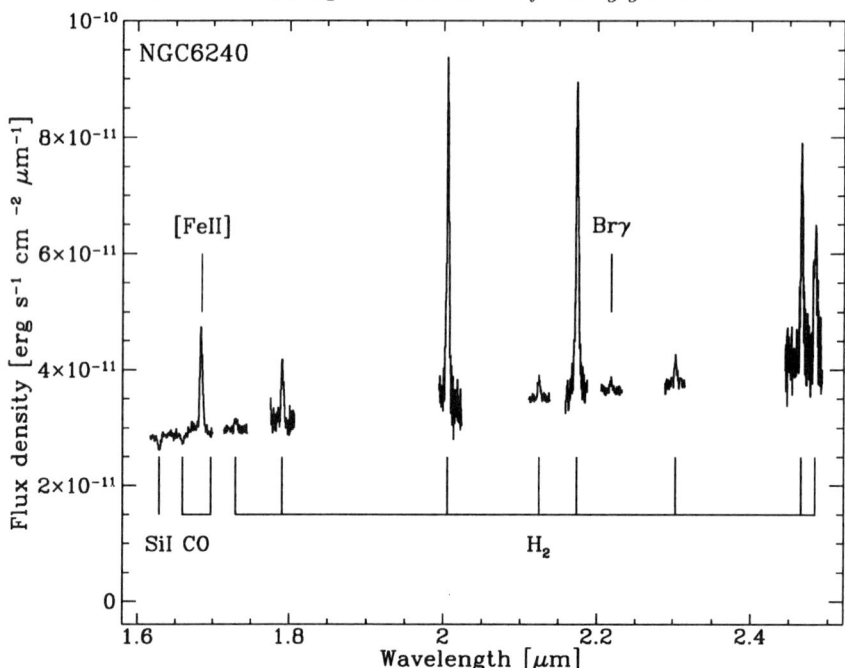

FIGURE 2. $H+K$-band spectrum of NGC 6240 showing H_2 lines up to the $v = 1 \to 0$ S(9) line (Van der Werf et al. 2000).

and concentrations associated with the two nuclei, and a further concentration approximately (but not precisely) between the two nuclei. The relative brightness of the H_2 emission from the southern nucleus is deceptive, since this nucleus is much better centred in the filter that was used for these observations than the other emission components, in particular the northern nucleus. Taking this effect into account, it is found that 32% of the total H_2 flux is associated with the southern nucleus, 16% is associated with the northern nucleus, and 12% with the component between the two nuclei, the remaining 40% being associated with extended emission. Using inclination-corrected circular velocities (from Tecza 1999) of 270 and 360 km s^{-1} for the southern en northern nucleus respectively, and of 280 km s^{-1} for the central component (Tacconi et al. 1999), the mass infall rates derived using Eqs. (3.1) and (3.2) are 80 $M_\odot$ yr^{-1} for the southern nucleus, 22 $M_\odot$ yr^{-1} for the northern nucleus and 28 $M_\odot$ yr^{-1} for the central component.

The derived molecular gas inflow rate to the two nuclei is remarkably close to the mass consumption rate by star formation of approximately 60 $M_\odot$ yr^{-1}, indicating that the H_2 emission from the nuclei directly measures the fueling of the starbursts in these regions. The gas inflow towards the central component is more remarkable, since this component is not associated with a prominent stellar component. The gravitational potential at this position is therefore most likely dominated by the gas itself, which is also indicated by high resolution interferometry in the CO $J = 2 \to 1$ line (Tacconi et al. 1999). The absence of prominent star formation at this position shows that the central gas component is gravitationally stabilized, possibly by a high local shear. However, as the central gas column density increases, the shear must eventually be overcome and given the high gas density (and hence short free-fall time) an explosive starburst will result. In that stage NGC 6240 will become a true ultraluminous infrared galaxy.

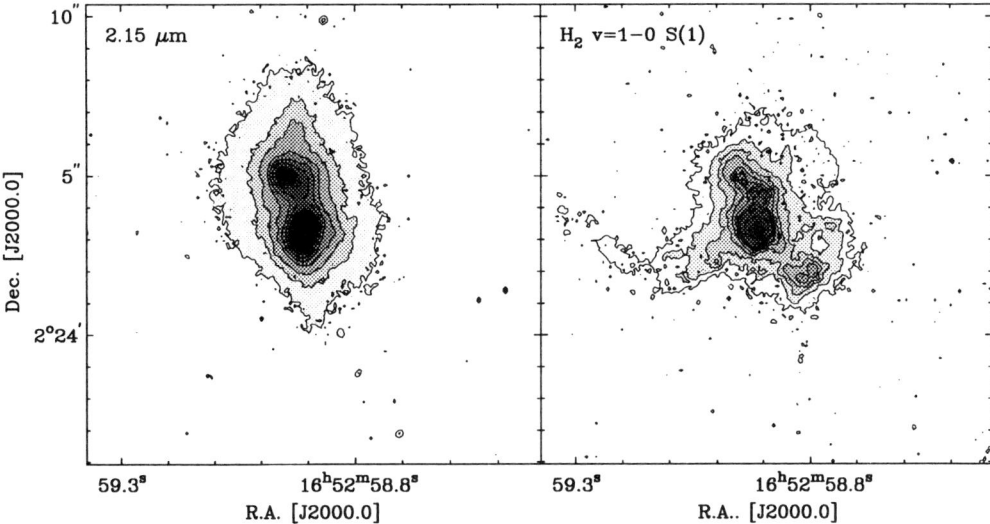

FIGURE 3. High resolution imaging of NGC 6240 in the H_2 $v = 1 \to 0$ S(1) line and the 2.15 μm continuum with NICMOS/HST (Van der Werf et al. 2000).

4. General implications

The two case studies discussed above illustrate the unique diagnostic power of H_2 emission lines provided that the excitation mechanism can be determined. The extended PDR emission detected in NGC 891 reveals the presence of a widespread fluorescent component, which is probably a common feature of star forming galaxies. For instance, Pak et al. (1996) have detected extended diffuse H_2 vibrational emission from the inner Milky Way and argue that this emission is UV-excited; similarly, Harrison et al. (1998) argue for purely fluorescent extended H_2 emission in the moderate luminosity starburst galaxy NGC 253.

The configuration of shocked H_2 emission from a pronounced molecular gas concentration between the two nuclei as found in NGC 6240 is not unique either: it is also found in Arp 220 (Van der Werf & Israel 2000), and several other mergers. Perhaps the best case in point is the well-known merging "Antennae" system (NGC 4038 – 4039) where pronounced CO emission is found in the interaction zone between the two nuclei (Stanford et al. 1990), which is the site of vigorous obscured star formation (Mirabel et al. 1998). These geometries may be due to the fact that the gas components are dissipative, and thus merge on a shorter time scale than the dissipationless stellar components, which merge by the slower process of dynamical friction. The H_2 emission thus provides unique insight into the physics of these gas-rich mergers.

I would like to thank my collaborators Frank Israel, Alan Moorwood, Tino Oliva, and Edwin Valentijn for discussions on this subject, and Guido Kosters for reducing the NICMOS data of NGC 6240.

REFERENCES

ANDERSSON, B.G., & WANNIER, P.G. 1993, ApJ, 402, 585
BRYANT, J.J., & HUNSTEAD, R.W. 1999, MNRAS, 308, 431
CHROMEY, F.R., ELMEGREEN, B.G., & ELMEGREEN, D.M. 1989, AJ, 98, 2203
DRAINE, B.T., & WOODS, D.T. 1990, ApJ, 363, 464

Draine, B.T., Roberge, W.G., & Dalgarno, A. 1983, ApJ, 264, 485

Engelbracht, C.W., Rieke, M.J., Rieke, G.H., Kelly, D.M., & Achtermann, J.M. 1998, ApJ, 505, 639

Forbes, D.A., Ward, M.J., Rotaciuc, V., Blietz, M., Genzel, R., Drapatz, S., Van der Werf, P.P., & Krabbe, A. 1993, ApJ, 406, L11

Garcia-Burillo, S., Guélin, M., Cernicharo, J., & Dahlem, M. 1992, A&A, 266, 21

Gérin, M., & Phillips, T.G. 1997, Neutral carbon in external galaxies. In: Wilson, A. (ed.), The far infrared and submillimetre universe, ESA SP-401, p. 105

Goldader, J.D., Joseph, R.D., Doyon, R., & Sanders, D.B. 1995, ApJ, 444, 97

Goldader, J.D., Joseph, R.D., Doyon, R., & Sanders, D.B. 1997, ApJS, 108, 449

Harrison, A., Puxley, P., Russell, A., & Brand, P. 1998, MNRAS, 297, 624

Israel, F.P., Van der Werf, P.P., & Tilanus, R.P.J. 1999, A&A, 344, L83

Jaffe, W., & Bremer, M.N. 1997, MNRAS, 284, L1

Madden, S.C., Geis, N., Genzel, R., Herrmann, F., Poglitsch, A., Stacey, G.J., & Townes, C.H. 1994, Infrared Phys. Technol., 35, 311

Mirabel, I.F., et al. 1998, A&A, 333, L1

Moorwood, A.F.M., & Oliva, E. 1988, A&A, 203, 278

Moorwood, A.F.M., & Oliva, E. 1990, A&A, 239, 78

Mouri, H., & Taniguchi, Y. 1995, ApJ, 449, 134

Pak, S., Jaffe, D.T., & Keller, L.D. 1996, ApJ, 457, L43

Puxley, P.J., Hawarden, T.G., & Mountain, C.M. 1988, MNRAS, 234, 29P

Puxley, P.J., Hawarden, T.G., & Mountain, C.M. 1990, ApJ, 364, 77 (erratum ApJ 372, 73 (1991))

Sakamoto, S., Handa, T., Sofue, Y., Honma, M., & Sorai, K. 1997, ApJ, 475, 134

Scoville, N.Z., Thakkar, D., Carlstrom, J.E., & Sargent, A.I. 1993, ApJ, 404, L59

Stanford, S.A., Sargent, A.I., Sanders, D.B., & Scoville, N.Z. 1990, ApJ, 349, 492

Sugai, H., Malkan, M.A., Ward, M.J., Davies, R.I., & McLean, I.S. 1997, ApJ, 481, 186

Tacconi, L.J., Genzel, R., Blietz, M., Cameron, M., Harris, A.I., & Madden, S. 1994, ApJ, 426, L77

Tacconi, L.J., Genzel, R., Tecza, M., Gallimore, J.F., Downes, D., & Scoville, N.Z. 1999, ApJ, 524, 732

Tanaka, M., Hasegawa, T., & Gatley, I. 1991, ApJ, 374, 516

Tecza, M., 1999, PhD thesis, Ludwig-Maximilians University, Munich, Germany

Valentijn, E.A., & Van der Werf, P.P. 1999a, ApJ, 522, L29

Valentijn, E., & Van der Werf, P., 1999b, The ISO SWS survey for molecular hydrogen in galaxies. In: Cox, P., & Kessler, M. (eds.), The universe as seen by ISO, ESA Sp-427 2, p. 821

Valentijn, E.A., Van der Werf, P.P., De Graauw, T., & De Jong, T. 1996, A&A, 315, L145

Van der Werf, P.P., & Israel, F.P. 2000, in preparation

Van der Werf, P.P., Goss, W.M., & Vanden Bout, P.A. 1988, A&A, 201, 311

Van der Werf, P.P., Dewdney, P.E., Goss, W.M., & Vanden Bout, P.A. 1989, A&A, 216, 215

Van der Werf, P.P., Genzel, R., Krabbe, A., Blietz, M., Lutz, D., Drapatz, S., Ward, M.J., & Forbes, D.A. 1993, ApJ, 405, 522

Van der Werf, P.P., Moorwood, A.F.M., & Israel, F.P. 2000, in preparation

Vanzi, L., Alonso-Herrero, A., & Rieke, G.H. 1998, ApJ, 504, 93

Wannier, P.G., Lichten, S.M., & Morris, M. 1983, ApJ, 268, 727

Ward, M.J., Blanco, P.R., Wilson, A.S., & Nishida, M. 1991, ApJ, 382, 115

5. Outlook

John Black and Craig Kulesa enjoying tropical drinks behind the palm tree.

H_2 in the Universe: Perspectives

By John H. BLACK

Onsala Space Observatory, Chalmers University of Technology, S-439 92 Onsala, Sweden

From the broadest perspective, the hydrogen molecule is found virtually everywhere in the Universe. Some issues concerning H_2 in space clearly deserve more attention. For example, can traces of the formation process of H_2 in the interstellar medium be observed? Is it possible for large quantities of very cold H_2 to escape detection? How can H_2 be used to probe gas at high redshift and in the centers of active galaxies?

1. Introduction

The hydrogen molecule plays myriad roles on the cosmic stage. During this conference, we have been reminded how the study of H_2 ranges widely in space, time, and energy: from the microcosm of molecular processes to the giant molecular clouds, from the origin of structure in the early universe to places where a star will form tomorrow. We marvel at speculations about clumpuscules of H_2 that might be as cold as 3 K yet contribute measurably to gamma radiation from the Galactic halo. In trying to offer a forward-looking perspective on H_2 in space, it seems best to concentrate on a few topics where rapid progress in observation is taking place and where the interpretation of existing results is inadequate.

2. On the interpretation of astronomical spectra of H_2

Dilute matter in space generally exists in a chemical and physical state far out of thermodynamic equilibrium. The state of such dilute matter reflects a competition among microscopic processes, which often operate in contact with several thermal reservoirs at different effective temperatures.

The spectrum of H_2 in an astronomical source results in principle from a great variety of processes, which govern the excitation by coalition. in particular, the microscopic rates of formation and destruction, of absorption and spontaneous emission of ultraviolet light, of inelastic collisions, and of reactive collisions can often have comparable values. Thus emission mechanisms may be mixed in the same sources, so that it is not instructive to seek a clean division between ultraviolet-excited and collisionally excited sources. The hydrogen molecule is a valuable spectroscopic diagnostic probe precisely because so many competing mechanisms contribute to its excitation.

It is now known that temperatures rivalling those of molecular shocks ($T \approx 10^3$ K) can be attained in photon-dominated regions (PDRs) in steady state (Hollenbach & Tielens 1997); therefore, one should not conclude that quasi-thermal excitation at $T \approx 10^3$ K implies shocked gas unless the kinematical signatures of such a shock are also present. There has also been a tendency to interpret the relative abundances of ortho and para levels in a few highly excited states as though they apply directly to the most abundant levels of the $v = 0$ ground vibrational state. This is dangerous, as pointed out recently by Sternberg and Neufeld (1999). In general, we expect the ortho/para ratio in H_2 to be controlled by the competition between (1) reactive processes that can alter the nuclear-spin state and (2) the processes of formation (which set the initial value) and destruction (which limit the time-scale over which conversion must be efficient).

3. H_2 is everywhere

It is worthwhile to summarize the range of environments where H_2 is important. On Earth, of course, the hydrogen molecule remains fascinating for spectroscopists and molecular physicists: it poses challenges for both accurate measurement and *ab initio* calculation.

3.1. *The solar system*

The atmospheres of the Jovian planets are composed mainly of H_2, which can be detected through its collision-induced absorption. In some cases, e.g. the thermosphere of Uranus (Trafton et al. 1999), the vibration-rotation lines of H_2 appear in emission and carry interesting information about the structure and thermal balance of the atmosphere. A good understanding of the opacity of H_2 in planetary atmospheres has natural extensions to the study of the atmospheres of brown dwarfs and of giant extrasolar planets. Indeed, giant planets lying closer than 1 AU to their parent stars are expected to have greatly distended atmospheres with more highly excited H_2 than Jupiter. Recent investigations (Seager & Sasselov 1998, Goukenleuque et al. 2000) have concentrated on the reflection and thermal emission spectra of such atmospheres; however, the fluorescence of H_2 that is driven by the stellar EUV emission may provide a superthermal signature of the planetary atmosphere. H_2 emission is a feature of the Jovian dayglow and aurora (e.g. Wolven & Feldman 1998, Liu & Dalgarno 1996ab). Fluorescent emission pumped by solar H I Lα radiation has been observed at a Comet Shoemaker-Levy 9 impact site on Jupiter (Wolven et al. 1997).

3.2. *Stars*

In the Sun's atmosphere, H_2 survives in small amounts and its UV spectrum probes an interesting part of the lower chromosphere overlying sunspots (Jordan et al. 1977ab, Bartoe et al. 1979, Sandlin et al. 1986). There are accidental resonances between H_2 lines of the Lyman ($B^1\Sigma_u^+ - X^1\Sigma_g^+$) and Werner ($C^1\Pi_u - X^1\Sigma_g^+$) systems and various atomic emission lines that dominate the spectra of stellar chromospheres and coronae, notably H I Lα. Through these transitions, the H_2 can be radiatively excited to produce a cascade of fluorescent emission lines, whose intensities probe the internal radiation field, density, and temperature where the molecules live. Fluorescent processes that excite H_2 line emission in the Sun also operate in the atmospheres of other late-type stars (e.g. McMurry, Jordan, & Carpenter 1999), in T Tauri (Brown, Ferraz, & Jordan 1984), and in Herbig-Haro objects (Schwartz 1983, Curiel et al. 1995). Sometimes the H_2 emission of H-H objects is purely collisionally excited (Gredel 1996), while in others the combination of radiative and collisional processes at work is not adequately explained by existing models (Raymond, Blair, & Long 1997). The fluorescent emission lines of H_2 in the spectrum of the giant star α Tau (McMurry et al. 1999) highlight a need of collisional quenching rates for molecules in levels of the $B^1\Sigma_u^+$ state.

3.3. *Interstellar Matter*

It is in the study of interstellar matter that H_2 has had its widest applications. Although expected to be the most abundant species in dense molecular clouds, H_2 was not directly observed in a cold, dense, quiescent cloud until Lacy et al. (1994) succeeded in measuring the weak, electric quadrupole vibration-rotation lines in absorption toward highly obscured infrared sources. This has made it possible to determine directly the fractional abundances of other molecules, like CO, which are used more widely as surrogate tracers of hydrogen. The hydrogen molecule offers a superb spectroscopic diagnostic tool for determining density, temperature, intensity of ultraviolet radiation, cosmic ray flux, and the interstellar deuterium abundance. It is interesting to consider whether H_2 can exist in solid form in the interstellar medium. An earlier report of the identification of solid-phase H_2 (Sandford, Allamandola, & Geballe 1993) has since been retracted (Geballe 1999); however, such observations raise very interesting questions about the chemistry of cold solids in the interstellar medium (Buch & Devlin 1994; Dissly, Allen, & Anicich 1994) and certainly deserve to be pursued further.

3.4. External Galaxies

Since its first detection in the Seyfert galaxy NGC 1068 (Thompson, Lebofsky, & Rieke 1978), H_2 has been a recognized feature in the infrared spectra of active galaxies. Indeed, emission lines of the $v = 1 \to 0$ band are often as intense as the H I Brγ line from ionized gas in the K-band spectra of active galaxies (cf. Veilleux, Goodrich, & Hill 1997). Broadly speaking there are two kinds of activity in centers of galaxies: the true active galactic nuclei (AGN) are thought to derive their power from accretion onto supermassive black holes, while central starbursts represent regions where the luminosity arises from massive episodes of star-forming activity. Current theories and observations of AGN suggest that a torus or disk of molecular gas may be an essential element of these systems, providing both the reservoir of gas to fuel the power source and the obscuring matter that hides the center when viewed more nearly edge-on than face-on. Although much of the luminosity of a starburst may originate in the ultraviolet radiation of massive stars, the formation of massive stars is likely to be attended by shock waves and to be followed soon (i.e. within $\approx 10^7$ years) by supernova explosions, which also inject mechanical energy into the surrounding molecular gas. In some systems, a true AGN may be surrounded by a starburst region on larger scales—a good example is NGC 1068. Thus it is not surprising that H_2 emission lines alone do not cleanly distinguish the different kinds of activity. H_2 lines are characteristic features of the infrared spectra of ultra-luminous galaxies that show no evidence of AGN (Murphy et al. 1999); yet they also appear in the spectra of AGN such as Cygnus A (Thornton, Stockton, & Ridgway 1999). Moreover, H_2 line emission near 1.87 μm wavelength can sometimes confuse the search for H I Pα, which is otherwise a valuable probe of an obscured broad-line emission region of a buried AGN (Veilleux, Sanders, & Kim 1999).

The hydrogen molecule is energetically predisposed to (1) control the cooling of shock waves that pass through molecular gas, (2) reprocess a significant fraction of the ultraviolet flux incident on a molecular gas at wavelengths $\lambda \approx 91 - 110$ nm into a few infrared lines at $\lambda \approx 2$ μm, and (3) respond with collisionally excited infrared radiation to the the passage of X-rays through molecular gas. Thus it is very difficult to assign a single excitation mechanism to the spectra that emerge from the unresolved central regions of galaxies. At high densities, the appearance of a UV-excited spectrum (Black & van Dishoeck 1987) becomes difficult to distinguish from that of a collisionally excited spectrum (cf. Sternberg & Dalgarno 1989).

Despite the difficulties of interpretation of H_2 line emission in centers of galaxies, it remains a very sensitive way to detect the presence of molecular gas in these hostile environments, where other commonly used molecular tracers may be difficult to measure (Black 1998). In relatively nearby starburst galaxies, H_2 line emission can be used to measure how the radiant and mechanical energy associated with star formation is put back into the molecular gas.

3.5. Quasi-stellar Objects and the Universe at High Redshift

Although H_2 is rarely observed through its redshifted ultraviolet absorption lines in the spectra of distant (i.e. redshift $z \gtrsim 2$) quasi-stellar objects (QSOs), there are three instances where the identification seems secure (Foltz et al. 1987; Ge & Bechtold 1997, 1999; Srianand & Petitjean 1998). In all three cases, the H_2 absorption resembles that observed in the diffuse molecular clouds within our Galaxy. There are several selection effects that may operate against detection of redshifted UV absorption lines of H_2 in QSO spectra. The absorption spectrum of H_2 will invariably overlap the numerous lines of the H I Lα forest that are present in the spectra of all high-z QSOs. If effective conversion of H into H_2 requires surfaces of dust particles, then the dust that accompanies detectable amounts of H_2 will extinguish the light of the background QSO; therefore, such QSOs are less likely to be identified in surveys carried out at visible wavelengths in the observer's frame. A more subtle spectroscopic effect can arise through absorption in a foreground galaxy that contains a large amount of H_2: a rotating disk of molecular clouds might show kinematical broadening of the absorption lines over the full range of rotational velocities. With increasing linewidth δv, the spectrum eventually becomes indistinguishable in practice from broad continuous absorption. Figure 1 illustrates the effect with synthetic spectra of H and H_2 where three values of linewidth have been applied to modest column densities, $N(H) = N(H_2) = 10^{20}$ cm^{-2}. In all three cases, the rotational excitation temperature of the H_2 is $T_{rot} = 300$ K. The spectrum with greatest broadening in Figure 1 would look very much like an atomic H Lyman-limit absorption system at an anomalous redshift.

It is gratifying that cosmologists now appreciate the physical properties of H_2 that must be included in numerical simulations of the origin of structure in an expanding Universe (cf. Abel et al. 1998; Corbelli, Galli, & Palla 1998).

4. Formation of H_2

Two major questions remain about the association of H atoms on grain surfaces to form H_2: what is the effective rate of the process in the interstellar medium and what is the distribution of nascent molecules among excited vibration-rotation states. Recent experimental studies and theoretical treatments of H-atom association on astrophysically relevant surfaces represent important progress and help to clarify crucial issues. Pirronello et al. (1997a,b; 1999) present measurements of molecular hydrogen formation on olivine and carbonaceous substrates at low temperatures (< 30 K). They conclude that adsorbed H atoms are not nearly as mobile as assumed by Hollenbach & Salpeter (1971) and that the effective formation rate in the interstellar medium may be approximately an order of magnitude lower than found in that classic study. Katz et al. (1999) model these laboratory measurements in a way that permits direct extension to interstellar conditions, while Biham et al. (1998) have already discussed the behavior of two limiting cases in **steady state**. Several points deserve further investigation. First, does the steady state formulation apply when the time-dependences of the grain surface coverage and desorption processes are important? Second, when a grain surface is sparsely covered by H atoms, the effective formation rate **per grain** seems to have a quadratic dependence on the surface area (Biham et al. 1998), although other factors favoring the smallest grains may be buried in the coefficients of diffusion and desorption. The effective H_2 formation rate should be described for a realistic distribution of grain sizes. It is of particular interest to understand whether the very small grains (sizes $\lesssim 0.01$ μm) contribute to the formation of H_2. If so, they may compensate partly for a reduced formation efficiency on classical dust grains compared with the model of Hollenbach & Salpeter (1971). If large molecules (e.g. polycyclic aromatic hydrocarbons = PAHs) act like very small grains, might they be important sites of H-atom association for form H_2?

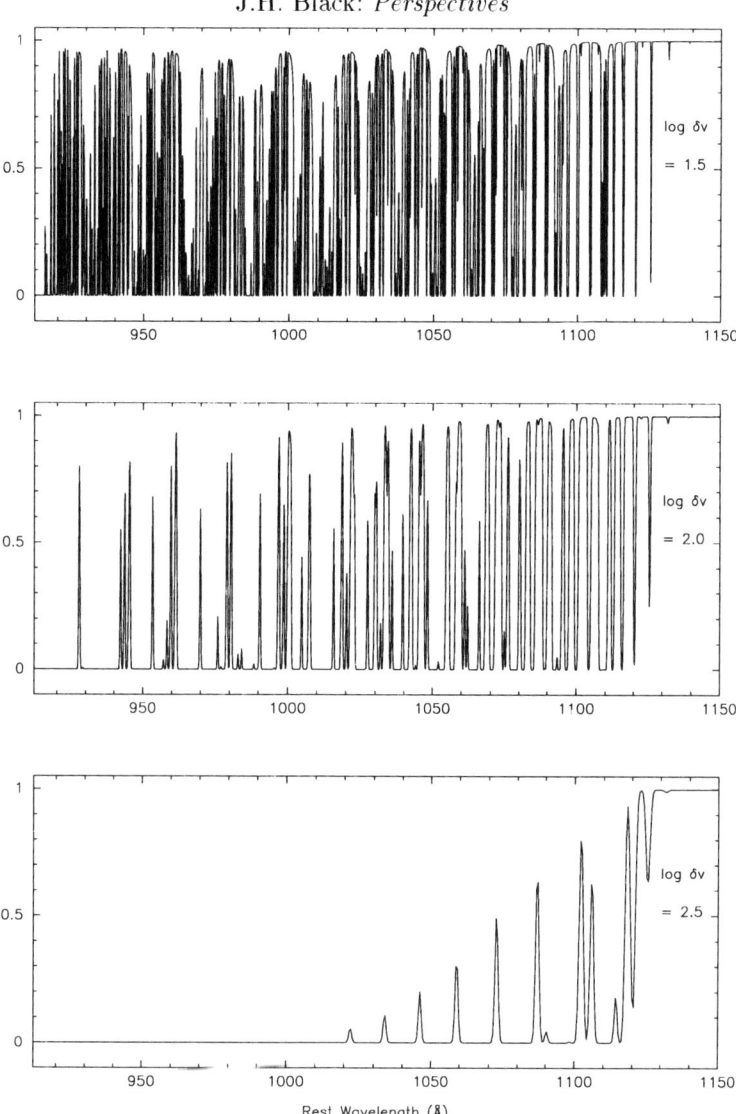

FIGURE 1. Absorption spectra of H and H_2 broadened with a gaussian linewidth δv. In each panel the linewidth (in km s^{-1}) is indicated and the vertical axis displays intensity relative to the true continuum.

Recent discussions of H-atom association on PAH cations offer conflicting results (Herbst and Le Page 1999, Bauschlicher 1998).

5. Cold H_2 in interstellar clouds

A commonly repeated falsehood in astrophysics has asserted that cold H_2 in dense and thick molecular clouds is not directly observable because the molecule lacks the allowed rotational spectrum possessed by its surrogate tracers like CO. Dramatic evidence to the contrary was supplied by Lacy and collaborators (§3.3) with their measurement of quadrupole lines in absorption toward obscured infrared sources. With recent improvements in the sensitivity and resolution of infrared spectrometers, such measurements are now much easier, and with the new generation of large telescopes, they will become almost routine. Coordinated infrared absorption observations of H_2, CO, ^{13}CO, and H_3^+ have recently been made for a small sample of molecular clouds (Kulesa & Black 2000). Preliminary results can be summarized for three sources with fully reduced data (AFGL 2591, AFGL 490, and NGC 2024 IRS2). Observed column densities of H_2 lie in the range of $(2-4) \times 10^{22}$ cm^{-2}. Upper limits on H_2 in $J = 1$ and measurements of H_3^+ are consistent with ortho/para ratios in both molecules that are thermalized at the same temperature described by the CO populations. The **directly measured** abundance ratio, CO/H_2 by number of molecules, lies in the range $(2-5) \times 10^{-4}$. The abundances of H_3^+ are in harmony with ionization rates due to penetrating cosmic rays of the order of a few times 10^{-17} s^{-1}. The measured H_2 absorption lines are typically 1% deep at a resolution of approximately 5 km s^{-1}. Such observations are difficult, but not impossible. Although the absorption measurements are restricted to directions where background infrared sources are found, such observations will soon be possible toward many more, fainter sources.

A broader question concerns the possibility that a large mass of baryons might remain virtually undetectable in the form of extremely cold molecular hydrogen in small, dense clouds or clumpuscules (Combes & Pfenniger 1997, and references therein; see also Kalberla et al. 1999 and Wardle & Walker 1999). Whether such clouds might be distributed in an extended disk or a flattened halo, it is important to consider whether they can, in fact, remain stable and cold. If the Galactic cosmic ray spectrum inferred near Earth also applies throughout the volume occupied by such clouds, then the heating by cosmic rays may prevent their temperatures from approaching that of the cosmic background radiation. In the general interstellar medium, a cosmic ray nucleus loses a minuscule fraction of its total energy to ionization and heating in each encounter with a giant molecular cloud, which is typically transparent to it. In contrast, the central column density of an "invisible" clumpuscule might be at least as large as 10^{24} molecules cm^{-2}, which corresponds to the range of a 1 MeV proton. The range implies the length over which the cosmic ray will be stopped and its energy completely degraded. As a result, the heating efficiency of cosmic rays on dense clumpuscules must be considerably larger than in the general interstellar medium: the total heating rate can be estimated from the geometrical cross section of a clumpuscule, the cosmic ray energy spectrum, and an average heating efficiency. Simple estimates suggest steady-state temperatures that might approach 100 K. If the clumpuscules are distributed within the reach of the observed Galactic cosmic rays and if they account for a dynamically significant total mass, then they would likely affect the propagation of cosmic rays. More work is needed.

6. H_2 in diffuse molecular clouds

As we await the publication of new data from the *Far Ultraviolet Spectroscopic Explorer* (FUSE) on ultraviolet absorption lines of H_2 in diffuse molecular clouds, it is important to recall some unanswered questions left over from the pre-FUSE era. The large body of data from the *Copernicus* observatory revealed interstellar lines of H_2 with inferred (but unresolved) Doppler broadening parameters $b \approx 3$ to 5 km s^{-1}. These values are rather larger than the observed line widths of CH, CN, and C_2 ($b \approx 1$ km s^{-1}) that characterize the molecule-containing component of many of the same lines of sight. In a few cases, H_2 has been observed with sufficient resolution that the line broadening can be measured directly. For example, Jenkins & Peimbert (1997) noted that b increases with rotational quantum number J toward ζ Ori A. It remains an interesting question whether the broadening of H_2 lines reflects the kinematical distribution (i.e. discrete

Doppler components) of the absorbing gas or whether it might indicate something profoundly unthermalized about the molecular speed distributions on the microscopic scale. If molecules in levels $J \geq 5$, for example, derive part of their excitation from the formation process itself, then their excess kinetic energy at birth cannot be fully thermalized during their radiative lifetimes ($\lesssim 10^8$ s) by elastic collisions at the densities of diffuse clouds. In this way, at least part of the large line widths may be microscopic in nature. Approaching such questions with improved data will eventually make it possible to assess the total amount of highly excited H_2 in diffuse molecular clouds.

7. H_2 in the Crab Nebula

The discovery of infrared line emission of H_2 in filaments of the Crab Nebula (Graham, Wright, & Longmore 1990) is remarkable, not least for the fact that it has been so widely overlooked. These molecule-containing filaments are exposed to intense X-radiation with a well measured, continuous spectrum. They are also bombarded by relativistic electrons that are responsible for the synchrotron radiation of the nebula. Thus the filaments might offer an excellent laboratory for studying the response of molecular gas to X-rays and energetic charged particles. Such X-ray dominated regions (XDRs) have attracted considerable theoretical interest (Lepp & Tiné 1998; Maloney, Hollenbach, & Tielens 1996; Dalgarno, Yan, & Liu 1999), yet clear comparisons between theoretical models and observation are lacking.

REFERENCES

ABEL, T., ANNINOS, P., NORMAN, M.L., & ZHANG, Y. 1998 *ApJ* **508**, 518–529.

BARTOE, J.-D.F., BRUECKNER, G.E., NICOLAS, K.R., SANDLIN, G.D., VANHOOISER, M.E., & JORDAN, C. 1979 *MNRAS* **187**, 463–471.

BAUSCHLICHER, C.W., JR. 1998 *ApJ* **509**, L125–L127; erratum, 1999, *ApJ* **517**, L67–L67.

BIHAM, O., FURMAN, I., KATZ, N., PIRRONELLO, V., & VIDALI, G. 1998 *MNRAS* **296**, 869–872.

BLACK, J.H. 1998, "Excitation and detectability of molecules in active galactic nuclei", Chap. 21 of *The Molecular Astrophysics of Stars and Galaxies*, T.W. Hartquist & D.A. Williams, editors, (Oxford: Clarendon Press), pp. 469–487.

BLACK, J.H., & VAN DISHOECK, E.F. 1987 *ApJ* **322**, 412–449.

BROWN, A., FERRAZ, M.C. DE M., & JORDAN, C. 1984 *MNRAS* **207**, 831–859.

BUCH, V., & DEVLIN, J.P. 1994 *ApJ* **431**, L135–L138.

COMBES, F., & PFENNIGER, D. 1997 *A&A* **327**, 453–466.

CORBELLI, E., GALLI, D., & PALLA, F. 1998 "Molecular hydrogen in the early Universe", *Mem. Soc. Astronomica Italiana*, volume 69, number 2.

CURIEL, S., RAYMOND, J.C., WOLFIRE, M., HARTIGAN, P., MORSE, J., SCHWARTZ, R.D., & NISENSON, P. 1995 *ApJ* **453**, 322–331.

DALGARNO, A., YAN, M., & LIU, W. 1999 *ApJS* **125**, 237–256.

DISSLY, R.W., ALLEN, M., & ANICICH, V.G. 1994 *ApJ* **435**, 685–692.

FOLTZ, C.B., CHAFFEE, JR., F.H., & BLACK, J.H. 1987 *ApJ* **324**, 267–278.

GE, J., & BECHTOLD, J. 1997 *ApJ* **477**, L73–L77.

GE, J., & BECHTOLD, J. 1999 in *Highly Redshifted Radio Lines*, C.L. Carilli, S.J.E. Radford, K.M. Menten, & G.I. Langston, editors, (San Francisco: Astronomical Society of the Pacific), p. 121.

GEBALLE, T.R. 1999 remarks at IAU Symposium 197, "Astrochemistry: From Molecular Clouds to Planetary Systems".

GOUKENLEUQUE, C., BÉZARD, B., JOGUET, B., LELLOUCH, E., & FREEDMAN, R. 2000 *Icarus* **143**, 308–323.

GRAHAM, J.R., WRIGHT, G.S., & LONGMORE, A.J. 1990 *ApJ* **352**, 172–183.

GREDEL, R. 1996 *A&A* **305**, 582–591.

HERBST, E. & LE PAGE, V. 1999 *A&A* **344**, 310–316.

HOLLENBACH, D.J. & SALPETER, E.E. 1971 *ApJ* **163**, 155–164.

HOLLENBACH, D.J. & TIELENS, A.G.G.M. 1997 *Annu. Rev. Astron. Astrophys.* **35**, 179–215.

JENKINS, E.B., & PEIMBERT, A. 1997 *ApJ* **477**, 265–280.

JORDAN, C., BRUECKNER, G.E., BARTOE, J.-D.F., SANDLIN, G.D., & VANHOOSIER, M.E. 1977a *Nature* **270**, 326–327.

JORDAN, C., BRUECKNER, G.E., BARTOE, J.-D.F., SANDLIN, G.D., & VANHOOSIER, M.E. 1977b *ApJ* **226**, 687–697.

KALBERLA, P.M.W., SHCHEKINOV, YU.A., & DETTMAR, R.-J. 1999 *A&A* **350**, L9–L12.

KATZ, N., FURMAN, I., BIHAM, O., PIRRONELLO, V., & VIDALI, G. 1999 *ApJ* **522**, 305–312.

KULESA, C.A., & BLACK, J.H. 2000 in preparation.

LACY, J.H., KNACKE, R., GEBALLE, T.R., & TOKUNAGA, A.T. 1994 *ApJ* **428**, L69–L72.

LEPP, S., & TINÉ, S. 1998, "X-ray Dominated Regions", in *The Molecular Astrophysics of Stars and Galaxies*, T.W. Hartquist & D.A. Williams, editors, (Oxford: Clarendon Press), pp. 489–505.

LIU, W., & DALGARNO, A. 1996a *ApJ* **467**, 446–453.

LIU, W., & DALGARNO, A. 1996b *ApJ* **471**, 480–484.

MALONEY, P.R., HOLLENBACH, D.J., & TIELENS, A.G.G.M. 1996 *ApJ* **466**, 561–584.

MCMURRY, A.D., JORDAN, C., & CARPENTER, K.G. 1999 *MNRAS* **302**, 48-58.

MURPHY, JR., T.W., SOIFER, B.T., MATTHEWS, K., KIGER, J.R., & ARMUS, L. 1999 *ApJ* **525**, L85–L88.

PIRRONELLO, V., BIHAM, O., LIU, C., SHEN, L., & VIDALI, G. 1997a *ApJ* **483**, L131–L134.

PIRRONELLO, V., LIU, C., SHEN, L., & VIDALI, G. 1997b *ApJ* **475**, L69–L72.

PIRRONELLO, V., LIU, C., ROSER, J.E., & VIDALI, G. 1999 *A&A* **344**, 681–686.

RAYMOND, J.C., BLAIR, W.P., & LONG, K.S. 1997 *ApJ* **489**, 314–318.

SANDFORD, S.A., ALLAMANDOLA, L.J., & GEBALLE, T.R. 1993 *Science*, **262**, 400–402.

SANDLIN, G.D., BARTOE, J.-D.F., BRUECKNER, G.E., TOUSEY, R., & VANHOOSIER, M.E. 1986 *ApJS* **61**, 801–898.

SCHWARTZ, R.D. 1983 *ApJ* **268**, L37–L40.

SEAGER, S., & SASSELOV, D.D. 1998 *ApJ* **502**, L157–L161.

SRIANAND, R., & PETITJEAN, P. 1998 *A&A* **335**, 33–40.

STERNBERG, A., & DALGARNO, A. 1989 *ApJ* **338**, 197–233.

STERNBERG, A., & NEUFELD, D.A. 1999 *ApJ* **516**, 371–380.

THOMPSON, R.I., LEBOFSKY, M.J., & RIEKE, G.H. 1978 *ApJ* **222**, L49–L53.

THORNTON, R.J., STOCKTON, A., & RIDGWAY, S.E. 1999 *AJ* **118**, 1461–1467.

TRAFTON, L.M., MILLER, S., GEBALLE, T.R., TENNYSON, J., & BALLESTER, G.E. 1999 *ApJ* **524**, 1059–1083.

VEILLEUX, S., GOODRICH, R.W., & HILL, G.J. 1997 *ApJ* **477**, 631–660.

VEILLEUX, S., SANDERS, D.B., & KIM, D.C. 1999 *ApJ* **522**, 139–156.

WARDLE, M., & WALKER, M. 1999 *ApJ* **527**, L109–L112.

WOLVEN, B.C., & FELDMAN, P.D. 1998 *Geophys. Res. Letters* **25**, 1537–1540.

WOLVEN, B.C., FELDMAN, P.D., STROBEL, D.F., & MCGRATH, M.A. 1997 *ApJ* **475**, 835–842.

Author Index

Abel, T., **237**
Abergel, A., 211
Abgrall, H., 13
Alves, J., **217**
André, M., 179

Bechtold, J., **301**
Bertoldi, F., 131, 231
Bessega, M. C., 263
Biham, O., 71
Black, J., **317**
Bluhm, H., 165
Boissé, P., **221**
Boissel, P., 107
Borisov, A. G., 89
Boulanger, F., **211**
Brand, P., 143, 197
Burton, M., 143, 197

Cabrit, S., 123
Cernicharo, J., 151
Cesarsky, D., 205
Chrysostomou, A., 143
Ciardi, B., 269
Clary, D., 99
Combes, F., **275**
Cox, P., **205**
Cuillandre, J. C., 221
Cutri, R. M., 201

Dalgarno, A., **3**
De Boer, K., 165
Dettmar, R. J., 57
Deutscher, S. A., 89
Draine, B. T., **131**
Drapatz, S., 231
Duvert, G., 221

Falgarone, E., 155, 211, **225**
Farebrother, A., 99
Ferlet, R., **179**
Ferrara, A., **269**
Field, D., **155**
Fisher, A., 99
Flower, D., **23**, 117, 123

Galli, D., 247
Gerin, M., 155
Gerlich, D., **63**
Gingell, J., 99

Hébrard, G., 179
Habart, E., 211
Haiman, Z., 237
Haud, U., 297
Heras, A., 151
Herbst, E., **85**
Herpin, F., **151**
Hily-Blant, P., 225

Ibanez, M. H., **263**
Illemann, J., 63
Israel, F. P., **293**

Jackman, R., 99
Jarrett, T., 201
Jeloaica, L., 89
Joblin, C., **107**
Jungen, Ch., **31**
Jura, M., **161**

Kalberla, P. M., 57, **297**
Kerp, J., 297

Lada, C., 217
Lada, E., 217

Le Bourlot, J., 23
Leach, S., 155
Lecavelier, A., 179
Lemaire, J. L., 155
Lemoine, M., 179
Liu, X., 13

Maillard, J. P., 107, 155
Manico, G., 71
Marggraf, O., 165
Mason, N., 99
Meijer, A., 99

Pagani, L., 221
Palla, F., **247**
Pech, C., 107
Perry, J., 99
Pfenniger, D., **285**
Pijpers, F. P., 155
Pineau des Forêts, G., 23, **117**, 123, 155, 179, 205, 211, 225
Pirronello, V., **71**
Price, S., 99
Puxley, P., 143
Puy, D., **259**

Ramsay Howat, S., **143**
Rawlings, J., 99
Reach, W. T., **193**, 201
Rho, J., 193, **201**
Richter, P., **165**
Rosenthal, D., **231**
Roser, J. E., 71
Ross, S. C., 31
Rostas, F., 155
Rouan, D., 155
Roueff, E., **13**, 23, 179

Schaefer, J., **47**
Schlemmer, S., 63
Shchekinov, Y. A., **57**

Shemansky, D., 13
Sidis, V., **89**
Snow, T. P., **171**

Tedds, J. A., **197**
Thoraval, S., 221
Tielens, A. G., 139
Todini, P., 269

Ubachs, W., **39**

Van Dyk, S., 201
Van den Ancker, M., **139**
Van der Werf, P., **307**
Vannier, L., 155
Vauglin, I., 107
Verstraete, L., 211, 225
Vidal-Madjar, A., 179
Vidali, G., 71

Wesselius, P. R., 139
Wilgenbus, D., **123**
Williams, D. A., **99**
Williams, D. E., 99
Wright, C. M., **189**